ELEMENTS OF
QUANTUM THEORY
Second Edition

ELEMENTS OF QUANTUM THEORY

Second Edition

Revised and enlarged

FRANK J. BOCKHOFF
Professor of Chemistry
Cleveland State University

1976

Addison-Wesley Publishing Company

Advanced Book Program
Reading, Massachusetts

London · Amsterdam · Don Mills, Ontario · Sydney · Tokyo

Frank J. Bockhoff, Elements of Quantum Theory
First edition, 1969
Second edition, revised and enlarged, 1976

This book is in the
ADDISON-WESLEY SERIES IN CHEMISTRY

Francis T. Bonner, Consulting Editor

Library of Congress Cataloging in Publication Data

Bockhoff, Frank J
    Elements of quantum theory.

    (Addison-Wesley series in chemistry)
    Bibliography: p.
    Includes index.
    1. Quantum theory. 2. Wave mechanics. I. Title.
QC174.12.B6 1976   530.1'2   76-41769
ISBN 0-201-00799-1

Printed in the United States of America

ABCDEFGHIJ-HA-79876

To Esther

CONTENTS

This text is intended to be a first introduction for chemists to the concepts, postulates, and applications of quantum theory, using the wave mechanical approach. Although separate undergraduate courses in quantum mechanics are now commonplace in the curricula of chemists and physicists, the introductory treatments given in standard undergraduate texts are often too superficial to provide either real understanding or an adequate base for advanced study. On the other hand, many more comprehensive texts assume a background in mathematics and classical physics which the average student simply lacks, thus presenting what seems to be an almost insurmountable barrier to real understanding. The primary aim of this text is to provide within a self-contained and conceptually ordered structure a basic understanding and appreciation of what quantum mechanics is all about. In particular, we'll try to identify those factors which necessitate and justify the concepts and postulates of quantum mechanics, which so often appear bewildering, even occult, to the novice.

This book is intended to be neither an introductory survey nor a complete treatise of quantum mechanics. Rather, I have chosen to consider with relative rigor and thoroughness a limited number of concepts and applications which seem to provide a coherent framework for further development. Blackbody radiation, the photoelectric effect, and the Bohr theory of the hydrogen atom are discussed in detail in the first three chapters. Wherever profitable in subsequent chapters, analogies are drawn between quantum mechanics and classical wave theory. In order to develop an appreciation of wave analogies, I have devoted Chapters 4 and 5 to important aspects of classical wave theory.

In Chapter 6, the postulates of quantum mechanics are introduced and developed for a single particle and are then illustrated in Chapters 7, 8, and 9 for simple systems such as the free particle, particles in potential energy wells and boxes, and the linear harmonic oscillator. Because an understanding of the quantum mechanical model of the hydrogen atom is vital to an understanding of the behavior and properties of electrons in more complex atoms and molecules, the treatment of the hydrogen atom and hydrogen-like ions in Chapters 10 and 11 is relatively detailed. In a similar fashion, the study of the helium atom seems to provide an ideal medium for introducing approximation methods and symmetry properties of wave functions for identical particles; therefore, its treatment in Chapters 12 and 13 is more extensive than is found in most introductory texts. Chapters 14 and 15 introduce more complex atoms and molecules, and serve as transitions to more detailed texts in these areas.

It is assumed that the student will have had one year of introductory physics and at least one year of calculus. Although a course in differential equations would certainly be helpful, such a course is not required in order to follow the mathematical development of the text, since advanced mathematical techniques (in addition to those which might require review) are introduced and discussed at those points in the text where they are used. For this reason, and because derivations are usually presented in detail, the book is also well suited for those who desire to study quantum mechanics independently.

It is a pleasure to acknowledge the generous assistance of Professor Joseph R. Crook of Western Washington State College and Professor Steinar S. H. Huang of Cleveland State University, each of whom read the entire initial manuscript and offered many helpful comments and suggestions. I am also indebted to Professors Francis T. Bonner and Robert F. Schneider, of the State University of New York at Stony Brook, for the helpful advice offered in their reviews of the final manuscript, as well as to Professors Kerro Knox, Earl M. Mortensen, and Bruce F. Turnbull, of Cleveland State University, and Professor Walter Kauzmann of Princeton University for suggestions now incorporated into the revised edition. Finally, I appreciate the competent assistance of Kathleen Kennedy and Janet Casaregola, who typed and proofread the manuscript.

F. J. B.

ELEMENTS OF
QUANTUM THEORY
Second Edition

1 INTRODUCTION

1-1 THE FIRST THIRTY YEARS

On December 14, 1900, before a meeting of the German Physical Society, Max Planck suggested that the spectral distribution of radiation emanating from hot objects may be accounted for *only* if one assumes that matter absorbs and emits energy in discrete bundles called *quanta*. On that day the *quantum theory* was born. Although the discrete division or *quantization* of *matter* into indivisible packets called atoms had been accepted since the beginning of the nineteenth century, the transmission and absorption of *energy* was still firmly believed to occur in the continuous fashion demanded by classical electromagnetic wave theory. Planck's revolutionary proposal was in effect parallel to similar theories concerning the indivisibility of matter which had begun with the first ideas of Leucippus in 450 B.C. and which were ultimately extended to the atomic theory of John Dalton in 1808. We might expect that such a bold proposal involving the quantization of energy would have caused a major stir in the scientific world. Surprisingly, it did not! In fact, a routine search through the scientific journals reveals that practically nothing was said about Planck in the years 1900 through 1904. The reluctance of scientists to accept Planck's proposal is understandable, since the acceptance of such a radical concept involved an inherent rejection of the fundamental framework of classical theory which had been developed to such a magnificent extent in the nineteenth century. Planck himself, a firmly grounded classical physicist, was very disturbed with the significance of his own proposal and spent many years in a futile attempt to reconcile his quantum restriction with classical physics.

Einstein in 1905 provided the first major support for Planck's quantum hypothesis when he studied the theoretical aspects of radiation balance in a cavity and suggested, as a means of supporting his conclusions, that the photo-electric effect (the emission of electrons from the illuminated surfaces of certain metals) could be explained only in terms of quantization of light itself. He proposed that light is *propagated* in the form of *corpuscles* or packets in which the amount of energy is directly proportional to the frequency. In Einstein's original words:*

> "... the energy in the light propagated by rays from a point is not smeared out continuously over larger and larger volumes, but rather consists of a finite number of energy quanta localized at space points, which move without breaking up and which can be absorbed or emitted only as wholes."

These small packets or *quanta* of light were later to be called *photons* by G. N. Lewis in 1926.

In 1907 J. J. Thomson proposed a model for the atom in which discrete electrons were assumed to be distributed randomly within a cloud of positive charge such that the entire atom was neutral. Thomson's model has often been called the "plum pudding" model, the plums representing the electrons and the pudding representing the positive charge. This model, however, was short-lived. As a result of alpha-scattering experiments in 1911 Rutherford concluded that the positive charge must be concentrated in a very small, dense *nucleus* at the center of the atom and first introduced what we now know as the *nuclear atom*.

But there were serious problems in the stability of the Rutherford atom which could not be explained in terms of classical theory. In an attempt to solve these problems, the Danish physicist Niels Bohr in 1913 boldly proposed a *quantum* model for the hydrogen atom which provided remarkable agreement with the observed spectrum of hydrogen. Bohr's model was based on the postulate of quantized energy levels for the lone electron in the hydrogen atom, a postulate which had absolutely no justification in classical physics. But then again, neither were the earlier quantum postulates of Planck and Einstein justifiable in a classical sense. For the third time in 13 years scientists had encountered experimental observations which could be explained *only* in terms of quantized energy.

The period between 1913 and 1923 was spent largely in extending and patching the Bohr theory to fit the interpretation of more complex spectra, and the tools were becoming most unwieldy. However, a major breakthrough appeared in 1924 in the doctoral thesis of the French physicist Louis de Broglie, who proposed that the motion of electrons is intimately associated with *pilot waves*. In fact, he proposed a specific mathematical relationship between the momentum of a particle and the wavelength of its associated wave which supported Bohr's earlier model of the hydrogen atom.

Late in 1925 Werner Heisenberg, a German physicist, with the help of Max Born and Pascual Jordan, published his first papers on *matrix mechanics*, a technique which allows one to determine mathematically the possible transitions

---

* A. Einstein, *Ann. Physik*, **17**, 132 (1905).

between different states of an atom and further enables one to determine the frequencies associated with such transitions. Matrix calculations led to results which were in closer agreement with experimental values than the most refined methods based on the older Bohr theory.

Almost simultaneously with the work of Heisenberg, Erwin Schrödinger, an Austrian physicist, developed the formalism known as *wave mechanics*, in which he combined earlier aspects of classical wave theory with de Broglie's wave-particle relationship to postulate an entirely new technique for determining the properties of atomic systems. Schrödinger's methods are based on differential equations and are thus more easily comprehended by the average student than are the matrix calculations of Heisenberg.

The simultaneous appearance of the papers of Heisenberg and Schrödinger astounded the scientific world. They looked entirely different in their approaches and yet led to the same results in the interpretation of atomic structure and spectra. In fact, for about a year they were thought of as two rival and distinct theories until Schrödinger and Carl Eckart, an American physicist, independently recognized the mathematical identity of the two approaches. What is referred to today as *quantum mechanics* is usually a fusion of certain aspects of each of the above approaches plus contributions from still other methods.

A remaining major refinement in quantum theory was made in 1929 by P. A. M. Dirac, a British physicist, who united the theory of relativity with quantum theory to produce the relativistic wave equation, which describes an atomic electron at velocities approaching the velocity of light. It also provides mathematically acceptable interpretations for such properties as intrinsic magnetic moments of electrons.

The bulk of the theoretical basis of quantum mechanics was laid by 1930 and much experimental work has been completed since that time. The validity of the formalisms of quantum mechanics has been confirmed repeatedly in measurements involving atomic structure, spectra, covalent bonds, free radicals, mechanisms of chemical reactions, magnetic properties of materials, stereochemistry, semiconductors, and in other areas.

We must, however, appreciate that quantum mechanics is a *model* which we do not fully understand. It is a formalism based on postulates which have no foundation in classical physics and which can be justified only in terms of its correctness in predicting experimental behavior. Thus, we must be prepared to cope with ideas and concepts which will appear not only strange but often quite unreasonable in terms of classical theory.

1–2 SOME INSIGHTS AND IMPLICATIONS

The bewildering sequence of events beginning with Planck's quantum hypothesis in 1900 and culminating in the quantum mechanical formalisms of Heisenberg and Schrödinger in 1926 represented a total reorganization of the fundamental philosophy of science. The development of early quantum theory, that is, up to 1925, led to a clear recognition of the impossibility of forming a coherent causal description of atomic phenomena in terms of ordinary mechanical concepts. Radical new ideas were required in order to reduce experimental observations to some semblance of order.

In particular, a renewed discussion of the principles of causality and determinism has emerged from the quantum theory which has forced an essential change in our attitude toward experimental observation. The concept of *determinism* implies that all natural phenomena are interconnected in an invariable order, such that a given set of conditions will always lead us to the same result. According to quantum theory, however, we'll find that we must accept the impossibility of experimentally determining or predicting the exact properties of an individual particle at a given moment in time. Rather, we must deal with the statistical regularities which characterize measurements involving large numbers of particles or which characterize a large number of measurements on an individual particle. In a way, quantum mechanics, which deals with probabilities rather than certainties, introduces a margin of uncertainty into our ability to measure and to know nature.

Throughout our study of quantum theory, we must appreciate that all new experiences arise within the framework of customary or traditional forms of perception and expression. For this reason, we shall attempt, insofar as possible, to reinterpret classical concepts by giving them suitable quantum mechanical meaning.

The story of the quantum is the story of a confused and groping search for knowledge conducted by scientists of many lands on a front wider than the world of physics had ever seen before, illumined by flashes of insight, aided by accidents and guesses, and enlivened by coincidences such as one would expect to find only in fiction.

    —BANESH HOFFMANN, *The Strange Story of the Quantum*, Dover Publications, Inc., New York, 1959. Reprinted through permission of the publisher.

ENERGY PACKETS AND THE
2   QUANTUM HYPOTHESIS

If we are to understand where and how classical mechanics failed to provide satisfactory explanations for experimental observations, we must consider two developments in much greater detail: Planck's quantum hypothesis of 1900 and Einstein's photoelectric theory of 1905. Planck and Einstein each found it necessary to postulate quantized energy levels in order to explain satisfactorily two apparently separate experimental phenomena involving the absorption of electromagnetic radiation.

2-1 WAVE PROPERTIES OF ELECTROMAGNETIC RADIATION

Visible light, heat rays, radio waves, and gamma radiation are different forms of the same fundamental radiation known as electromagnetic radiation. It is convenient to think of electromagnetic radiation as the propagation of a wave in which electrical and magnetic fields at right angles to one another periodically change magnitude and direction (see Fig. 2-1). Furthermore, electromagnetic waves have energy and are able to exchange this energy with matter. It may be shown that the energy of electromagnetic radiation at any point is proportional to the *square* of either the electrical or the magnetic vector.

The nature of a particular radiation is more precisely defined in terms of three variables: *wavelength, frequency*, and *speed of propagation*. The wavelength, λ, is the distance between two adjacent wavecrests or wavefronts and is often expressed in centimeters per cycle or simply in centimeters. The frequency, v, represents the number of wavecrests or wavefronts which pass a given point in a unit time an is usually expressed in cycles/second or simply second^{-1}. The reciprocal second s^{-1} is called a *Hertz* (Hz). Thus, it follows that the speed of propagation, c, of the wave may be obtained by multiplying the frequency by the wavelength:

$$c = \lambda v. \tag{2-1}$$

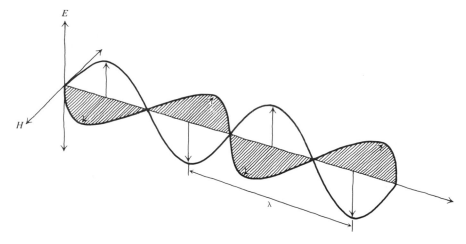

FIG. 2–1   The fluctuation of electrical (E) and magnetic (H) fields in a propagating electro-magnetic wave (plane polarized).

For all forms of electromagnetic radiation the speed c in a vacuum is constant and is approximately equal to 3.0×10^{10} cm/sec. Thus, electromagnetic radiation is classified conveniently either in terms of its wavelength, λ, or its frequency, v, which is inversely proportional to the wavelength. The electro-magnetic spectrum, shown in Fig. 2–2, includes radio waves several miles long at one end and very short, highly penetrating gamma rays at the other. Visible light constitutes only a very small portion of the total spectrum.

One of the most interesting and important aspects of electromagnetic radiation is the periodic electrical and magnetic disturbance produced at a given point in space as a wave passes by. An electron, for example, experiences a passing electromagnetic wave as a periodically fluctuating push and pull which causes it to oscillate at the same frequency as that of the wave. Any other charged particle would experience this same effect. Even more interestingly, the

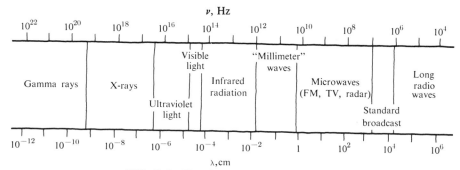

FIG. 2–2   The electromagnetic spectrum.

entire process may be reversed. That is, if a charged particle is forced to oscillate, electromagnetic radiation is generated. A floating cork subject to the effect of a ripple wave on the surface of a pond is a fair analogy to the electron subject to an electromagnetic wave, even though the ripple wave is quite complex in form. As the wave passes the position of the cork, the periodic fluctuation in the surface causes the cork to bob up and down at the same frequency as that of the wave. The forces on the cork operate primarily in a vertical or *transverse* direction, that is, perpendicular to the direction of wave propagation, just as the electrical and magnetic forces in an electromagnetic wave are oriented perpendicularly to the direction of wave propagation. Conversely, if one begins with a smooth surface on the pond and mechanically forces the cork to oscillate up and down on the surface, a ripple wave is generated.

A basic appreciation of the reversibility of absorption and emission of electromagnetic energy by an oscillating charge is important to an understanding of the next topic: blackbody radiation.

2–2 THE NATURE OF BLACKBODY RADIATION

All heated objects emit electromagnetic radiation. At relatively low temperatures a metal or ceramic rod emits mostly long-wavelength electromagnetic radiation known as infrared radiation, which is not visible but may be felt by the skin as heat. As the temperature is increased the predominant wavelength of the emitted radiation steadily decreases until a dull red visible radiation is observed at about 600°C. As the temperature is raised higher the color of the visible radiation gradually changes from red to yellow and eventually appears as a bright white incandescence at about 2000°C. At still higher temperatures significant ultraviolet radiation is also emitted.

The exact nature of the radiation, particularly the variation of intensity with wavelength, seems to depend to some extent on the solid being heated. For convenience, we may imagine a system in which the emitting solid is in complete equilibrium with the radiation so that the solid is hypothetically both a perfect absorber of radiation and also a perfect emitter. That is, it completely absorbs radiation of all wavelengths and in turn freely emits radiation of all wavelengths. Such a hypothetical solid is called a *blackbody radiator*. A perfectly dull black surface at least absorbs all of the *visible* radiation which impinges upon it. This is why it appears to be black. A perfect blackbody would totally absorb *all* electromagnetic radiation from gamma rays to radio waves.

The hypothetical "blackbody" may be approximated experimentally by constructing a hollow container (for example, a sphere) using a single pinhole to admit and release radiation. Radiation entering through the pinhole would have an extremely low probability of being reflected immediately back (see Fig. 2–3). In effect it eventually would be completely absorbed. On the other

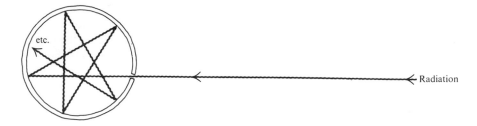

FIG. 2–3   A perfect blackbody radiator.

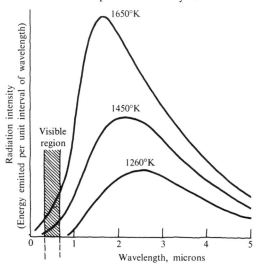

FIG. 2–4   Experimental curves of intensity versus wavelength for blackbody radiation.

hand, if we heated the sphere to incandescence, radiation leaving through the pinhole would also have had a very good chance to equilibrate with the incandescent surface before escaping. It would in effect represent *equilibrium radiation*.

It is found experimentally that the equilibrium radiation emitted from such a blackbody is identical for all solids and that the distribution of intensity of radiation as a function of wavelength is dependent *only on the temperature* of the incandescent solid. Typical experimental curves are shown in Fig. 2–4. Note that the wavelength of the most intense radiation (the peak) is lowered as the temperature is increased.

2–3   ALLOWED AND NONALLOWED VIBRATIONS

One of the major problems facing classical physicists at the end of the nineteenth century was that of explaining theoretically the nature of experimentally observed curves of intensity versus wavelength for blackbody radiation.

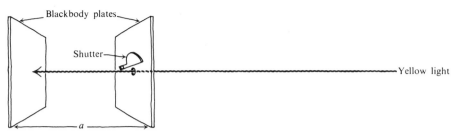

FIG. 2-5   A hypothetical one-dimensional blackbody cavity.

Classical physics attributes the origin of electromagnetic radiation in heated solids to the random vibrations and oscillations of charged nuclei and electrons. It is a well-known classical principle that charged particles emit electromagnetic radiation when accelerated. For example, the periodic to-and-fro acceleration of electrons in a radio transmitter antenna results in the emission of radio waves having the same frequency as the oscillating electrons. In a heated solid the myriad of random frequencies involved in the motion of the nuclei and electrons results in a random distribution of wavelengths in the emitted radiation.

However, in the case of the hypothetical blackbody, in which the radiation is forced to equilibrate within an enclosed cavity, classical wave theory dictates that only those wavelengths may persist (and may eventually be emitted) which may be reflected (or absorbed and immediately re-emitted) without destructive interference.

We may more easily understand the equilibration process in a blackbody cavity if we imagine the *one-dimensional* enclosure shown in Fig. 2-5. Such an imaginary device may be built by placing two parallel blackbody surfaces at an arbitrary separation distance, a. We may drill a pinhole in one plate and cover it with a moveable shutter. Now imagine that we open the shutter and allow electromagnetic radiation of *any* frequency distribution, for example, yellow light, to enter the cavity, and then immediately close the shutter. If the plates were considered to be merely parallel mirrors (which they are not), the radiation would be reflected back and forth in the single dimension. As we'll show more rigorously later, in such a reflecting case only those wavelengths are allowed and persist for which the distance between the plates is an integral multiple of one-half the wavelength, or for which

$$a = n\left(\frac{\lambda}{2}\right), \tag{2-2}$$

where n = an integer = 1, 2, 3, . . . , ∞.

Such persistent waves are called *standing* or *stationary waves. All other wavelengths disappear through destructive interference.* Thus, if the plates were merely reflecting surfaces, the yellow light would in effect be "filtered" to leave only those wavelengths for which $\lambda = 2a/n$. If we were to now open the shutter

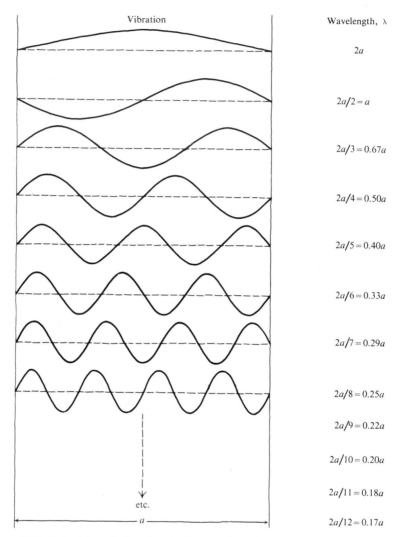

FIG. 2–6   Allowed vibrations in the one-dimensional blackbody cavity.

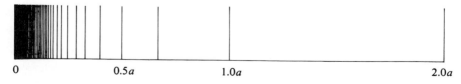

FIG. 2–7   Spectrum of allowed modes of vibration for the one-dimensional blackbody.

after equilibrium had been reached, a distribution of such selected wavelengths would emerge through the pinhole.

Blackbody surfaces, however, have the special characteristic of absorbing *all* wavelengths and emitting *all* wavelengths. For example, radiation of wavelength λ_1 might be absorbed by a blackbody surface and might be immediately re-emitted at the same wavelength λ_1, or it might be stored briefly and then be re-emitted by the same or an adjacent oscillator without loss of phase relationship. In either case the process is equivalent to reflection and would be expected to be a common occurrence, since a particular oscillator in the solid which absorbs radiation of wavelength λ_1 also emits radiation of wavelength λ_1. On the other hand, it is also possible that radiation of wavelength λ_1 might be absorbed, stored briefly as energy in the blackbody surface, and then might be re-emitted at a *different* wavelength λ_2, *which may be higher or lower than* λ_1, since the blackbody is capable of emitting radiation of all wavelengths. This latter process is equivalent to an *exchange* of frequencies. The eventual result of a very large number of "reflections" and wavelength exchanges would be an equilibrium distribution which contains *all* of the allowed or persistent wavelengths.

Thus, if yellow light (or any other light for that matter) were introduced into the hypothetical one-dimensional blackbody cavity, it would be converted through eventual equilibration to *all* of the allowed wavelengths. The cavity would contain all sorts of radiation from γ-rays increasing in wavelength to the maximum allowed wavelength, $2a$ (in which $n = 1$).

Then the introduction of *any* electromagnetic radiation into the cavity or the generation of *any* electromagnetic radiation within the cavity through incandescence should lead to the same eventual classical distribution of wavelengths, which at equilibrium would contain all vibrations for which

$$\lambda = \frac{2a}{n}, \qquad n = 1, 2, 3, \ldots, \infty.$$

All other vibrations would disappear through destructive interference.

For the one-dimensional model, the persistent or "allowed" vibrations are graphically depicted in Fig. 2–6. The allowed vibrations as a function of a wavelength are depicted in Fig. 2–7. Observe that the points representing allowed vibrations merge together as λ approaches zero. That is, the number of allowed vibrations approaches infinity as the wavelength of the vibration approaches zero.

For our purpose it is especially instructive to plot the number of vibrations *per unit interval of wavelength* versus the wavelength. We may use the data from Fig. 2–6 and choose intervals in such a way that each interval contains a convenient number of vibrations (see Table 2–1).

As we continue to choose smaller and smaller intervals, the number of allowed vibrations per unit interval rapidly becomes larger and, in fact,

TABLE 2–1

Distribution of Allowed Vibrations in the One-Dimensional
Blackbody Cavity

Interval, units of a	Number of allowed vibrations	Number of allowed vibrations per unit wavelength interval
1.1 to 2.1	1	$1/1 = 1.0$
0.8 to 1.1	1	$1/0.3 = 3.3$
0.6 to 0.8	1	$1/0.2 = 5.0$
0.45 to 0.60	1	$1/0.15 = 6.7$
0.30 to 0.45	2	$2/0.15 = 13.3$
0.23 to 0.30	2	$2/0.07 = 28.6$
0.19 to 0.23	2	$2/0.04 = 40.0$
0.16 to 0.19	2	$2/0.03 = 66.7$
·	·	·
·	·	·
·	·	·
0	∞	∞

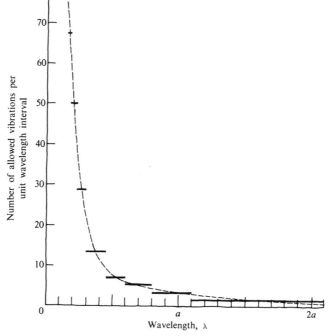

FIG. 2–8 Number of allowed vibrations per unit wavelength interval in the one-dimensional
blackbody cavity.

approaches infinity as the wavelength approaches zero. A plot of the data tabulated above is shown in Fig. 2–8. The steplike data may be represented by a curve (dashed line, Fig. 2–8) if the intervals are very small and if the spacings between adjacent wavelengths are even smaller.

2-4 EQUIPARTITION OF ENERGY AND THE ULTRAVIOLET CATASTROPHE

A very serious paradox arose when Lord Rayleigh (M. J. O. Strutt) and Sir James Jeans attempted to apply the classical *principle of equipartition of energy* to the problem of blackbody radiation. We can understand the complication more clearly if we return to the idealized one-dimensional model. According to the principle of equipartition of energy, there is no reason that any one allowed vibration should be favored over any other. In other words, at equilibrium all allowed vibrations within the cavity have equal *a priori* probabilities. Furthermore, when a given total amount of energy is introduced as radiation into the blackbody cavity, there is no reason to believe that any one vibration should have more energy associated with it than any other. In fact, *the classical principle of equipartition of energy predicts that the total radiation energy should be divided equally among all of the allowed vibrations.* Thus, for example, the vibration for which $\lambda = 2a$ should have the same energy as the vibration for which $\lambda = 2a/13$. The energy associated with a single vibration in the γ-ray region should be exactly equal to the energy associated with a single vibration in the infrared region. It then follows from the classical principle of equipartition of energy that the energy associated with a given interval of wavelength should be directly proportional to the number of allowed vibrations in that interval, since each vibration accepts an equal share of the total energy. On this basis, then, let's relabel Fig. 2–8 in terms of relative energy per unit wavelength interval (intensity) versus wavelength. Such a plot is shown in Fig. 2–9.

The Rayleigh-Jeans treatment for the *three-dimensional* blackbody cavity leads to the same shape of curve as Fig. 2–9 and predicts that for a given total quantity of energy almost all of the energy will eventually reside (after equilibration) in vibrations of infinitesimal wavelengths because the number of normal modes of vibration per unit wavelength interval corresponding to such wavelengths approaches infinity as the wavelength approaches zero. Thus, were classical predictions correct, if we were to open the shutter in our one-dimensional blackbody cavity after a suitable period for equilibration, we would be bombarded with extremely short radiation—*an ultraviolet catastrophe* to say the least—more appropriately a γ-ray catastrophe! The longer wavelengths would have disappeared in favor of infinitesimally short radiation. This, of course, is not experimentally observed. If classical predictions were correct, one would not dare to open an ordinary muffle furnace for fear of being showered by γ-radiation!

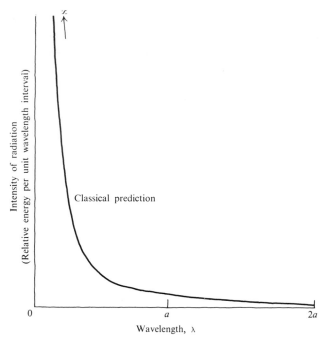

FIG. 2–9   Intensity of radiation as a function of wavelength in the one-dimensional black-body cavity.

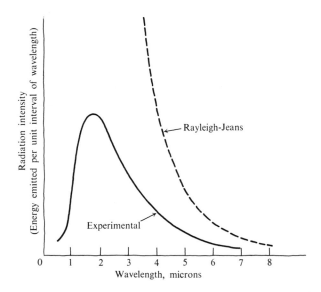

FIG. 2–10   Comparison of the Rayleigh-Jeans classical prediction with the experimental spectrum for blackbody radiation at 1650° K.

In Fig. 2–10, the experimental curve at $1650°K$ is compared with the predicted Rayleigh-Jeans spectrum at the same temperature. In the limit of long wavelengths, the Rayleigh-Jeans curve approaches the experimental curve, but it is "catastrophically" in error as the wavelength becomes shorter.

2–5 MAX PLANCK AND THE QUANTUM HYPOTHESIS

A solution to the Rayleigh-Jeans paradox was proposed by Max Planck* at a meeting of the German Physical Society on December 14, 1900. Planck suggested that the ultraviolet catastrophe could be avoided by assuming that the oscillators (nuclei and electrons) in the blackbody solid are allowed only certain *discrete energy levels*, $\epsilon_1, \epsilon_2, \epsilon_3, \ldots$, and that the energy absorbed in passing from one level to the next adjacent level is directly proportional to the frequency, ν, of the oscillator; that is,

$$(\epsilon_2 - \epsilon_1) \propto \nu. \tag{2–3}$$

Insertion of a proportionality constant yields

$$\epsilon_2 - \epsilon_1 = h\nu. \tag{2–4}$$

The constant h is called *Planck's constant* and was later shown to have the approximate value 6.6×10^{-27} erg-sec. Allowed energy levels for an oscillator of frequency ν are graphically depicted in Fig. 2–11. The discrete energy jumps, $h\nu$, separating the levels are called *quanta*.

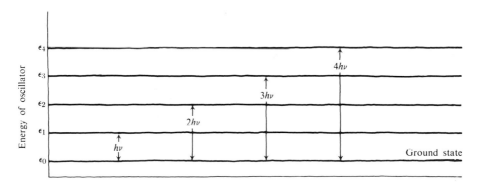

FIG. 2–11 Allowed energy levels for an oscillator having a frequency, ν.

Now it is well known from statistical mechanics that the greater the energy of a given state above a reference level, the lower is the probability that a system has that energy. Specifically, referring to Fig. 2–11, we may show from a consideration of the statistics governing thermal equilibrium that the probability

---

* M. Planck, *Ann. Physik*, **4**, 553 (1901).

p_1 that an oscillator of frequency ν will have an energy ϵ_1 compared to the probability p_0 that it will have energy ϵ_0 is given by

$$\frac{p_1}{p_0} = \exp\left[-\frac{(\epsilon_1 - \epsilon_0)}{kT}\right] = \exp\left(-\frac{h\nu}{kT}\right), \qquad (2\text{–}5)$$

where k, the *Boltzmann constant*, is the gas constant R divided by Avogadro's number, T is the absolute temperature, and the symbol exp (x) represents the quantity e^x. Thus the larger the value of the quantum jump in energy, $h\nu$, between two energy levels of an oscillator, the lower is the probability that the energy jump will occur. In fact, according to Eq. (2–5), as λ approaches zero (as ν approaches infinity), the probability that an energy jump will occur approaches zero. We may infer then that it is much easier for oscillators of low frequency to acquire energy in small jumps having relatively large probabilities than for oscillators of high frequency to acquire energy in large jumps having relatively low probabilities.

The principle of equipartition of energy is based on the assumption that energy may be absorbed continuously by an oscillator and that in an equilibrium system oscillators of all frequencies will each have the same total average energy. On the other hand, Planck's quantum hypothesis requires that oscillators must absorb or emit energy in quanta whose size is directly proportional to the frequency of oscillation, which results, through statistical mechanics, in the lowering of the probability that oscillators of high frequency (low wavelength) will gain energy.

By restricting high-frequency oscillators to large quantum jumps, we seriously limit the total amount of energy they may acquire in an equilibrium distribution.

It then follows, since electromagnetic radiation is emitted at the frequency of the oscillator when the oscillator drops from a higher to a lower energy level, that low-frequency oscillators will emit more radiation than high-frequency oscillators. This is because at equilibrium the number of oscillators in energy states above ground level is larger for low-frequency oscillators than for high-frequency oscillators.

The net effect of Planck's quantum hypothesis on the spectral-distribution curve of Rayleigh and Jeans (dashed line of Fig. 2–10) is that the radiation intensity is lowered at an exponential rate as the wavelength approaches zero, *which is exactly what is required by experiment*! In fact, Planck's correction results in a final spectral distribution which agrees with the observed curve for blackbody radiation within experimental error!

As a final thought, it is important for us to note that Planck's quantum hypothesis was *not* a logical extension of then current physical theory but represented a radical departure necessary to obtain agreement between theory and experiment. Actually, the hypothesis was considered so radical that most scientists preferred to ignore the whole idea of quantized oscillators in the conviction that a more satisfactory classical interpretation would be forth-coming. Their hopes were not to be fulfilled.

2–6 THE PHOTOELECTRIC EFFECT

Hertz first noted in 1887 that electrons are emitted if light is allowed to fall on a clean metal plate in a vacuum. A simple device for illustrating this phenomenon, which is called the *photoelectric effect*, is shown in Fig. 2–12. A metal plate, from which electrons are to be photoemitted, is placed in an evacuated tube and a voltage is applied between it and another more positive electrode so that a current may be detected in the galvanometer, G, whenever electrons are emitted. Not all metals readily show the photoelectric effect. Metals which hold their electrons more loosely, such as the alkali metals, seem to work better. For most metals, ultraviolet radiation is necessary, but potassium and cesium release electrons easily when bombarded by visible light.

The following three observations in particular are impossible to explain in terms of classical theory:

1. Although the *number* of electrons emitted is dependent on the intensity of radiation, the *energy* of the average emitted electron is *not dependent* on the intensity of incident radiation. According to classical physics, if radiation were absorbed continuously, electrons should be able to absorb more energy from a more intense beam before escaping.

2. The maximum energy of escaping electrons is shown to depend on the *frequency* of the incident radiation. In fact, there is observed to be a *threshold frequency*, ν_0, different for each metal, below which photoemission does not occur.

3. There is no apparent lag between the time the light strikes the plate and the time electrons are emitted. Classical physics requires a time during which the evenly distributed, continuous radiant energy accumulates to a level sufficient for the electron to overcome surface forces in the metal and escape. For

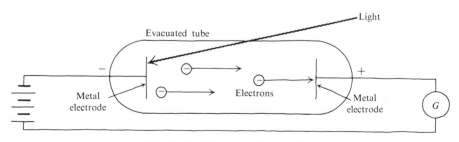

FIG. 2–12   The photoelectric effect.

example, even for a light level as low as 10^{-10} watt/meter² (W/m²), the intensity at a point 200 miles away from a 100-W light source, delay times no longer than 10^{-9} sec have been observed. A conservative classical estimate (See Prob. 2–8), however, shows that no photoemission may be expected until a lapse of several hundred years!

In explaining the observed data, Einstein proposed in 1905 that *electromagnetic radiation itself is quantized*, just as Planck had previously postulated that energy levels in a blackbody oscillator are quantized. Einstein visualized the elementary act of photoemission as resulting from the collision of one corpuscle or quantum of electromagnetic radiation (later to be called a *photon*) with one electron in the surface of the metal. He further postulated that *each photon has an energy hv*. He proposed in effect that Planck's quantized energy levels in the blackbody oscillator are not really fundamental properties of the oscillator alone but result from *quantized energy levels in electromagnetic radiation itself.*

There is a minimum energy, ϕ, called the *work function*, which must be expended in order to remove an electron from the surface of a metal. The value of ϕ varies from metal to metal and depends on the tenacity with which a given metal holds its electrons. The work function for alkali metals is relatively low compared to that for transition metals. Assume that one photon of energy hv is absorbed by one electron in the surface of the metal:

1. If $hv < \phi$, the electron cannot be emitted, since it does not acquire sufficient energy to escape.

2. If $hv > \phi$, the electron may escape and the excess energy appears as kinetic energy in the emitted electron. That is,

$$hv = \phi + \text{kinetic energy.} \qquad (2\text{–}6)$$

Since the kinetic energy of an electron of mass m_e and velocity v is given as $\frac{1}{2}m_ev^2$,

$$hv = \phi + \tfrac{1}{2}m_ev^2, \qquad (2\text{–}7)$$

or

$$\tfrac{1}{2}m_ev^2 = hv - \phi. \qquad (2\text{–}8)$$

Equation (2–8) predicts that the kinetic energy of an emitted electron should depend only on v, the frequency of the incident radiation. An increase in *intensity* of radiation only increases the *number* of electrons emitted, not the kinetic energy. Furthermore, if v is too low, that is, if $hv < \phi$, the electron cannot escape. Apparently, then, the threshold frequency v_0 is given by

$$hv_0 = \phi$$

or

$$v_0 = \phi/h \qquad (2\text{–}9)$$

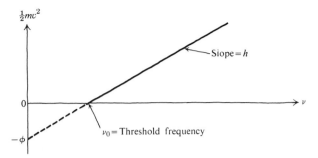

FIG. 2–13   Kinetic energy of a photoemitted electron as a function of the frequency of the incident radiation.

A helpful graphical picture is provided if we note that Eq. (2–8) is the equation of a straight line where $\frac{1}{2}m_e v^2$ is the ordinate and v is the abscissa. Then h is the slope and $-\phi$ is the ordinate intercept. The linear plot is shown in Fig. 2–13.

Equation (2–8) also predicts that all metals should show the same slope on a plot of kinetic energy versus frequency of incident radiation. It was not until 1916 that Millikan experimentally verified Einstein's prediction by making careful measurements conducted in an attempt to *disprove* Einstein's photo-electric theory. The classical concepts of Maxwellian waves were so deeply ingrained in physicists that even Planck remained reluctant to accept quanti-zation of radiation.

After all, what now is electromagnetic radiation? Is it wavelike or is it corpuscular? It seems impossible that it can be both simultaneously. Yet, we must consider light to be wavelike if we are to explain diffraction and inter-ference and we must consider light to be corpuscular if we are to explain the photoelectric effect. The general concept of *wave-particle dualism* made its debut with Einstein's photoelectric hypothesis in 1905. There seemed to be no alternative choice. For a second time in five years scientists were forced to resort to quantization of energy to explain experimental observations.

The relationship of the two images (wave and particle) in general is known with great clarity and surprising detail. No one doubts its correctness and general validity. With regard to the combination into a single, concrete, obvious image, opinions are so divided that many consider it entirely impossible. . . . Do not count on forming a uniform, concrete picture; and do not blame my lack of skill in representation nor your slowness in understanding for your failure—for up to now no one has been successful.

—ERWIN SCHRÖDINGER, in a lecture on September 4, 1952*

* W. Heisenberg, M. Born, E. Schrödinger, and P. Auger, *On Modern Physics*, Clark-son N. Potter, Inc., New York, 1960, p. 48.

They could but make the best of it, and went around with woebegone faces sadly complaining that on Mondays, Wednesdays, and Fridays they must look on light as a wave: on Tuesdays, Thursdays and Saturdays as a particle. On Sundays they simply prayed.

 —BANESH HOFFMANN, *The Strange Story of the Quantum.**

PROBLEMS

2–1 According to Planck's theory, the wavelength at which the emission of energy by a blackbody is a maximum is given as $\lambda_{max} = ch/4.97kT$. What is the approximate surface temperature of the sun, for which λ_{max} is 4650 Å?

2–2 a) Assume that for a given oscillator the frequency is $v = kT/h$. Using equations similar to Eq. (2–5), calculate the probability, in terms of p_0, for each of the first four quantum states above the ground state, in which the energy is ϵ_0.

 b) Repeat the calculations of part (a), assuming that the oscillator frequency is ten times as great, that is $10kT/h$.

2–3 Calculate the frequency and the energy per photon (the *quantum energy*) for electromagnetic radiation whose wavelength is:

 a) 10^6 cm (long radio wave).

 b) 4×10^{-5} cm (4000 Å; visible radiation).

 c) 10^{-10} cm (γ-ray).

2–4 The threshold wavelength for photoemission of electrons from lithium is about 520 mμ. What is the velocity of an electron emitted as the result of absorption of a photon whose wavelength is 300 mμ?

2–5 Calculate the maximum kinetic energy of an electron which is emitted from a clean potassium surface which is illuminated with light having a wavelength of 3000 Å. For potassium, $\phi = 2.0$ electron volts. (An electron volt, abbreviated eV, is the amount of energy transferred when a single electronic charge is moved through a potential-field change of 1 volt.)

2–6 For a monochromatic beam of light which has an intensity of 1 W/cm^2 (10^7 erg-s^{-1}-cm^{-2}) calculate the average number of photons per cubic centimeter in

 a) 10 kHz radio waves,

 b) 10 MeV (million electron volt) gamma rays.

2–7 The work function for sodium is about 2 eV. Assuming that classical theory is correct (that is, assuming that light is a wave and that its energy may be absorbed continuously) and assuming that a single sodium atom may collect its energy from a target area of the surface equivalent to ten atomic diameters in radius (10^{-7} cm), how long would it take for a single electron to absorb enough energy to be photoemitted if it were illuminated by a 25 W (joule/second) light source placed 1000 cm away? How does your answer compare with observed time lags?

---

* From *The Strange Story of the Quantum*, Dover Publications, Inc., New York, 1959. Reprinted through permission of the publisher.

2–8 According to the text, the observed delay time for photoemission of electrons from a metal target located 200 miles (3.2×10^7 cm) from a 100 watt light source is about 10^{-9} seconds. Assuming that the metal has a work function of 2 eV and an effective absorption radius of 10^{-7} cm, calculate the delay time in years which would be expected classically.

2–9 A stainless steel table knife having a work function of 4.7 eV and an effective area of 30.0 cm^2 is place on a table 200 cm from a 100 watt lamp. Assuming the lamp to be a point source and all parts of the knife to be equidistant from the source, calculate classically the maximum number of electrons released from the metal each second. Why isn't such an electron release actually observed?

LINE SPECTRA,
THE BOHR MODEL,
3   AND DE BROGLIE WAVES

The model of the nuclear atom prevalent in 1912 presented many problems to the classical physicist. Alpha-particle scattering experiments conducted in 1911 led Rutherford* to propose that the atom was composed mostly of empty space with electrons revolving at relatively great distances around a small, very dense, positive nucleus. In many respects the Rutherford model was similar to a miniature solar system.

Classical electromagnetic theory requires that an accelerated charged particle (one which is changing either speed or direction or both) should emit electromagnetic radiation. The planetary electron of the Rutherford model, in traveling in a circular orbit, undergoes constant acceleration toward the nucleus and thus should continuously emit radiation. But should the electron emit radiation, its energy would decrease, its velocity would drop and it would begin to spiral toward the nucleus, constantly losing energy as its orbital radius steadily decreased. The ultimate result would be *complete collapse of the electrons* into the nucleus! Classical calculations show that the entire collapse should occur in one hundred-millionth of a second (10^{-8} s)! Of course, such a collapse does not occur. Neither do we observe the continuous emission of radiation. Rather, excited atoms emit radiation only at selected discrete wavelengths in what is called a *line spectrum*.

3–1 LINE SPECTRA OF ATOMS: SOME EXPERIMENTAL OBSERVATIONS

It is well known that salts of certain metals, when heated to high temperatures in a flame or an electric arc, impart characteristic colors to the emitted light. Thus, sodium salts impart a yellow color to a flame, potassium salts impart a violet color, and barium salts show a light green color. In addition, when an

---

* E. Rutherford, *Phil. Mag.* **21**, 669 (1911).

electrical discharge is passed through a gas at a sufficiently low pressure the gas emits a characteristically colored glow discharge. The orange-red discharge of neon, for example, is commonly seen in commercial neon signs. Mercury vapor emits a greenish-blue glow discharge, and hydrogen emits a violet-blue discharge.

We may more completely characterize such flame colors or glow discharges by dispersing the emitted light through a prism so that the various wavelengths are spread out into a spectral pattern. An appropriate experimental arrangement is shown in Fig. 3–1. The slits serve to collimate the beam into a narrow band of light which enters the prism. Were visible white light used as the source, the narrow band would be dispersed into a continuous spectrum from red light at the long-wavelength end to violet light at the short-wavelength end. When, however, the flame colored by a metal salt or the glow discharge of a gas is used as the source, the spectrum consists of a series of characteristic sharp lines, which reproducibly appear at the same wavelengths for a given atom. Foucault, for example, accurately identified in 1848 the principal sharp line in the emission spectrum of sodium which is responsible for its yellow color in a flame. Further work by Kirchhoff and Bunsen in 1861 identified the principal sodium line to be actually a pair of closely spaced lines (a doublet) at a wavelength of about 5890 Å.

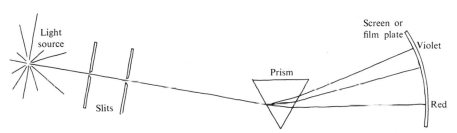

FIG. 3–1 Dispersion of light by a prism.

Because of the association characteristic of discrete line spectra with certain elements and because of the sharpness of the lines, it soon became convenient to use emission spectroscopy as a tool for the analysis of complex mixtures of elements. For this pragmatic reason alone, an enormous amount of empirical data on line spectra was gathered quickly for most elements.

3–2 THE LINE SPECTRUM OF HYDROGEN

Balmer in 1885 carefully examined the series of about 35 resolvable lines appearing in the visible and near-ultraviolet regions of the emission spectrum of atomic hydrogen. The spectral distribution of these lines is shown in Fig. 3–2. Observe that the lines are spaced more and more closely together as their wavelengths decrease until a continuum finally appears at 3646.6 Å.

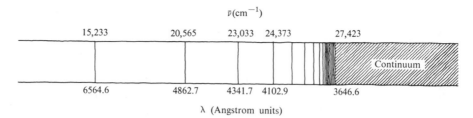

FIG. 3-2   The Balmer series for hydrogen. Although spectroscopic measurements are often made in air, the values of λ and $\bar{\nu}$ shown above have been reduced to *vacuo*, that is, to values which would be obtained in a vacuum*.

Balmer found that the wavelengths in vacuo of the spectral lines may be related mathematically in a purely empirical way which may be expressed as

$$\bar{\nu} \equiv \frac{1}{\lambda} = R_{\mathrm{H}}\left(\frac{1}{2^2} - \frac{1}{n^2}\right), \tag{3-1}$$

where $\bar{\nu}$ is called the *wave number* and is defined as $1/\lambda$, R_{H} is a constant called the Rydberg constant, and n may have the integral values 3, 4, 5, 6, . . . , ∞. The Rydberg constant is determined experimentally to be $109{,}677.586 \pm 0.090$ cm^{-1}. Since wavelengths may be measured with great precision, it is one of the most accurately known physical constants.

In 1906 Lyman discovered another similar spectral series for hydrogen in the ultraviolet region. Paschen detected a third series in the infrared region in 1909. It was soon found that all of the spectral emission lines for hydrogen can be expressed by a more generalized version of Eq. (3-1):

$$\bar{\nu} = R_{\mathrm{H}}\left(\frac{1}{n_f^2} - \frac{1}{n_i^2}\right), \tag{3-2}$$

where n_f is allowed the values 1, 2, 3, . . . , ∞, and n_i is allowed the values $(n_f + 1)$, $(n_f + 2)$, . . . , ∞.

Using Eq. (3-2), one can generate mathematically any of the observed series by appropriate choice of n_f, as shown in the accompanying table.

Series	n_f	n_i	Region
Lyman	1	2, 3, 4, . . .	Ultraviolet
Balmer	2	3, 4, 5, . . .	Visible and near ultraviolet
Paschen	3	4, 5, 6, . . .	Infrared
Brackett	4	5, 6, 7, . . .	Infrared
Pfund	5	6, 7, 8, . . .	Infrared

* Tables for correcting wavelengths and wavenumbers to *vacuo* are given in *Handbook of Chemistry and Physics*, 45th ed., Chemical Rubber Co., 1964, p. E–106.

It is important to note that the relationship given by Eq. (3–2) was deduced entirely from empirical data. Until 1913 there was no theoretical explanation for the origin of discrete spectral lines.

3-3 BOHR'S MODEL FOR THE HYDROGEN ATOM

In 1913, Niels Bohr* postulated a model for the hydrogen atom with which he attempted to solve several of the perplexing conflicts between classical theory and experiment. It is evident, for example, that atoms do not collapse as would be predicted by the application of classical mechanics to the Rutherford model. Is it possible that the energy of the electron in its orbit is quantized in the same way that the energy of Planck's oscillator and the energy of Einstein's photon are quantized? Furthermore, an adequate atomic model must explain a mechanism for the origin of line spectra and the precise reproducibility of the characteristic wavelengths for a given element.

The postulates for Bohr's model of the hydrogen atom were:

1. The single electron in the hydrogen atom moves about a stationary nucleus in circular orbits in which the electrostatic attraction between the negatively charged electron and the positively charged nucleus is exactly balanced by the centrifugal force due to the velocity of the electron in its orbit.

2. While in a given orbit, the energy of the electron is constant. Furthermore, the electron remains in a given orbit unless a quantum of energy of exactly the right amount is absorbed or emitted. (This postulate, often called the *stationary-state assumption*, is contrary to the ordinary laws of electrodynamics, since the accelerating electron should constantly lose energy. Bohr neatly handled this problem by assuming that it did not exist!)

3. Only certain orbits are allowed. Such orbits are restricted to those for which the magnitude of angular momentum L of the electron is an integral multiple of $h/2\pi$ or

$$L = m_e v r = nh/2\pi, \tag{3-3}$$

where m_e is the mass of the electron, v is its linear velocity, r is the radius of the orbit, and n may assume the values $1, 2, 3, \ldots, \infty$. (The best argument for this bold postulate, known as the *quantum restriction postulate*, is provided by the final result to which it leads. This precise postulate is required if theory is to account for experimental results.)

4. When an electron makes a transition from an initial stationary energy state E_i to a lower final stationary energy state E_f the energy difference is *emitted* as a single photon such that

$$E_i - E_f = h\nu, \tag{3-4}$$

where ν is the frequency of the corresponding spectral line. Conversely, if energy is *absorbed* by an electron, the same relationship must apply. That is, the

---

* N. Bohr, *Phil. Mag.*, **26**, 1 (1913).

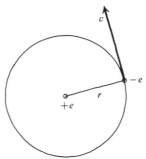

FIG. 3–3   Orbital motion of the electron about the nucleus.

wavelength of the absorbed photon is given by Eq. (3–4). (This postulate is in accord with Einstein's quantum hypothesis for electromagnetic radiation.)

5.   One must renounce all attempts to physically describe the electron when it is between stationary states!

Consider a single electron of charge $- e$ revolving with a linear velocity v in a stable circular orbit or radius r about the single proton (charge $= + e$), which is the nucleus of hydrogen. The relationship is illustrated in Fig. 3–3. According to the first postulate, the force of electrostatic attraction F_A between the electron and the nucleus must be equal to the centrifugal force F_C which tends to separate them:

$$F_A \text{ (force of attraction)} = F_C \text{ (centrifugal force).} \qquad (3\text{–}5)$$

Coulomb's law states that the magnitude F of the force acting between two charges q_1 and q_2 at a separation r is given by

$$F \propto \frac{q_1 q_2}{r^2}.$$

Usually, absolute values for q_1 and q_2 are used, in which case like charges result in a force of repulsion while unlike charges result in a force of attraction. Ordinarily a proportionality is converted to an equation by inserting a constant on one side. In the cgs system we assign unit value to the constant by defining the electrostatic unit of charge (esu or statcoulomb) as that quantity of electricity which, when placed in a vacuum 1 cm from an equal charge, will be acted on by a force of 1 dyne. Thus, in a vacuum, by definition,*

$$F = \frac{q_1 q_2}{r^2}, \qquad (3\text{–}6)$$

---

* In the rationalized mks system, Coulomb's law is written as $F = (1/4\pi\epsilon_0)q_1 q_2/r^2$, where F is expressed in newtons, r in meters, q_1 and q_2 in coulombs (ampere-seconds) and ϵ_0 is called the permittivity constant, which in a vacuum is equal to 8.85419×10^{-12} coulomb^2-newton^{-1}-meter^{-2}.

where F is expressed in dynes (g-cm-sec^{-2}), r is expressed in centimeters, and q is expressed in esu or statcoulombs (g$^{1/2}$-cm$^{3/2}$-sec^{-1}). It is important to note the correct charge units to use in Eq. (3–6). The charge on the electron, e, is 4.803242×10^{-10} esu. Of course, for hydrogen this is also equal to the charge on the nucleus.

Noting that centrifugal force F_c is given by $m_e v^2/r$, where m_e is the mass of the electron, we may write Eq. (3–5) as

$$\frac{q_{\text{nucleus}} \, q_{\text{electron}}}{r^2} = \frac{m_e v^2}{r},$$

or, using an absolute value of e for the charge on both the electron and the nucleus and noting that the force is an attractive force, as

$$e^2/r = m_e v^2. \tag{3–7}$$

The total energy E of the electron in its orbit is given as the sum of its kinetic energy T and its potential energy V:

$$E = T + V. \tag{3–8}$$

The kinetic energy T is given as $\frac{1}{2} m_e v^2$. The potential energy, however, is a relative value which depends on the state we define as having zero potential energy. *It is convenient to arbitrarily define the system to have a potential energy of zero when the electron is infinitely removed from the nucleus.* As the electron is brought from a distance ∞ to a distance r, the system could be made to do work on the surroundings due to the attractive force between the nucleus and the electron. Thus, the potential energy decreases as r decreases. The total amount of work the electron does in moving from ∞ to a distance r is thus equal to the potential energy V at a distance r and is given by

$$V = \int_{\infty}^{r} F_A \, dr = \int_{\infty}^{r} \frac{e^2}{r^2} \, dr = e^2 \int_{\infty}^{r} \frac{dr}{r^2},$$

or

$$V = -e^2/r.$$

Substitution of the values for T and V into Eq. (3–8) yields

$$E = m_e v^2/2 - e^2/r. \tag{3–9}$$

We may now substitute Eq. (3–7) into Eq. (3–9) to obtain

$$E = -m_e v^2/2 = -e^2/2r. \tag{3–10}$$

Substitution of the quantum restriction equation,

$$m_e v r = nh/2\pi, \tag{3–3'}$$

into Eq. (3–10) yields

$$v = 2\pi e^2/nh, \tag{3–11}$$

where $n = 1, 2, 3, 4, \ldots, \infty$.

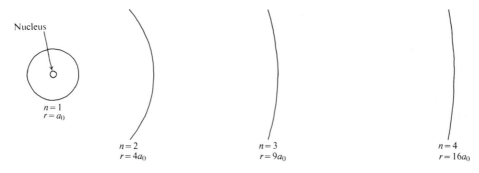

FIG. 3-4 Allowed radii for the first four Bohr orbits of hydrogen (a_0 = the radius of the first Bohr orbit).

The integer n is called the *principal quantum number*. Substitution of various values of n into Eq. (3-11) generates the discrete linear velocities of the electron in the allowed Bohr orbits. Allowed energies in each of the stationary states or orbits may be generated by inserting Eq. (3-11) into Eq. 3-10 to yield

$$E = -2\pi^2 m_e e^4/n^2 h^2, \tag{3-12}$$

where $n = 1, 2, 3, \ldots, \infty$.

It is also convenient to combine Eq. (3-12) with Eq. (3-10) to eliminate E:

$$e^2/2r = 2\pi^2 m_e e^4/n^2 h^2,$$

from which the allowed radii for the Bohr orbits are given as

$$r = n^2 h^2/4\pi^2 m_e e^2, \tag{3-13}$$

where $n = 1, 2, 3, \ldots, \infty$.

Equations (3-12) and (3-13) can be further simplified by grouping constants. Thus

$$E = -R(1/n^2), \tag{3-14}$$

where

$$R = 2\pi^2 m_e e^4/h^2, \tag{3-15}$$

and

$$r = a_0 n^2, \tag{3-16}$$

where

$$a_0 = h^2/4\pi^2 m_e e^2. \tag{3-17}$$

The allowed orbital radii for the Bohr model obtained by inserting successive integers into Eq. (3-16), are

$$a_0, 4a_0, 9a_0, 16a_0, 25a_0, \ldots$$

Relative orbit sizes are shown in Fig. 3-4. When $n = 1$ in Eq. (3–16), a_0 is shown to be the radius of the first Bohr orbit for hydrogen. The calculated value of a_0 from Eq. (3–17) is 0.5291770 Å, which is in good agreement with the estimated size of the hydrogen atom in the closely packed solid.

The real triumph for the Bohr model, however, is to be found in the precision with which it predicts the spectral lines for hydrogen. Allowed energy levels are obtained by substituting values for the principal quantum number n into Eq. (3–14). Thus:

$$E = -R, \ -R/4, \ -R/9, \ -R/16, \ -R/25, \ \ldots, \ 0.$$

Note that as n gets larger the energy *algebraically* increases. Energies of the electron in allowed orbits are graphically depicted in Fig. 3–5. According to the fourth postulate of Bohr's theory, the energy *emitted* in the transition of a single electron from an initial stationary state in which the energy is E_i to a final

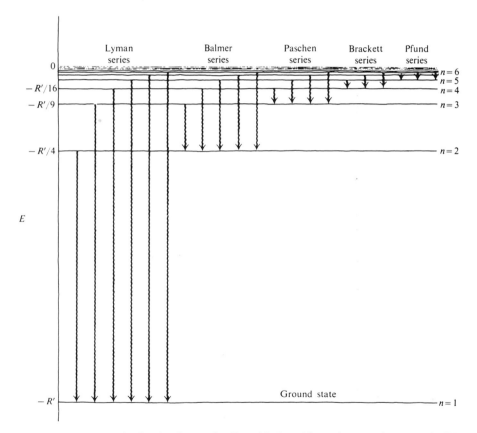

FIG. 3–5 Total energies for the electron in allowed Bohr orbits and energy drops responsible for observed spectral lines ($R' = 2\pi^2 \mu e^4 / h^2$).

stationary state in which the energy is E_f is given by

$$\epsilon = h\nu = E_i - E_f, \tag{3-18}$$

where ν is the frequency of the corresponding line observed in the spectrum. Substitution of Eq. (3-14) into Eq. (3-18) yields

$$h\nu = -R\left(\frac{1}{n_i^2} - \frac{1}{n_f^2}\right) = R\left(\frac{1}{n_f^2} - \frac{1}{n_i^2}\right),$$

where n_i and n_f are the initial and final quantum numbers respectively. But $\nu = c/\lambda = c\bar{\nu}$, so that

$$\bar{\nu} = \left(\frac{R}{ch}\right)\left(\frac{1}{n_f^2} - \frac{1}{n_i^2}\right), \tag{3-19}$$

or

$$\bar{\nu} = \left(\frac{2\pi^2 m_e e^4}{ch^3}\right)\left(\frac{1}{n_f^2} - \frac{1}{n_i^2}\right). \tag{3-20}$$

The form of Eq. (3-20) is identical to that of Eq. (3-2), the experimentally deduced Rydberg equation, which was

$$\bar{\nu} = R_{\mathrm{H}}\left(\frac{1}{n_f^2} - \frac{1}{n_i^2}\right). \tag{3-2'}$$

A major test of the validity of the Bohr model may thus be made by comparing the calculated value of $2\pi^2 m_e e^4/ch^3$, which is given in Eq. (3-20), with the experimental value for R_{H}. However, because the Rydberg constant may be measured experimentally with such great precision, it is important that we first make a minor correction in the mathematical treatment in order to account for the finite mass of the nucleus.

3-4 CORRECTION FOR THE FINITE MASS OF THE NUCLEUS

Up to now we have assumed that the single electron in the hydrogen atom revolves about a *stationary* nucleus. Such an assumption introduces only a small error because the mass of the electron is such a small fraction (about 1/2000) of that of the nucleus. However, the assumption of a stationary nucleus is entirely valid only if the mass of the nucleus is *infinite*. If the nucleus has a finite mass, the electron *and* the nucleus must *each* revolve about a common center of mass on a line connecting the two particles, as shown in Fig. 3-6. The radius of revolution, x, for the nucleus is much smaller than the radius of revolution, $r - x$, for the electron, and approaches zero as the mass of the nucleus approaches infinity.

We have indicated that the magnitude of orbital angular momentum of a single particle is given as mvr, or $mr^2\omega$, where $\omega = v/r$ is the angular velocity of the particle in radians per second. Thus the total orbital angular momentum L for the hydrogen atom is the sum of the orbital angular momenta for

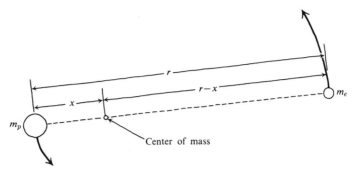

FIG. 3-6 The motion of the nucleus (a proton) of the hydrogen atom and the single electron about a common center of mass. The mass of the electron is m_e and that of the protron is m_p.

the electron of mass m_e and for the proton (nuclcus) of mass m_p:

$$L = m_e(r - x)^2\omega + m_p x^2\omega, \tag{3-21}$$

where the angular velocity ω is the *same* for the electron as for the proton. Since the center of mass is defined by the relationship

$$m_p x = m_e(r - x),$$

we may write

$$x = \frac{m_e r}{m_p + m_e} \tag{3-22}$$

and

$$r - x = \frac{m_p x}{m_e} = \frac{m_p r}{m_p + m_e}. \tag{3-23}$$

Substitution of Eqs. (3-22) and (3-23) into Eq. (3-21) yields

$$L = \frac{m_e m_p^2 r^2 \omega}{(m_p + m_e)^2} + \frac{m_p m_e^2 r^2 \omega}{(m_p + m_e)^2}$$

$$= \frac{m_e m_p (m_p + m_e) r^2 \omega}{(m_p + m_e)^2} = \left(\frac{m_e m_p}{m_p + m_e}\right) r^2 \omega, \tag{3-24}$$

or,

$$L = \mu r^2 \omega = \mu r(r\omega), \tag{3-25}$$

where $\mu = m_e m_p/(m_e + m_p)$ is called the *reduced mass*. In a similar way, it can be shown that the kinetic-energy and centrifugal-force expressions are

$$T = \mu(r\omega)^2/2 \tag{3-26}$$

and

$$F_c = \mu(r\omega)^2/r. \tag{3-27}$$

The above three expressions suggest that we can correctly calculate L, T, and F_c if we adopt a hypothetical, but *mathematically equivalent*, model of the hydrogen atom in which we imagine an electron of *assigned mass* μ to revolve at a distance r about a stationary nucleus and to have a linear velocity $v = r\omega$. Then

$$L = \mu r v, \quad T = \mu v^2/2, \quad \text{and} \quad F_c = \mu v^2/r,$$

and Bohr's quantum postulate must be revised to $\mu v r = nh/2\pi$. The only difference in the above expressions as compared to the original expressions in Sec. 3-3 is that we have substituted the reduced mass μ for the electron mass m_e, which in turn will result in the substitution of μ for m_e in the expressions for E, r, a_0, and $\bar{\nu}$. The corrected equations are then

$$E = \frac{-2\pi^2\mu e^4}{n^2 h^2} = -R'\left(\frac{1}{n^2}\right), \tag{3-28}$$

where

$$R' = \frac{2\pi^2\mu e^4}{h^2},$$

and

$$r = \frac{n^2 h^2}{4\pi^2\mu e^2} = a_0 n^2, \tag{3-29}$$

$$a_0 = \frac{h^2}{4\pi^2\mu e^2}, \tag{3-30}$$

$$\bar{\nu} = \left(\frac{2\pi^2\mu e^4}{ch^3}\right)\left(\frac{1}{n_f^2} - \frac{1}{n_i^2}\right). \tag{3-31}$$

3-5 THE BOHR THEORY COMPARED WITH EXPERIMENT

Comparison of Eq. (3-31) with Eq. (3-2) leads to the theoretical expression for the Rydberg constant for hydrogen:

$$R_{\mathrm{H}} = \frac{2\pi^2\mu e^4}{ch^3}. \tag{3-32}$$

Let's now calculate R_{H} from the values of the constants given in Appendix A. First, we'll calculate the reduced mass μ:

$$\mu = \frac{m_e m_p}{m_p + m_e} = \frac{(9.109534 \times 10^{-28}\text{ g})(1.6726485 \times 10^{-24}\text{ g})}{(9.109534 \times 10^{-28}\text{ g}) + (1.6726485 \times 10^{-24}\text{ g})}$$

$$= 9.104575 \times 10^{-28}\text{ g}.$$

The calculated value of R_{H} is then

$$R_{\mathrm{H}} = \frac{2\pi^2\mu e^4}{ch^3} = \frac{2\pi^2(9.104575 \times 10^{-28}\text{ g})(4.8032424 \times 10^{-10}\text{ cm}^{3/2}\text{ g}^{1/2}\text{ sec}^{-1})^4}{(2.99792458 \times 10^{10}\text{ cm-sec}^{-1})(6.626176 \times 10^{-27}\text{ g-cm}^2\text{ sec}^{-1})^3}$$

$$R_{\mathrm{H}} = 109{,}677.6\text{ cm}^{-1} \quad \text{(calculated)}.$$

Compare this value with the most recent experimentally observed value

$$R_H = 109{,}677.586 \pm 0.090 \text{ cm}^{-1} \quad \text{(observed)}.$$

The remarkable agreement between the theoretical and experimental values of the Rydberg constant for hydrogen represented a major victory for the Bohr model.

It is interesting to note that, had we not corrected for the finite mass of the hydrogen nucleus, that is, had we used m_e instead of μ in calculating the Rydberg constant, the resultant value would have been

$$R_\infty = \frac{2\pi^2 m_e e^4}{ch^3} = 109{,}737.3 \text{ cm}^{-1}.$$

The subscript ∞ denotes the assumption of infinite nuclear mass.

Through substitution of appropriate values for n_f into Eq. (3–31) we may calculate wave numbers for the lines in all five of the major spectral series for hydrogen. That is, when $n_f = 1$, the Lyman series is generated by using the allowed integers for n_i; when $n_f = 2$, the Balmer series is generated, and so forth. Energy drops responsible for the lines in the various spectral series are schematically shown in Fig. 3–5. We'll see later that the *continuum* at the limit of a given spectral series, as shown for example at the end of the Balmer series in Fig. 3–2, can be explained in terms of absorption and emission of energy by a *free* or *ionized* electron. Wave mechanics implies that free electrons, like free waves, are not restricted to discrete energy levels but may change energy in a continuous fashion, thus giving rise to a continuous variation in wavelength for emitted photons, which results in a continuum in the spectrum.

The Bohr theory was certainly a great advance in the development of the atomic model and was soon followed by modifications considered necessary to explain fine-line structure observed under varying conditions. Sommerfeld,* in particular, extended the Bohr model to accommodate elliptical orbits having quantized degrees of eccentricity, which required the introduction of a second quantum number k having the allowed integral values $1, 2, \ldots, n$. The shapes of the corresponding orbits are given by

$$\frac{n}{k} = \frac{a}{b},$$

where a is the length of the major axis and b is the length of the minor axis. Thus, when k is equal to n the Bohr-Sommerfeld orbit is a circle. Sommerfeld also showed that if one considers relativity, the energy $E_{n,k}$ for an electron in a given stationary elliptical orbit depends not only on n but also to a very slight degree on k, that is,

$$E_{n,k} = \frac{-2\pi^2 \mu e^4}{n^2 h^2}\left[1 + \frac{\alpha^2}{n}\left(\frac{1}{k} - \frac{3}{4n}\right)\right],$$

where $\alpha = 2\pi e^2/hc = 7.29735 \times 10^{-3}$ is called the *Sommerfeld fine structure*

---

* A. Sommerfeld, *Ann. Physik*, **51**, 1 (1916).

constant. The use of Sommerfeld corrected energies rather than Bohr energies in the calculation of spectral wavelengths results in very slight shifts from the values given by Eq. (3–2). In particular, the use of the n different allowed k values associated with a given single value of the principal quantum number leads to a group of very closely spaced lines in the spectrum at those positions for which the uncorrected Bohr theory predicts single lines. In spite of several important modifications of the Bohr model which were to follow in the years between 1913 and 1925, even the revised model was not able successfully to interpret the spectra of atoms containing more than one electron. Nor was it able to account for relative intensities of different spectral lines, or to explain the binding of atoms in molecules. Clearly, some fundamental changes in the model of the atom were in order.

3–6 SOME BRIEF COMMENTS ON RELATIVITY

Although we'll not examine the detailed arguments of the theory of relativity as postulated by Einstein in 1905, it will help us to be familiar with a few of the more important results which relate directly to the development of the concepts of quantum mechanics. Specifically, according to the theory of relativity, the *effective* or *relativistic mass*, m, of a particle having a velocity, v, is given as

$$m = \frac{m_0}{\sqrt{1 - (v^2/c^2)}}, \tag{3–33}$$

where c, the speed of light, represents the *maximum possible* speed of the particle, and m_0, the *rest mass*, is the mass of the particle when $v = 0$. In addition, according to relativity theory, the *total relativistic energy* E of the particle is given as

$$E = mc^2. \tag{3–34}$$

Substitution of Eq. (3–33) into Eq. (3–34) yields

$$E = mc^2 = m_0 c^2 \left(\frac{1}{\sqrt{1 - (v^2/c^2)}} \right), \tag{3–35}$$

and expansion in a Maclaurin series of the term in brackets leads to

$$E = mc^2 = m_0 c^2 \left(1 + \frac{v^2}{2c^2} + \frac{3v^4}{8c^4} + \cdots \right).$$

Since v is usually much smaller than c, we may ignore, to a very close approximation, all of the terms in the series beyond the second, so that

$$E = mc^2 \cong m_0 c^2 + \tfrac{1}{2} m_0 v^2. \tag{3–36}$$

But the classical nonrelativistic kinetic energy T of the particle is given as $T = \tfrac{1}{2} m_0 v^2$, so that Eq. (3–36) may be written as

$$E = mc^2 \cong m_0 c^2 + T. \tag{3–37}$$

If the particle also has a potential energy V, it is possible to extend the argument to show that

$$E = mc^2 \cong m_0c^2 + T + V. \tag{3–38}$$

That is, to a close approximation, the total relativistic energy of a particle is the sum of its *rest-mass energy* m_0c^2, *which is a constant*, and its classical total energy $T + V$. It is important to emphasize that the total relativistic energy E and the total classical or *nonrelativistic* energy $T + V$ differ by a constant. In those cases in which we are primarily interested in *differences* in energies between two separate energy states of a particle, we may ignore the *rest-mass energy* m_0c^2, since it always cancels out in the subtraction of total-energy terms. Later on, for example, in our initial investigation of energy levels for the electron in the hydrogen atom, we'll ignore the rest-mass energy when we write the total energy of the atom in its various energy states. We refer to a treatment as a *non-relativistic treatment* when we ignore the rest-mass energy and use nonrelativistic rather than relativistic kinetic-energy terms.

It is sometimes also convenient to express the relativistic energy in terms of the *relativistic momentum p*, which is defined as

$$p = mv.$$

Elimination of the relativistic mass m between the above equation and Eq. (3–34) yields

$$v = \frac{pc^2}{E},$$

and substitution of the above value for v into Eq. (3–35), followed by squaring and rearrangement, gives

$$E^2 = c^2p^2 + m_0^2c^4. \tag{3–39}$$

We'll find Eq. (3–39) to be particularly useful in our later development of the concept of wave packets in Chapter 7.

3–7 THE PILOT WAVES OF LOUIS DE BROGLIE

In the mathematical treatments of blackbody radiation, the photoelectric effect, and the Bohr model for the hydrogen atom, we have observed that integers appear as necessary parts of the solutions. We have referred to the restriction of solutions to values involving integers as quantum restrictions and have observed that only certain solutions are allowed.

There is another well-known branch of physics in which integers appear naturally in solutions and in which only certain discrete values of physical properties are allowed, that is, classical wave theory. In our discussion of blackbody radiation we have stated that persistent waves in a one-dimensional enclosure are restricted to those for which the distance between the boundary walls is an integral multiple of one-half the wavelength. The audio frequencies emitted by violin strings are not continuous but are limited to certain frequencies referred to as fundamentals and overtones. *Is it possible that the quantum restrictions we have observed are caused by a fundamental wave nature associated with all particles?*

The photon apparently has both particle and wave properties associated with it. At least, in order to explain diffraction we find it convenient to think of electromagnetic radiation as a wave, and in order to explain the photoelectric effect we find it convenient to think of radiation as a stream of discrete particles.

The energy ϵ of a photon is related to the frequency of the associated electromagnetic wave by the Planck-Einstein relationship

$$\epsilon = h\nu. \tag{3-40}$$

But we also note that according to the theory of relativity, the relativistic energy of the photon is

$$\epsilon = mc^2, \tag{3-41}$$

where m is the *relativistic* mass of the photon (the *rest* mass of the photon can be shown to be equal to zero) and c is the speed of light. Combination of Eq. (3–40) and Eq. (3–41) leads to

$$mc^2 = h\nu = hc/\lambda,$$

from which

$$\lambda = h/mc. \tag{3-42}$$

The product mc is the momentum in a vacuum of the photon considered as a *particle*, whereas λ is the wavelength of the photon considered as a *wave*.

In a bold move, Louis de Broglie[*] in 1924 intuitively extended the particle-wave relationship for photons to *all particles* (including electrons) and hypothesized that all particles are guided by associated symbolic waves which he called *pilot waves*. Specifically, de Broglie postulated that equations of the forms of Eqs. (3–40) and (3–42) are applicable to *all particles* and that the frequency ν *and the wavelength λ of the pilot wave associated with a particle of relativistic mass m, velocity v, and total relativistic energy E are given by the equations*

$$\nu = E/h \tag{3-43}$$

and

$$\lambda = h/mv. \tag{3-44}$$

---

[*] L. V. de Broglie, "Researches on the Quantum Theory," Thesis, Paris University, 1924.

It is important to understand that v is the *velocity of the particle* and *is not* (except for photons) equal to the *phase velocity u of the pilot wave*, which in turn may be written as

$$u = \lambda v. \tag{3–45}$$

In fact, substitution of Eqs. (3–43) and (3–44) in Eq. (3–45) yields

$$u = E/mv = mc^2/mv = c^2/v.$$

Thus the relationship between the velocity v of the particle and the phase velocity u of its associated pilot wave is

$$uv = c^2. \tag{3–46}$$

For all particles except photons traveling in a vacuum, v must be smaller than c, and u must therefore be greater than c. Such a requirement is, at first, disturbing for two reasons: (1) It appears that particles cannot keep up with their own pilot waves, and (2) the phase velocities of the pilot waves themselves exceed the speed of light. However, we'll show in Chapter 7 that we are able to associate mathematically the position of a free particle with a *packet* or *group* of superposed simple pilot waves in such a way that the *group velocity* of the packet is less than the phase velocities u of the constituent pilot waves. Furthermore, we'll also show that the group velocity of the packet is equal to the velocity of the particle. Since the speed v of the wave packet itself, which is the speed at which mass and energy are transferred, is always less than c, there is no conflict with the theory of relativity.

3–8 WAVELENGTHS OF ELECTRONS AND OTHER PARTICLES

Let's now consider an electron in an allowed orbit. If the properties of the electron are governed by a wave, the only way to avoid destructive interference is to require that the circumference of the orbit contain an integral number of wavelengths:

$$2\pi r = n\lambda. \tag{3–47}$$

Only under such conditions are standing waves produced in a circular path. Wavelengths which do not satisfy this condition will eliminate themselves by destructive interference. Examples of allowed and nonallowed electron waves are shown in Fig. 3–7. The problem of fitting standing electron waves into orbits is the same as that of fitting standing acoustical waves into a metal hoop which is struck with a hammer. In the case of the metal hoop, only those audio frequencies are heard for which standing waves may persist. As in the case of the plucked violin string, which we'll consider in detail in the next chapter, persistent vibrations consist of discrete allowed frequencies referred to as *fundamental modes*.

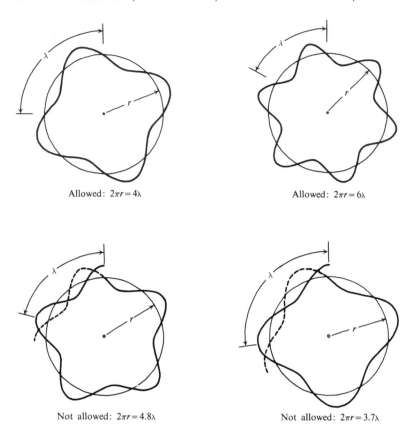

FIG. 3-7   Allowed and nonallowed de Broglie orbits.

A surprising and important consequence of de Broglie's postulate results when we combine Eqs. (3–44) and (3–47) to eliminate λ, assuming that the rest mass m_e of the electron may be substituted for the relativistic mass at ordinary orbital velocities:

$$m_e v r = nh/2\pi. \qquad (3\text{–}48)$$

Equation (3–48) is *identical to Bohr's quantum restriction postulate*! By attributing wave properties to the electron, de Broglie was able to justify the bold empirical postulate made by Bohr a decade earlier.

The fact that electrons do indeed have associated wavelengths was experimentally verified in 1927 by Davisson and Germer,* who computed the wavelength for electrons from the observed diffraction pattern from a nickel surface.

---

* C. Davisson and L. H. Germer, *Phys. Rev.*, **31**, 705 (1927).

The calculated wavelength for electrons at the experimental velocity exactly agreed with that predicted by the de Broglie relationship, Eq. (3–44).

The de Broglie equations are also applicable to particles other than electrons. Knowing the mass and velocity of *any* particle, we may directly calculate its associated wavelength from Eq. (3–44). For example, an average hydrogen molecule at 200°C has a velocity of 2.4×10^5 cm/sec. Since the mass of the hydrogen molecule is 3.3×10^{-24} g, its associated wavelength is calculated to be

$$\lambda = \frac{h}{mv} = \frac{6.62 \times 10^{-27} \text{ g-cm}^2\text{-sec}^{-1}}{(3.3 \times 10^{-24} \text{ g})(2.4 \times 10^5 \text{ cm-sec}^{-1})} = 0.83 \times 10^{-8} \text{ cm} = 0.83 \text{ Å}.$$

In like manner, we may calculate that the wavelength of a 2-ton automobile traveling 20 mph is 4.1×10^{-28} Å! This is a ridiculously small wavelength which cannot be measured by any known technique! A few simple calculations of this type immediately disclose that de Broglie wavelengths are experimentally detectable *only* for *very small particles*.

A curious thing about the atom-model work of Bohr, prior to 1923 or 1924, was that if you look at the then-current papers you get the impression that everybody in the world was terrifically excited about the Bohr model and believed in it hook, line, and sinker, including the electron orbits as they are used in the ads for the atomic age nowadays. Bohr, on the other hand, was constantly making remarks, speeches, and admonitions to the effect that this is temporary and we ought to be looking for a way to do it right.

> —EDWARD U. CONDON, to the Philosophical Society of Washington, December 2, 1960.*

PROBLEMS

3–1 a) Calculate the wavelength to which each of the following series converges: Lyman series, Paschen series, Brackett series, Pfund series.

  b) In the spectral region immediately below each of the wavelengths calculated in part (a), the observed spectrum for hydrogen is continuous rather than composed of discrete lines. What does the presence of the continuum allow you to say concerning the energies of free or ionized electrons?

3–2 a) Calculate the reduced mass of the deuterium atom.

  b) Calculate the Rydberg constant for deuterium.

  c) What is the calculated wavelength of the first line in the Balmer series of the deuterium spectrum as compared to that in the hydrogen spectrum?

---

* Complete address reprinted in *Physics Today*, October 1962, 37–46. Used with permission of Professor Condon and the American Institute of Physics.

d) What further correction must be made in order that the value calculated from part (c) for the first line in the Balmer series for hydrogen will agree *exactly* with the experimental value given in Fig. 3–2?

3–3 Assuming a finite nuclear mass for the hydrogen atom according to the model of Fig. 3–6, prove that

a) the kinetic energy of the atom is given by $\mu(r\omega)^2/2$, and
b) the centrifugal force is given by $\mu(r\omega)^2/r$.

3–4 Calculate the energy required to completely remove an electron from the ground state of the hydrogen atom (from $n = 1$ to $n = \infty$). How does the calculated energy compare with the observed energy, as given by the ionization potential of hydrogen in Table 14–5?

3–5 Using the Bohr theory, derive equations similar to Eq. (3–12) and Eq. (3–13) giving the allowed energies and radii of the single electron in a monoelectronic ion whose nuclear charge is Ze.

3–6 Calculate the wavelengths of the first two lines in the Balmer series of He^+. What is the minimum wavelength predicted for an electron transition *within* the He^+ ion?

3–7 a) Calculate the energy in ergs of the least energetic photon emitted in the Lyman series of the hydrogen spectrum.
b) Does the above photon have sufficient energy to cause photoemission of an electron from the surface of lithium (work function = 2.3 eV)? If so, calculate the energy in ergs of the photoemitted electron.

3–8 Maclaurin's series for the expansion of $f(x)$ is given as

$$f(x) = f(0) + xf'(0) + \frac{x^2}{2!} f''(0) + \frac{x^3}{3!} f'''(0) + \cdots,$$

where $f'(0)$ represents the first derivative of the function with zero substituted for x, etc. Using a Maclaurin series expansion, prove that

$$\frac{1}{\sqrt{1 - (v^2/c^2)}}$$

may be expanded to

$$\left(1 + \frac{v^2}{2c^2} + \frac{3v^4}{8c^4} + \cdots\right).$$

3–9 Neglecting relativistic effects (that is, assuming that the mass of the electron remains constant), what is the wavelength for an electron which is accelerated through a potential of 1000 V? 10,000 V?

3–10 What is the de Broglie wavelength in angstroms for an electron

a) In the ground state of the hydrogen atom?
b) In the first excited state ($n = 2$)?

3–11 The relativistic formula for kinetic energy is *not* $T = mv^2/2$ but rather can be shown to be

$$T = m_0 c^2 \left[\left(1 - \frac{v^2}{c^2} \right)^{-\frac{1}{2}} - 1 \right]$$

Using the results of problem 3–8, show that at low velocities the relativistic formula reduces to the classical or nonrelativistic form for T.

The success of the de Broglie equation in deriving the Bohr quantum restriction postulate, coupled with the experimental fact that electrons can be diffracted, seems to present convincing evidence that the behavior of particles is governed by associated waves. It is thus both desirable and expedient that we review several of the basic mathematical properties of waves and oscillations.

4-1 SIMPLE HARMONIC MOTION

Consider a point P which revolves counterclockwise on the circumference of a circle of radius A. Such an arrangement is shown in Fig. 4-1. Further assume that the point starts at zero time from position S and revolves with a *constant*

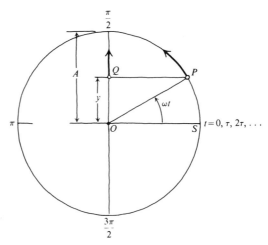

FIG. 4-1   Simple harmonic motion as a projection of circular motion on the diameter of the circle.

angular velocity ω (radians/second). We'll designate point Q as the horizontal projection of point P on the y-axis passing through the origin. As point P revolves, point Q oscillates up and down in what is called *simple harmonic motion. We may define simple harmonic motion as the projection of circular motion on the diameter of the circle.*

The motion of point Q repeats itself during each successive revolution. The period of time required for one revolution or *cycle* is called the *period* τ, and is given by

$$\tau = \frac{2\pi}{\omega} \frac{\text{rad/cycle}}{\text{rad/s}} = \frac{2\pi}{\omega} \frac{\text{s}}{\text{cycle}} \qquad (4\text{--}1)$$

It also follows that the *frequency* of revolution, ν, in cycles/second or Hz is given by

$$\nu = 1/\tau = \omega/2\pi,$$

or

$$\omega = 2\pi\nu. \qquad (4\text{--}2)$$

In order to define mathematically the motion of point Q, which is undergoing simple harmonic oscillation, it is convenient to define the *displacement* y as the distance of point Q from the origin or center of oscillation. Thus, from Fig. 4–1,

$$y = A \sin(\omega t). \qquad (4\text{--}3)$$

Equation (4–3) mathematically defines simple harmonic motion and is graphically plotted in Fig. 4–2. The maximum displacement A is called the *amplitude*, and the quantity ωt is called the *phase*.

If, instead of starting the circular motion of point P at S, we start at some displaced angle δ, as shown in Fig. 4–3, the displacement is given by

$$y = A \sin(\omega t + \delta), \qquad (4\text{--}4)$$

where the *phase* is now $\omega t + \delta$ and δ is referred to as the *initial phase* or *phase constant*. The displacement y is plotted as a function of t according to Eq. (4–4) in Fig. 4–4. At any point in time, the phases of the waves shown in Figs. 4–2 and 4–4 are different by the phase constant δ. The two waves are said to be *out of phase*.

The velocity of point Q at any time may be expressed as dy/dt. Differentiation of the more general form of the wave equation for harmonic motion, Eq. (4–4), yields

$$\frac{dy}{dt} = A\omega \cos(\omega t + \delta). \qquad (4\text{--}5)$$

The acceleration of point Q, which is undergoing simple harmonic motion, is

$$\frac{d^2y}{dt^2} = -A\omega^2 \sin(\omega t + \delta). \qquad (4\text{--}6)$$

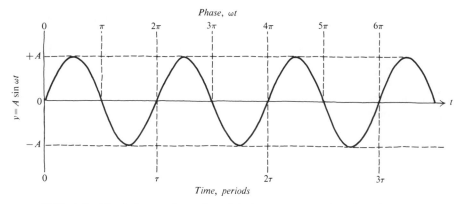

FIG. 4–2   Simple harmonic motion. Displacement y as a function of time t.

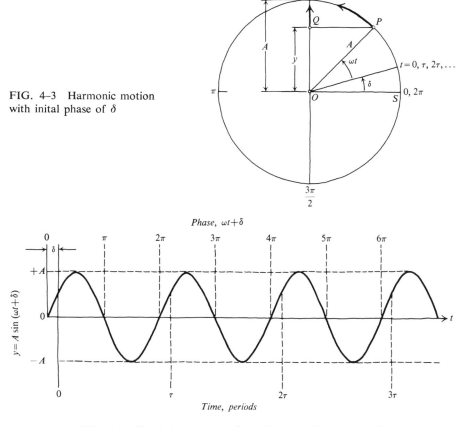

FIG. 4–3   Harmonic motion
with inital phase of δ

FIG. 4–4   Simple harmonic motion with phase displacement δ.

Combination of Eq. (4–4) and Eq. (4–6) to eliminate $A \sin (\omega t + \delta)$ yields

$$\frac{d^2y}{dt^2} = -\omega^2 y, \qquad (4\text{–}7)$$

which is the *differential equation of harmonic motion*. Equation (4–7) is a more general description of harmonic motion than Eq. (4–4) because it does not include dependence on the constants A and δ. Furthermore, we'll find that functions other than the sine function can satisfy this simple differential equation.

4–2 OSCILLATION OF A MASS ON A SPRING

In a physical sense, simple harmonic motion may also be defined as the *motion of a body which is constrained to move in a straight line, subject to a restoring force which is proportional to its displacement from the equilibrium position.* To show this, consider a mass which is oscillating in a vacuum and in the absence of gravity on a weightless spring (see Fig. 4–5). Further, assume that the spring is attached on the other end to a large immovable mass such as a wall. If y is the displacement from the equilibrium position, *Hooke's law* may be stated as

$$-f = \kappa y, \qquad (4\text{–}8)$$

where $-f$ is the restoring force and κ is a characteristic constant of the spring called the *force constant*. The negative sign in Eq. (4–8) indicates that the force acts in opposition to the displacement; that is, it is a *restoring force*.

The velocity of the mass at any point is dy/dt, and the acceleration is d^2y/dt^2. According to Newton's second law of motion, the restoring force is also given by

$$f = ma = m\frac{d^2y}{dt^2}. \qquad (4\text{–}9)$$

Combination of Eqs. (4–8) and (4–9) to eliminate f yields

$$\frac{d^2y}{dt^2} = \frac{-\kappa}{m} y, \qquad (4\text{–}10)$$

which is of the same form as Eq. (4–7), the differential equation of linear harmonic motion, thus revealing that the mass on the spring is a linear harmonic oscillator. A comparison of the constants in Eqs. (4–7) and (4–10) relates the angular frequency ω to κ and m such that

$$\omega^2 = \kappa/m. \qquad (4\text{–}11)$$

Furthermore, a satisfactory displacement equation for the mass must be given by Eq. (4–4), since Eq. (4–10) is derived through double differentiation of Eq. (4–4), so that

$$y = A \sin [(\kappa/m)^{1/2} t + \delta]. \qquad (4\text{–}12)$$

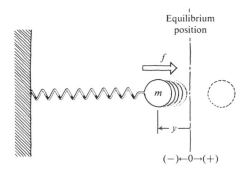

FIG. 4–5   Mass oscillating on a spring.

Let's now examine in detail how we might obtain in a *different* fashion the displacement equation $y(t)$ by the direct solution of Eq. (4–10). Even though we have already shown above, through comparison of the differential Eqs. (4–7) and (4–10), that a satisfactory displacement equation is given by Eq. (4–4), we'll perform this additional exercise in order (a) to introduce an important technique which we'll use often in solving linear differential equations such as Eq. (4–10), and (b) to show that the sine form given by Eq. (4–4) is not the only possible form for the displacement equation. But first we must review the mathematical concept of operators.

4–3 OPERATORS AND EIGENVALUE EQUATIONS

An *operator* is a symbol which directs one to perform an operation on the function which follows the symbol. Examples are shown in the table below.

Example	Operator	Operation
$\sqrt{3}$	$\sqrt{}$	Take the square root
$x\Phi$	$x \cdot$	Multiply by x
$\dfrac{d(y^2 - 6)}{dx}$	$\dfrac{d}{dx}$	Take the first derivative with respect to x

In quantum mechanics, operators are often (but not always) symbolized by a *circumflex* above a letter. Typical operator symbols are $\hat{p}$, $\hat{H}$, and $\hat{T}$. Consider two simple equations involving operators:

$$(d/dx)x^2 = 2x, \qquad (d/dx)e^{2x} = 2e^{2x}.$$

In each equation, the operator is d/dx. However, the second equation differs in an interesting way. That is, the operation of d/dx on the function e^{2x}

returns the original function multiplied by the constant 2. Any equation of the second type, in which an operation on a function yields the product of a constant times the original function, is called an *eigenvalue equation*. The function e^{2x} is said to be an *eigenfunction* of the operator d/dx, having an *eigenvalue* of 2. Eigenvalue equations occur frequently in quantum mechanics, and can be expressed generally as

$$\hat{g}\psi = g\psi.$$

An eigenvalue equation is solved by finding an eigenfunction ψ such that the operation of $\hat{g}$ on ψ yields the product of the original eigenfunction times a constant g. Usually more than one eigenfunction is found to satisfy a given operator, and more than one eigenvalue g results.

4-4 BACK TO THE MASS ON A SPRING

It is now apparent that Eq. (4–10) has the form of an eigenvalue equation where

$$\frac{d^2}{dt^2} \quad \text{is the operator,}$$

$$\frac{-\kappa}{m} \quad \text{is the eigenvalue,}$$

$$y \quad \text{is the eigenfunction.}$$

In order to solve Eq. (4–10) we must find a function $y(t)$ such that when the second derivative of $y(t)$ is taken with respect to t, the result is the original function $y(t)$ multiplied by $(-\kappa/m)$.

A common method for solving eigenvalue equations is trial and error, which becomes somewhat easier with practice. For example, we know that the *second* derivative of a sine function produces a negative sine function. We might, then, after a bit of experimentation, propose the general function

$$y = a \sin (bt + c) \tag{4–13}$$

as a possibly acceptable eigenfunction for the operator d^2/dt^2, where a, b, and c are constants. Let's perform the required operation in order to test the function;

$$\frac{dy}{dt} = ba \cos (bt + c), \qquad \frac{d^2y}{dt^2} = -b^2a \sin (bt + c),$$

or

$$\frac{d^2[a \sin (bt + c)]}{dt^2} = -b^2[a \sin (bt + c)].$$

That is,

$$\frac{d^2y}{dt^2} = -b^2y. \tag{4–14}$$

Comparison of Eq. (4–14) with Eq. (4–10) indicates that $y = a \sin (bt + c)$ is a satisfactory eigenfunction of the operator d^2/dt^2 if and only if

$$b^2 = \kappa/m.$$

Thus the displacement equation, Eq. (4–13), may be written

$$y = a \sin [(\kappa/m)^{1/2}t + c] \tag{4–15}$$

which is identical to Eq. (4–12) since a and c easily may be shown to be the amplitude A and initial phase δ respectively.

The frequency ν of vibration of the mass on the spring may be expressed by substituting $2\pi\nu$ for ω in Eq. (4–11), which yields:

$$\nu = \frac{1}{2\pi}\left(\frac{\kappa}{m}\right)^{1/2}. \tag{4–16}$$

The frequency given above is the only allowed *natural frequency*. Any other frequency forced upon the oscillating mass is said to be *anharmonic*.

Finally, it is important to note that functions other than the sine function may serve as satisfactory displacement equations for the linear harmonic oscillator. It can be shown easily, for example, that the functions

$$y = A \cos (\omega t + \delta) \tag{4–17}$$

and

$$y = A e^{\pm i(\omega t + \delta)}, \tag{4-18}$$

where $i^2 = -1$, are equally satisfactory eigenfunctions of the operator, d^2/dt^2 (prove), and are thus equally satisfactory displacement equations for the linear harmonic oscillator. In fact, it can be shown that *any linear combination* of the sine, cosine, and complex-exponential displacement functions also satisfies the general differential equation of harmonic motion (prove this; see Problem 4-10) and is, therefore, a valid description for simple harmonic motion.

4-5 ENERGY OF THE SIMPLE HARMONIC OSCILLATOR

Although kinetic energy T and potential energy V are periodically interconverted during the vibration of an isolated linear harmonic oscillator, their total remains constant and equal to E. That is,

$$E = T + V. \tag{4–19}$$

A comparison of Eq. (4–19) with Eq. (3–38) indicates that we have neglected the rest-mass energy m_0c^2. In general, in future references, the term E will refer to the *nonrelativistic* total energy rather than to the relativistic total energy unless specifically stated otherwise.

The potential energy V of the system is the amount of work done in displacing the mass from its equilibrium position to a displacement y. Since the restoring force is $-f$,

$$V = \int_0^y -f \, dy = \int_0^y \kappa y \, dy = \kappa y^2/2. \qquad (4\text{--}20)$$

Since $T = mv^2/2$, where v is the velocity of the mass, Eq. (4–19) may be written as

$$E = (mv^2/2) + (\kappa y^2/2). \qquad (4\text{--}21)$$

Although E is the same at any time during the oscillation, it is convenient to evaluate E at the point of maximum displacement, where $y = A$ and $v = 0$, whereupon Eq. (4–21) becomes

$$E = \tfrac{1}{2}\kappa A^2, \qquad (4\text{--}22)$$

and the *total energy of the simple harmonic oscillator is shown to be directly proportional to the square of its amplitude.* Recall that we have previously indicated that the total energy of electromagnetic radiation is also proportional to the square of either the electrical displacement vector or the magnetic displacement vector.

4–6 TRAVELING WAVES IN A STRETCHED STRING

In order to more completely understand the wave properties of a moving particle, which in quantum mechanics may be represented by the properties of a traveling wave, we'll examine in some detail the nature of the one-dimensional traveling wave which is produced in an infinitely long string under tension.

If one were to grasp one end of a long, taut rope and were to give the rope a single upward flip, a single pulse wave would be generated which would travel along the entire length of the rope (Fig. 4–6). If, however, one were to continually oscillate the rope end up and down at a fixed frequency, a traveling wave would be generated in the rope (Fig. 4–7). Note that although the waveform travels *longitudinally* (x-direction) along the rope, a given segment of rope undergoes a periodic *transverse* (y-direction) displacement which is a function of both time and longitudinal position in the rope.

In order to more completely analyze displacement as a function of time and longitudinal position, let's consider an infinitely long string which is under tension. Assume that

1. The string is uniform throughout its length.
2. The force of tension F acting axially in the string at any point is constant and is great enough so that it is not appreciably changed by the vibration.
3. The displacement y is small and is a smooth function of the longitudinal position x.

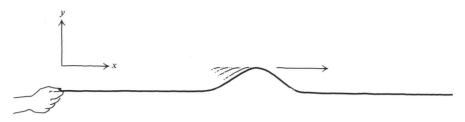

FIG. 4–6   A single pulse wave generated by a single impulse on the end of a taut rope.

FIG. 4–7   A traveling wave generated by oscillating the end of a taut rope.

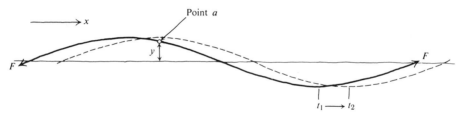

FIG. 4–8   Transverse displacement y as a function of time t and longitudinal position x in a traveling wave. The solid line shows the wave at time t_1, and the dashed line shows the wave at t_2, a short time later.

A section of such a taut string is shown in Fig. 4–8. The waveform is shown as moving from left to right at a *constant phase velocity* or *propagation velocity* in the x-direction, and the displacement y is a periodic function of both x and t.

Let's now analyze the forces acting on a differential length dL of the string at a specific moment in time, that is, *while holding time constant*. (A camera snapshot of the string represents a curve of displacement y as a function of longitudinal position x while t is constant.) Specifically let's look at the differential string segment dL shown at point a in Fig. 4–8. This section is magnified in Fig. 4–9. The force of tension F acts tangentially to the string at each end of

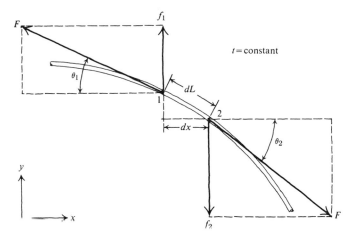

FIG. 4–9 The forces acting on a differential segment, dL, of a stretched string under vibration.

the segment dL. The component force f_2 acting *down* on the segment at point 2 is

$$f_2 = F \sin \theta_2, \tag{4–23}$$

whereas the component force f_1 acting *up* on the segment at point 1 is

$$f_1 = F \sin \theta_1. \tag{4–24}$$

The net downward component of force which tends to *restore* the string segment to its equilibrium position is

$$f = f_2 - f_1 = F(\sin \theta_2 - \sin \theta_1). \tag{4–25}$$

But if θ is small, which it is when the displacement is small, $\sin \theta \simeq \tan \theta$, so that

$$f = F(\tan \theta_2 - \tan \theta_1). \tag{4–26}$$

But $\tan \theta$ is the slope of the wave at constant time, $(\partial y/\partial x)_t$, so that

$$f = F\left[\left(\frac{\partial y}{\partial x}\right)_2 - \left(\frac{\partial y}{\partial x}\right)_1\right], \tag{4–27}$$

where $(\partial y/\partial x)_2$ is the slope of the wave at point 2, and $(\partial y/\partial x)_1$ is the slope of the wave at point 1. Note that both slopes are negative in our example and that $(\partial y/\partial x)_2$ is more negative than $(\partial y/\partial x)_1$. This results in a value for f which is also negative. That is, the force f is a *restoring* force which acts in opposition to the displacement vector.

Since the segment is a differential segment, we may write the change in slope with longitudinal position between points 1 and 2 as

$$\frac{(\partial y/\partial x)_2 - (\partial y/\partial x)_1}{dx} = \frac{\partial^2 y}{\partial x^2},$$

so that Eq. (4-27) becomes

$$f = F\left(\frac{\partial^2 y}{\partial x^2}\right) dx. \tag{4-28}$$

The term $\partial^2 y/\partial x^2$ is the *curvature*. Note that the restoring force is greatest where the curvature is greatest.

We may now apply Newton's second law to the segment, that is,

$$f = ma, \tag{4-29}$$

where m, the mass of the differential segment, is given as $m = \mu\, dL$, where μ is the *mass per unit length* of string. Since a, the acceleration of the differential segment, is given as $\partial^2 y/\partial t^2$ at a constant value of x, Eq. (4-29) may be written as

$$f = (\mu\, dL)\frac{\partial^2 y}{\partial t^2}. \tag{4-30}$$

Combination of Eq. (4-28) and Eq. (4-30) to eliminate f yields

$$F\frac{\partial^2 y}{\partial x^2}\, dx = (\mu\, dL)\frac{\partial^2 y}{\partial t^2}.$$

But, for small displacements, $dx \cong dL$, so that

$$\frac{F}{\mu}\left(\frac{\partial^2 y}{\partial x^2}\right) = \frac{\partial^2 y}{\partial t^2}. \tag{4-31}$$

It is now convenient arbitrarily to define the constant F/μ in terms of a new constant u such that $u^2 = F/\mu$. Equation (4-31) then becomes

$$\frac{\partial^2 y}{\partial x^2} = \frac{1}{u^2}\left(\frac{\partial^2 y}{\partial t^2}\right). \tag{4-32}$$

Equation (4-32) is the *general differential equation of wave motion in one dimension*.

The remaining problem is now to determine y as a function of x and t, that is, $y(x, t)$, such that the function satisfies Eq. (4-32). The resultant function $y(x, t)$ will then be a satisfactory equation for the traveling wave.

Once again, we'll resort to a trial and error solution, but we have now gained at least a little experience in choosing functions which will probably work. Since we know that y is a *periodic* function of both x and t, a possible general trial function might be

$$y = A \sin (ax + bt + c), \qquad (4\text{-}33)$$

where A, a, b, and c are constants. We must now evaluate the usefulness of this trial function by substituting it into Eq. (4–32). Thus

$$\frac{\partial y}{\partial x} = Aa \cos (ax + bt + c),$$

$$\frac{\partial^2 y}{\partial x^2} = -Aa^2 \sin (ax + bt + c), \qquad (4\text{-}34)$$

and

$$\frac{\partial y}{\partial t} = Ab \cos (ax + bt + c),$$

$$\frac{\partial^2 y}{\partial t^2} = -Ab^2 \sin (ax + bt + c). \qquad (4\text{-}35)$$

Substitution of Eqs. (4–34) and (4–35) into Eq. (4–32) yields

$$u^2 a^2 = b^2,$$

or

$$b/a = \pm u. \qquad (4\text{-}36)$$

Thus our trial solution is a satisfactory solution if, and only if, $b/a = \pm u$. Equation (4–33) may be rearranged to

$$y = A \sin a\left(x + \frac{b}{a}t + \frac{c}{a}\right). \qquad (4\text{-}37)$$

Substitution of Eq. (4–36) into Eq. (4–37), and definition of $c/a = c'$, a new constant, yields

$$y = A \sin a(x \pm ut + c'). \qquad (4\text{-}38)$$

Equation (4–38) is a satisfactory equation describing the displacement of a traveling wave as a function of position and time. We may easily deduce by inspection that the constant A is the amplitude, or maximum displacement, of the traveling wave, since the sine term must vary between $+1$ and -1, causing the displacement y to vary between $+A$ and $-A$.

Although we have shown Eq. (4–38) to be specifically valid for a taut, vibrating string of infinite length, it may be shown that it is applicable to all forms of smooth and uniform traveling waves, including sound waves and electromagnetic waves.

4-7 THE PHASE VELOCITY OF A TRAVELING WAVE

The *phase velocity* of a traveling wave is defined as $(\partial x/\partial t)_y$ and may be epresented graphically as the differential longitudinal motion dx of a point in he wave at some selected value of y in time interval dt (Fig. 4–10). But, from

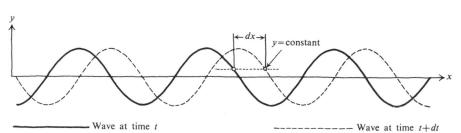

FIG. 4–10   The phase velocity of a wave is $(\partial x/\partial t)_y$.

Eq. (4–38), when y is constant,

$$A \sin a(x \pm ut + c') = \text{constant},$$

and since A and a are constants,

$$x \pm ut + c' = \text{constant}. \tag{4–39}$$

Taking the derivative of Eq. (4–39) with respect to t at constant y yields

$$\left(\frac{\partial x}{\partial t}\right)_y \pm u = 0,$$

or

$$\left(\frac{\partial x}{\partial t}\right)_y = \mp u = \text{phase velocity}.$$

The term u is thus the *magnitude* of the phase velocity.

The $\mp$ sign allows for propagation at the same speed in either direction. For a wave traveling in the $(+x)$-direction, that is, left to right, $(\partial x/\partial t)_y = +u$, which follows if a $(-)$ sign is used in front of ut in Eq. (4–38). Conversely, for a wave traveling in the $(-x)$-direction, $(\partial x/\partial t)_y = -u$, which follows if a $(+)$ sign is used in front of ut in Eq. (4–38).

From the original definition of u used in converting Eq. (4–31) into Eq. (4–32), the magnitude of the phase velocity may be expressed as

$$u = \sqrt{F/\mu}. \tag{4–40}$$

Thus, waves propagate more rapidly as the string becomes more tense or as it becomes lighter in mass.

4–8   FIXED ENDS ON A STRETCHED STRING: BOUNDARY CONDITIONS

Fixing both ends of a vibrating, stretched string at a *finite* length results in a wave motion which is quite different from the motion of the traveling wave in a taut string of *infinite* length. It is experimentally observed that when such a finitely confined string is plucked, evenly spaced *nodes* (points of zero displacement) and *antinodes* (points of maximum displacement) appear as illustrated for a typical mode in Fig. 4–11. The experimental conditions require, of course, that the fixed ends be nodes and undergo no transverse displacement. But in addition we observe, depending on how the string is energized, an integral number of nodes (or none) *between* the two ends, with the string wagging transversely between any two nodes. Such a wave pattern is called a *standing wave* or *stationary wave*. The most interesting feature of the stationary wave is that there is *no displacement at any time at the node positions*. Therefore, Eq. (4–38) for the traveling wave cannot directly apply to the stationary wave since it allows for transverse displacement y at *any* value of x. We may think conveniently of the stationary wave as being the additive wave produced by continued reflection of a traveling wave from fixed points at either end.

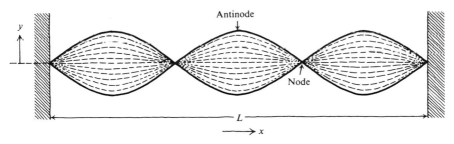

FIG. 4–11 A time exposure of a typical standing or stationary wave in a stretched string fixed at both ends. The loops wag transversely between the nodes.

Now that we are familiar with a few of the basic properties of stationary waves, let's attempt to develop a suitable function $y(x, t)$ to describe the variation of transverse displacement y with longitudinal position x and time t in a stretched string clamped on both ends at a distance L. In order to describe satisfactorily the behavior of a stationary wave, the function $y(x, t)$ must meet two general requirements:

1. It must satisfy the general differential equation of wave motion in one dimension,

$$u^2 \frac{\partial^2 y}{\partial x^2} = \frac{\partial^2 y}{\partial t^2} .$$ (4–32′)

2. It must satisfy the two *boundary conditions* imposed on the string by fixing the ends. These are

$$y = 0 \quad \text{when} \quad x = 0, \quad \text{and}$$
$$y = 0 \quad \text{when} \quad x = L,$$

for all values of t.

As a starting function let's begin with Eq. (4–38), which we have already shown to satisfy the general differential wave equation for a wave traveling in either direction in a taut string of infinite length. If the wave travels in the $(+x)$-direction (from left to right), the displacement y_+ is given as

$$y_+ = A \sin a(x - ut + c'), \tag{4-41}$$

whereas if the wave travels in the $(-x)$-direction (from right to left) the displacement y_- is given as

$$y_- = A \sin a(x + ut + c'). \tag{4-42}$$

The peculiar form of the stationary wave results from the combination of the displacements of traveling waves which are constantly reflected from the fixed ends of the string, so that they travel in both directions simultaneously.

There is a well-known principle of wave theory, called the *principle of superposition*, which applies to the combination of linear wave equations such as Eqs. (4–41) and (4–42), and it is this: *whenever two or more individual waves travel simultaneously in the same region, whether in the same or in different directions, the total displacement at any position is the sum of the displacements due to the individual waves.* Consider, for example, two independent waves of the same amplitude but of different wavelengths traveling in the same direction. At certain points where the wave crests of the two waves coincide, the total displacement is twice the amplitude of each individual wave, and we say that the waves *reinforce* one another. At other points, where the wave crest of one wave happens to coincide with the wave trough of the second wave, the two waves cancel one another completely so that the total displacement is *zero*, and we say that the waves *destroy* one another. The general phenomenon of partial or complete reinforcement and destruction is referred to as *interference*.

$$y = y_+ + y_-,$$

$$y = A[\sin a(x - ut + c') + \sin a(x + ut + c')]. \tag{4-43}*$$

But, using the trigonometric identity

$$\sin \alpha + \sin \beta = 2 \sin \left(\frac{\alpha + \beta}{2}\right) \cos \left(\frac{\alpha - \beta}{2}\right),$$

which holds for all values of α and β, we write Eq. (4–43) as

$$y = 2A \sin a(x + c') \cos aut,$$

noting that $\cos - \theta = \cos \theta$. More simply,

$$y = A' \sin a(x + c') \cos aut, \tag{4-44}$$

---

* Eq. 4–43 is a satisfactory solution to Eq. 4–32 because it is a linear combination of two separate satisfactory functions. See Problem 4–13.

where $A' = 2A$. Now let's determine what the constants a and c' must be in order for Eq. (4-44) to obey the two boundary conditions. At the left boundary of the string, $x = 0$ and $y = 0$. Substitution of zero for both x and y in Eq. (4-44) yields

$$0 = A' \sin ac' \cos aut,$$

which must be valid for all values of t. If neither A' nor a is restricted to zero, then it is necessary that

$$\sin ac' = 0,$$

which is true whenever the argument of the sine is an integral multiple of π, that is, when

$$ac' = n'\pi, \tag{4-45}$$

where $n' = 0, \pm1, \pm2, \pm3, \ldots, \pm\infty$. Substitution of c' from Eq. (4-45) back into Eq. (4-44) yields

$$y = A' \sin (ax + n'\pi) \cos aut. \tag{4-46}$$

At the right boundary of the string, $x = L$ and $y = 0$, and Eq. (4-46) becomes

$$0 = A' \sin (aL + n'\pi) \cos aut,$$

which must again be valid for all values of t. If, once more, neither A' nor a is restricted to zero, then

$$\sin (aL + n'\pi) = 0,$$

which is true whenever

$$aL + n'\pi = n''\pi, \tag{4-47}$$

where $n'' = 0, \pm1, \pm2, \pm3, \ldots, \pm\infty$. Rearrangement of Eq. (4-47) yields

$$a = \frac{(n'' - n')\pi}{L} = \pm\frac{n\pi}{L}, \tag{4-48}$$

where n, which we define as a *positive* integer, may have the values $n = 0, 1, 2, 3, \ldots, \infty$. Substitution of a from Eq. (4-48) back into Eq. (4-46) yields

$$y = A' \sin \left(\pm \frac{n\pi x}{L} + n'\pi\right) \cos \left(\pm \frac{n\pi ut}{L}\right). \tag{4-49}$$

But the shift in the first term of the argument of the sine by $n'\pi$ results either in no change in the sine term itself if n' is an even integer, or in a change of sign if n' is an odd integer. Either possibility may be accommodated if we write Eq. (4-49) as

$$y = \pm A' \sin \left(\pm \frac{n\pi x}{L}\right) \cos \left(\pm \frac{n\pi ut}{L}\right).$$

Finally, since $\sin (-\alpha) = -\sin \alpha$ and $\cos (-\alpha) = \cos \alpha$, we may write Eq.

(4–50) simply as

$$y = \pm A' \sin\left(\frac{n\pi x}{L}\right) \cos\left(\frac{n\pi ut}{L}\right), \qquad (4\text{–}50)$$

which represents a *series* of satisfactory functions for the stationary wave, including one positive function and one negative function for each value of n. For each of the positive functions, regardless of the value of n, the displacement y is a maximum and equal to A' whenever the sine term *and* the cosine term are each unity. Under these same conditions, the displacement is $-A'$ for each of the negative functions. In fact, for any given values of n, x, and t the displacements given by the positive and negative functions of Eq. (4–50) are always equal in magnitude but opposite in sign. Thus, it is apparent that any given negative function corresponding to a selected value of n is not really an independent function because it may be represented equally well by shifting the corresponding positive function by one-half period in time. Since the initial displacement of the wave at $t = 0$ is purely arbitrary, we'll choose to work only with the *positive* series of functions given by

$$y = A' \sin\left(\frac{n\pi x}{L}\right) \cos\left(\frac{n\pi ut}{L}\right), \qquad (4\text{–}51)$$

where $n = 0, 1, 2, 3, \ldots, \infty$. Let's now analyze the appearance of the string at some specific moment in time, which is equivalent to examining a snapshot photograph. With t constant, Eq. (4–51) becomes

$$y = A_t \sin\left(\frac{n\pi x}{L}\right), \qquad (4\text{–}52)$$

where $A_t = A' \cos(n\pi ut/L) = $ constant. Whenever nx/L is an integer, the argument of the sine term in Eq. (4–52) is an integral multiple of π, the sine term itself is zero, and the displacement y is zero. This is true regardless of the value of A_t, the amplitude of the fixed-time wave. Thus, *values of x for which nx/L is an integer define node positions*, which are positions of zero displacement at all times. Different node positions are generated for each value of n.
   When $n = 0$,

$$nx/L = 0x/L,$$

which may be an integer (0) for all values of x. Zero displacement for all values of x represents a flat string with no energy of vibration.
   When $n = 1$,

$$nx/L = x/L,$$

which may be an integer when $x = 0$ (integer $= 0$) or when $x = L$ (integer $= 1$). Since $x \leq L$, these are the only allowed values of x which result in nodes. The nodes occur at each end of the string, and the allowed waveform is shown in Fig. 4–12.

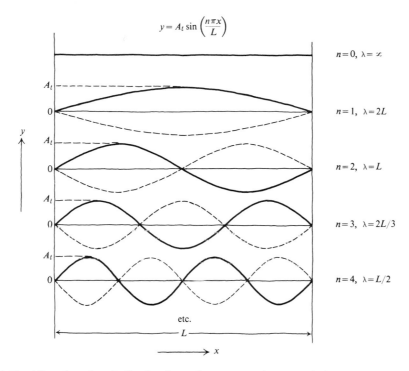

$$y = A_t \sin\left(\frac{n\pi x}{L}\right)$$

FIG. 4–12   Allowed modes of vibration for stationary waves in a stretched string fixed at both ends. The dashed curves show the displacements of the string at a time which is one-half period later.

When $n = 2$,

$$nx/L = 2x/L,$$

which may be an integer when $x = 0$, $L/2$, or L. For $n = 2$, nodes thus occur at each end and in the middle of the string (Fig. 4–12).

When $n = 3$,

$$nx/L = 3x/L,$$

which may be an integer when $x = 0$, $L/3$, $2L/3$, or L. For $n = 3$, nodes then occur at each end, at $L/3$ and at $2L/3$ (Fig. 4–12).

An examination of Fig. 4–12 indicates that *n is a direct measure of the number of antinodes in the stationary wave.* Corresponding allowed *wavelengths* are also listed in Fig. 4–12. Observe that *only those wavelengths are allowed or persist for which L, the interval of confinement of the wave, is an integral multiple of one-half the wavelength.* All other wavelengths are lost through destructive interference, and the final waveform for the string must be a linear combination of any or all of the allowed modes. That is, the string may be vibrating in a waveform which represents a superposition of several or all of the allowed modes, depending on how the string is plucked or bowed.

We have shown that the imposition of boundary conditions on (or the confinement of) the traveling wave results in solutions which are restricted to a series of wavelengths related by integers. Only certain restricted forms of motion are allowed. Although we have been concerned specifically with vibrating strings, it may be shown that the results we have obtained are applicable to all forms of waves, including sound waves and electromagnetic waves. Recall that we have subjected electromagnetic radiation to the standing-wave requirement in our previous treatment of blackbody radiation in Chapter 2, where we confined radiation within a cavity and required that the distance between blackbody walls be an integral multiple of one-half the wavelength for all persistent vibrations. In quantum mechanics we'll treat electrons and other particles as waves and will be much concerned with the stationary de Broglie waves which are formed when such particles are confined within the restricted spaces corresponding to atoms and molecules.

4–9 THE STATIONARY WAVE AS A SYSTEM OF COUPLED LINEAR HARMONIC OSCILLATORS

In order to gain some insight into the time behavior of a particular particle in the string at some given value of x, we may hold x constant in Eq. (4–51) so that the displacement y is then a function of time only. Thus Eq. (4–51) may be written as

$$y = A_x \cos \left(\frac{n \pi u t}{L} \right), \tag{4-53}$$

where $A_x = A' \sin (n\pi x / L) =$ constant. The vibration of such a particle is shown in Fig. 4–13. We may consider the wave for which $n = 2$ as an example. The position of the string is shown at various successive times: t_1, t_2, t_3, t_4. Times t_1 and t_4 correspond to the positions of maximum and minimum displacement for *any* value of x. According to Eq. (4–53), a point in

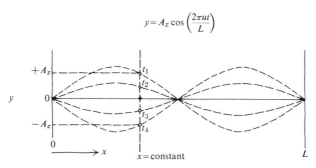

FIG. 4–13 The positions of a stationary wave at successive times. Any point in the string undergoes simple harmonic motion.

the string at any selected value of x will undergo simple harmonic motion, since the form of Eq. (4–53) is the same as that of Eq. (4–17), which defined simple harmonic motion, that is,

$$y = A \cos (\omega t + \delta). \qquad (4\text{–}17')$$

If Eqs. (4–17) and (4–53) are to represent the same motion, the initial phase δ must be zero and the angular frequency must be given as

$$\omega = n\pi u/L,$$

or, since $\omega = 2\pi v$,

$$2\pi v = n\pi u/L,$$

from which

$$v = nu/2L. \qquad (4\text{–}54)$$

But in a stretched string, according to Eq. (4–40),

$$u = \sqrt{F/\mu},$$

so that combination of the above relationship with Eq. (4–54) gives

$$v = (n/2L)\sqrt{F/\mu}. \qquad (4\text{–}55)$$

Equation (4–55) describes the allowed frequencies in a stretched string which is fixed at both ends. Note that regardless of the position of x, all particles in the string oscillate on their y-axes at the same frequency. The entire string might thus be imagined as a longitudinal series of an infinite number of coupled simple harmonic oscillators, all vibrating at the same frequency on parallel transverse axes (Fig. 4–14).

A violin string is a good example of a fixed stretched string. The allowed frequencies are

$$n = 1: \qquad v_1 = \frac{1}{2L}\sqrt{\frac{F}{\mu}}, \qquad \text{fundamental (first harmonic),}$$

$$n = 2: \qquad v_2 = \frac{2}{2L}\sqrt{\frac{F}{\mu}} = 2v_1, \qquad \text{first overtone (second harmonic),}$$

$$n = 3: \qquad v_3 = \frac{3}{2L}\sqrt{\frac{F}{\mu}} = 3v_1, \qquad \text{second overtone (third harmonic),}$$

and so forth. The overtone frequencies are observed as integral multiples of the fundamental frequency. In practice, the displacement of a violin string usually results in a waveform which is a superposed form of many of the allowed frequencies with different amplitudes. Among other things, the exact tonal quality of a violin note depends on the particular combination of fundamental and overtones and on their relative intensities, which in turn depends on the way in which the string is bowed.

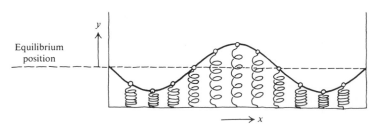

Equilibrium
position

FIG. 4–14 The stationary wave viewed as a series of coupled linear harmonic oscillators, all vibrating at the same frequency but with different amplitudes on parallel transverse axes.

One may generally observe, however, that for stationary waves having the same amplitudes at the antinodes, larger values of n represent higher energies in the string. We have shown in the case of the simple linear harmonic oscillator (the mass on a spring) that the energy of the oscillator is directly proportional to the square of the amplitude (Eq. 4–22). This same relationship is true for each differential segment of the string in a stationary wave, so that the energy per unit length of string or *energy density* at any point in the vibrating string is directly proportional to the square of the time-independent amplitude, A_x, at the point. That is,

$$\text{Energy density} \propto A_x^2.$$

Nodes thus represent positions of no energy and antinodes represent positions of maximum energy. It is clear, since A_x is a function of x, that the distribution of energy in the string between its end points is not uniform, but varies in such a way that the total energy density at any value of x remains constant with time. Furthermore, there are certain positions of x where the energy density is zero and certain positions where the energy density is a maximum.

By analogy, the ability to calculate variations in three-dimensional space of allowed time-independent amplitudes of deBroglie waves for confined particles such as electrons will be very helpful in our development of quantum mechanics. In particular, the *square* of the time-independent amplitude will be most important.

In the winter of 1821–2, one of us, while purifying some mercury by pouring it through a funnel from one flask to another, observed that at the surface of the mercury in the second flask the stream . . . caused an exceedingly complicated but regular pattern . . . which was seen to persist . . . unaltered as long as the inpouring mercury struck the surface at the same place and at the same velocity. But if the conditions varied, the pattern also changed. He recognized this figure as resulting from waves continually intercrossing each other at the same place . . . These results are best comprehended as due to the second mode, namely, stationary oscillations. These actually occur more frequently than progressive waves but so far, they have been almost entirely overlooked. It is to standing waves that we wish to direct the attention of physicists and mathematicians . . .

—ERNST and WILHELM WEBER, *Die Wellenlehre*, Leipzig (1825).

PROBLEMS

4–1 A particle oscillates with simple harmonic motion in which the displacement in centimeters is given by $y = 2 \cos (3\pi t + \pi/4)$ cm.
a) What is the amplitude in centimeters?
b) What is the initial phase?
c) What is the angular frequency?
d) What is the period of oscillation?
e) Calculate the displacement, velocity, and acceleration of the particle at the time $t = 1$ second.

4–2 What is the natural frequency of vibration of a 20-g mass oscillating on the end of a spring whose other end is fixed when the force constant is

a) 100 dyne-cm^{-1},   b) 1000 dyne-cm^{-1}.

4–3 At room temperature the vibrational frequency of a single silver atom in the solid state is about 10^{13} Hz. Assuming that the remaining atoms in the solid are at rest, and that the oscillating silver atom behaves like a mass on a single "spring," calculate the force constant of the "spring" (the atomic mass of silver is 107.9 amu).

4–4 A single mass of 1 g oscillates on a spring whose force constant is 2000 dyne-cm^{-1} with an amplitude of 1 cm. Assuming that the energy is quantized according to Planck's equation,

$$E = (n + \tfrac{1}{2})h\nu,$$

calculate the value of the quantum number n. What is the fractional energy increase when the oscillator is excited to the next highest quantum state (when n increases by unity)?

4–5 Show that $a \cos (bt + c)$ and e^{ibt} are eigenfunctions of the operator d^2/dt^2 ($i^2 = -1$, and b is a constant). What is the eigenvalue in each case?

4–6 Show that $z \exp (-z^2/2)$ is an eigenfunction of the operator, $-d^2/dz^2 + z^2$. What is the eigenvalue?

4–7 Prove that the function $c \exp (-ax)$ is an eigenfunction of the operator d^2/dx^2 and determine the eigenvalue (c and a are constants).

4–8 Show that $\psi(x, y) = A \exp[B \tan^{-1}(y/x)]$ is an eigenfunction of the operator

$$\left(x\frac{\partial}{\partial y} - y\frac{\partial}{\partial x} \right).$$

What is the eigenvalue? Hint:

$$\frac{d}{dx} \tan^{-1}u = \frac{1}{1 + u^2}\frac{du}{dx}$$

4–9 The function $\exp(-ax^2)$ is an eigenfunction of the operator $d^2/dx^2 - bx^2$ only under certain conditions. What are these conditions and what is the eigenvalue under these conditions? (a and b are constants.)

4–10 Prove that the linear-combination function

$$y = A \sin(\omega t + \delta) + B \cos(\omega t + \delta') + C \exp[i(\omega t + \delta'')],$$

where A, B, and C are arbitrary constants, is a satisfactory description of simple harmonic motion, that is, that the function satisfies Eq. (4–7).

4–11 Prove, by substitution in Eq. (4–32), that

$$y = A \cos a(x \pm ut + c')$$

is a satisfactory description of the traveling wave in one dimension.

4–12 Show that the linear-combination function

$$y = A \cos a(x \pm ut + c) + B \sin a(x \pm ut + c')$$

is a satisfactory equation for a wave traveling in one dimension.

4–13 Prove, by substitution into Eq. (4–32), that Eq. (4–51) satisfies the general differential equation of wave motion.

4–14 Show whether or not the function

$$y = A' \cos\left(\frac{n\pi x}{L}\right) \sin\left(\frac{n\pi ut}{L}\right)$$

is also a satisfactory equation for the stationary wave. That is, does it satisfy both Eq. (4–32) and the appropriate boundary conditions?

4–15 Consider two different standing waves in the same string:

$$y_n = A \sin\left(\frac{n\pi x}{L}\right) \cos\left(\frac{n\pi ut}{L}\right)$$

$$y_m = B \sin\left(\frac{m\pi x}{L}\right) \cos\left(\frac{m\pi ut}{L}\right)$$

where n and m are fixed integers.

a) Show that $y = y_n + y_m$ is a satisfactory equation for the motion of the string.
b) What relationship must exist between n and m for nodes to occur in the above string other than at the ends?

4–16 The equation for the displacement in a transverse wave traveling in a rope is

$$y = 0.5 \sin \pi(3.5x - 2.0t),$$

where x and y are in centimeters and t is in seconds.

  a) What is the amplitude of the wave?
  b) What is the frequency of the wave?
  c) What is the wavelength?

4–17 a) Consider the equation for displacement in a traveling wave:

$$y = b \sin (cx + dt + f)$$

  where b, c, d, and f are real positive constants. What is the amplitude of the wave?

  b) What is the phase velocity? In which direction is the wave traveling?
  c) Prove that the above displacement equation satisfies the differential equation for a one-dimensional traveling wave.

4–18 The transverse displacement of an element of string in a stationary wave is given by Eq. (4–51).

  a) Derive an equation for the *transverse* velocity w of a particle in the string at a given value of x, that is, $(\partial y/\partial t)_x$.
  b) The total energy of a given particle is the sum of its potential and kinetic energies. However, when $y = 0$, w is a maximum, and the total energy of the particle is given by the kinetic energy only. Show that for a given value of x,

$$w_{max} = \frac{\pm A' n \pi u}{L} \sin \left(\frac{n\pi x}{L}\right).$$

  c) Consider an element of string whose mass per unit length is given by μ. Show that the energy of the string per unit length is

$$E = \left[\frac{\mu n^2 \pi^2 u^2}{2L^2} \sin^2 \left(\frac{n\pi x}{L}\right)\right] A'^2.$$

  d) What is the energy per unit length at the nodes?
  e) What is the expression for energy per unit length at the antinodes?

Up to now we have been concerned primarily with one-dimensional wave motion, which is relatively simple to picture in a physical sense. Waves which propagate in more than one dimension are somewhat more difficult to visualize.

5-1 TWO-DIMENSIONAL WAVES

When a pebble is thrown into a pond, the resulting ripple wave travels in two dimensions. Similarly, waves travel in two dimensions in a stretched membrane in a drumhead. In fact, a vibrating drumhead represents a good two-dimensional analog of the one-dimensional stretched string fixed at both ends. We may calculate that only certain allowed standing two-dimensional waves may persist when the membrane is struck. Furthermore, two-dimensional nodes and antinodes occur. For the two-dimensional membrane the displacement D is a function of two coordinate positions x and y and time t. As for the stretched string fixed on both ends, only certain allowed types or *modes* of vibration persist, and the solutions for the allowed modes directly involve integers. As an example, several of the allowed *normal modes* of vibration for a uniformly stretched and clamped square membrane are shown in Fig. 5–1. The boundary conditions, which require that the displacement be zero along the entire periphery, leads to the superposition of stationary waves in both the x- and y-directions. In each direction the allowed or persistent modes are those for which the distance across the membrane is an integral number of half-wavelengths. It can be shown that at any time the total displacement D is proportional to the *product* of *two* sine terms, each analogous to the single sine term previously given in Eq. 4–52 for the one-dimensional standing wave.

It is important to recognize that the restriction of motion to discrete allowed modes of vibration of the membrane results from the requirement that the displacement function $D(x, y, t)$ simultaneously obey the *differential wave*

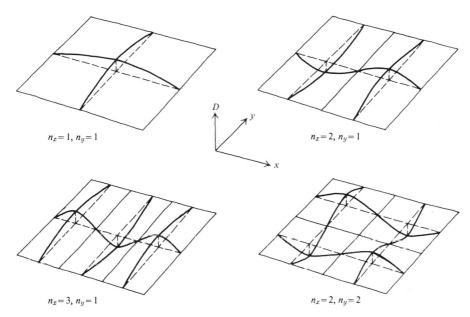

$n_x = 1, n_y = 1$

$n_x = 2, n_y = 1$

$n_x = 3, n_y = 1$

$n_x = 2, n_y = 2$

FIG. 5–1 Several of the allowed stationary modes of vibration for a square membrane uniformly stretched and clamped on the periphery. n_x is the number of half wavelengths in the x-direction and n_y is the number of half wavelengths in the y-direction. The displacements are exaggerated. Solid lines in the plane of the square represent nodes.

equation in two dimensions,

$$\frac{\partial^2 D}{\partial x^2} + \frac{\partial^2 D}{\partial y^2} = \frac{1}{u^2}\left(\frac{\partial^2 D}{\partial t^2}\right), \qquad (5\text{--}1)$$

and also obey the boundary condition, which requires that the entire periphery be a node. *If the two-dimensional waves were not confined, that is, if the surface of the membrane were infinite, nodes and antinodes would not arise.* Equation (5–1) is simply the two-dimensional analog of Eq. (4–32), which was the differential equation for the one-dimensional wave. The displacement D in two dimensions is analogous to the displacement y in one dimension.

5–2 THE GENERAL DIFFERENTIAL WAVE EQUATION IN THREE DIMENSIONS

Most of the problems of interest in quantum mechanics involve the confinement of particles having wave properties within three-dimensional enclosures. For example, the wave associated with the electron in the hydrogen atom must obey certain boundary conditions. It must also obey an appropriate differential wave equation. We would expect that only certain allowed modes of vibration involving integers may persist, but we do not as yet know exactly what an allowed mode of vibration for an electron really signifies.

The travel of an ordinary sound wave in air is a good example of a three-dimensional wave. If the sound wave is not confined, the wave motion may be represented as the emanation of spherical traveling waves which radiate outward in all directions from the source. If, however, we confine a three-dimensional sound wave within an enclosure such as a spherical cavity or a box (that is, apply boundary conditions), we cause interference through reflection of the sound waves from the boundaries, with the result that only certain allowed modes of vibration persist. That is, three-dimensional stationary waves are produced which consist of only certain allowed combinations of nodes and antinodes. Displacement in this three-dimensional system may be represented in terms of the concentration or density of gas molecules. At those positions of x, y, and z at which nodes occur in the standing wave, the concentration of gas molecules remains constant. At the antinode positions the extent of periodic fluctuation in gas density, that is, alternate compression and rarefaction, is maximum, and the sound intensity is greatest. At intermediate space positions between nodes and antinodes, the displacement, which is the difference between the concentration of gas molecules at a given position and the concentration of gas molecules at the node positions, periodically fluctuates with an amplitude somewhere between zero and the maximum amplitude given at the antinode positions.

In addition, at the antinode positions, the greatest number of gas molecules rush in and out of a given volume in a unit time so that the energy density associated with the wave motion is greatest at the antinodes. In fact, just as for the stationary wave in one dimension, the energy density at a given position in the coordinate system is directly proportional to the square of the space-dependent amplitude at that point.

If we now define Φ as the displacement of a three-dimensional wave, where Φ is a function of the three coordinate positions x, y, and z and is also a function of time t, we may, by analogy to Eqs. (4–32) and (5–1), write the *general differential equation of wave motion in three dimensions* as

$$\frac{\partial^2 \Phi}{\partial x^2} + \frac{\partial^2 \Phi}{\partial y^2} + \frac{\partial^2 \Phi}{\partial z^2} = \frac{1}{u^2}\frac{\partial^2 \Phi}{\partial t^2}. \tag{5-2}$$

We may then define the operator ∇^2 such that

$$\nabla^2 = \frac{\partial^2}{\partial x^2} + \frac{\partial^2}{\partial y^2} + \frac{\partial^2}{\partial z^2}, \tag{5-3}$$

so that Eq. (5–2) may be written more simply as

$$\nabla^2 \Phi = \frac{1}{u^2}\frac{\partial^2 \Phi}{\partial t^2}. \tag{5-4}$$

The operator ∇^2 ("del squared") is called the *Laplacian* operator.

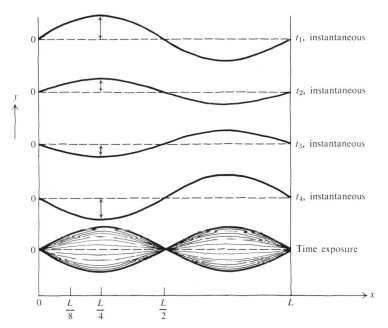

FIG. 5–2   Camera exposures of a vibrating string in the second allowed mode of vibration.

5–3   TIME-AVERAGE DISPLACEMENT IN STATIONARY WAVES

Equation (5–4) applies to all wave motion, including electromagnetic waves, and expresses the displacement Φ as a function of both *position* and *time*. If our purpose is to locate the positions of the nodes and antinodes in *stationary* waves produced by the imposition of boundary conditions, we are really interested in the *time-average* value of some displacement function at a particular point in space.

Let's briefly return to a consideration of one of the allowed modes of vibration for the violin string (Fig. 5–2). If we consider any given point in the string, for example, that point at which $x = L/4$, the value of the displacement y is seen to vary with time. We may imagine that we are snapping instantaneous camera exposures of the string at various times. If we leave the shutter open over at least a half period of vibration, we will produce a representative time exposure of the vibration. An examination of the time exposure photograph reveals the positions of nodes and antinodes in the stationary wave. It also reveals the size of the x-dependent amplitude of the vibration at any point in the string. Furthermore, the time exposure pattern of the wave is *time independent*. It may be thought of as an average picture of the wave over a representative period of time. In comparison, the time exposure photograph of a *traveling* wave in a free string (where no boundary conditions are imposed) appears as shown in Fig. 5–3, and no nodes or antinodes are detectable.

FIG. 5–3. Time exposures of the displacement in a free string or in a free traveling wave.

In the case of the equation for the one-dimensional stationary wave (Eq. 4–51), in which the magnitude u of the phase velocity was constant and in which solutions were restricted through boundary conditions to a set of privileged constant values for v, we were able to separate the complete displacement function into the product of a *space-dependent* part and a *time-dependent* part. It may be shown that when u and v are constant, we may make a similar separation in the equation for displacement of the *three-dimensional stationary* wave. We'll then try to relate the *space-dependent part* (time-independent part) of the function to a measure of the average displacement at any point, which is also time-independent. In other words, we'll try to define a function, dependent on space coordinates only, which will allow us to locate the positions of nodes and antinodes in a three-dimensional standing-wave system.

Let's begin by expressing $\Phi(x, y, z, t)$, the displacement function which must satisfy Eq. (5–4), as the product of $\psi(x, y, z)$, which depends only on position, multiplied by a periodic function which depends only on time. We thus write

$$\Phi(x, y, z, t) = \psi(x, y, z)e^{-i2\pi vt},$$

where $i^2 = -1$ and v is the frequency of oscillation of displacement in time at any space position. Or, more simply,

$$\Phi = \psi e^{-i2\pi vt}. \tag{5–5}$$

The time-dependent portion, $e^{-i2\pi vt}$, is arbitrary but will prove to be well chosen. We do know that Φ must be a periodic function of t at any given point in space, and we could, therefore, just as well choose $e^{+i2\pi vt}$, $\sin 2\pi vt$, or $\cos 2\pi vt$ as the time-dependent portion.* Recall that if v is constant (that is, independent of time) all these functions (the exponential function, the sine function, the cosine function, and any linear combinations thereof) are eigenfunctions of the operator $\partial^2/\partial t^2$, a condition required for simple harmonic motion. The exact nature of the space-dependent function, ψ, will depend on the physical parameters of the system and on the boundary conditions imposed. Before we continue with the problem of removing time dependence from the wave equation, we'll briefly review some of the characteristics of complex numbers and functions.

---

* Our sole reason for choosing the negative exponential form is that it will be consistent with the time-dependence relationship for *particle* waves which we will develop in Chapter 6.

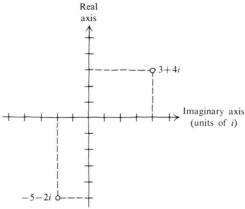

FIG. 5-4   The representation of complex numbers as points in the complex plane.

5-4 COMPLEX NUMBERS AND EULER'S FORMULA

The square root of a negative number is historically expressed in terms of the imaginary quantity i whose square is defined as -1, that is,

$$i^2 \equiv -1$$

so that, for example,

$$(2i)^2 = -4.$$

Thus, a number of the form

$$c = a + bi \qquad (5\text{-}6)$$

contains both a *real part*, a, and an *imaginary part*, bi, and is called a *complex number*. The *complex conjugate* of a complex number is defined by replacing i with $(-i)$ wherever it occurs. Thus the complex conjugate of c in Eq. (5-6) is

$$c^* = a - bi. \qquad (5\text{-}7)$$

It is important to note that the *product of a complex number and its complex conjugate is a real number;* thus

$$cc^* = (a + bi)(a - bi) = a^2 + b^2.$$

We express the *magnitude* or the *absolute value* of a complex number as

$$|c| = (cc^*)^{1/2} = (a^2 + b^2)^{1/2}, \qquad (5\text{-}8)$$

where the absolute value, $|c|$, is always real. As an aid in the mathematical handling of complex numbers, it is convenient to picture a complex number as a point on a two-dimensional graph having a *real axis* and an *imaginary axis* (Fig. 5-4). The plane created by the intersection of these two axes is called the *complex plane*. In terms of such a graphical representation, it can be shown

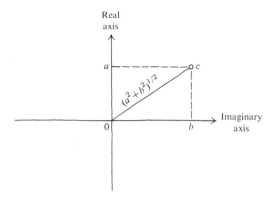

FIG. 5–5   The absolute value of the complex number $a + bi$ is the distance $0c$ which is equal to $(a^2 + b^2)^{1/2}$.

easily that the absolute value of a complex number is nothing more than the length of the line between the origin and the point representing the complex number (Fig. 5–5).

Complex numbers containing e^{ix} are particularly useful in wave mechanics. We have, for example, shown that $e^{i a x}$ is a satisfactory eigenfunction of the important operator $\partial^2/\partial x^2$. In order to construct a plot showing the variation of e^{ix} with x, it is convenient to express e^{ix} as a combination of a real cosine wave and an imaginary sine wave. *Euler's formula*, which is

$$e^{ix} = \cos x + i \sin x, \tag{5–9}$$

may be proved easily by expanding each of the terms in Eq. (5–9) in a Maclaurin series. The $\cos x$ series and the $i \sin x$ series generate alternate terms in the e^{ix} series. A graph of e^{ix} versus x is shown in Fig. 5–6. The real part of e^{ix} is plotted in the vertical plane as $\cos x$ versus x, while the imaginary part of e^{ix} is plotted in the horizontal plane as $i \sin x$ versus x. The *complex plane* is then perpendicular to the x-axis so that the absolute value of the complex number, graphically represented by a unit vector, is imagined to rotate counterclockwise as x increases. The function e^{ix} is shown to repeat itself exactly at the end of each period τ.

5–5 BACK TO THE TIME-AVERAGE DISPLACEMENT

Recall that our present objective is to locate node and antinode positions in the three-dimensional stationary wave. Let's, for the time being, imagine that we have chosen to express the time dependence of the displacement Φ in terms of a sine or cosine function, that is,

$$\Phi = \psi \sin 2\pi \nu t \qquad \text{or} \qquad \Phi = \psi \cos 2\pi \nu t.$$

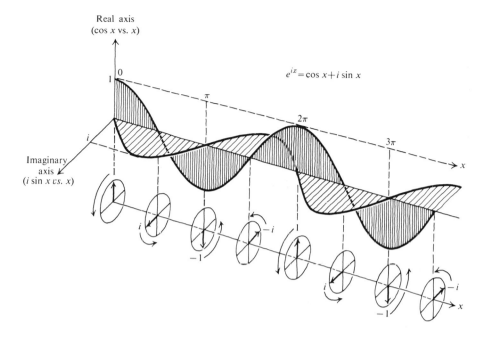

FIG. 5-6   A plot of e^{ix} as $(\cos x + i \sin x)$ versus x. $\cos x$ is a real number and is plotted as the vertical cosine wave, whereas $i \sin x$ is an imaginary number and is plotted as the horizontal sine wave on the imaginary axis. The complex planes shown below the curve are perpendicular to the x axis. At $x = 0, \pi, 2\pi, 3\pi, \ldots$, the function e^{ix} is entirely real. At $x = \pi/2, 3\pi/2, 5\pi/2, \ldots$, the function e^{ix} is entirely imaginary. At all intermediate values of x, e^{ix} is complex and consists of a real and an imaginary part.

Our first inclination might be to define antinode positions as those positions at which the average displacement with time is the greatest. However, a bit of reflection leads us to conclude that the average displacement at any point in space over a period of time is zero, since in uniform periodic motion positive displacement is exactly canceled by negative displacement. A better time-average displacement function would be Φ^2, since this is always positive even when Φ itself is negative. The time-average value of Φ^2 over a long period of time is then equal to the average of Φ^2 over a single representative period, since the waveform completely repeats itself in one period, that is, a time equal to τ or $1/\nu$.

Thus

$$\overline{\Phi^2} = \frac{1}{\tau} \int_0^\tau \Phi^2 \, dt, \qquad (5\text{--}10)$$

where $\overline{\Phi^2}$ is the *time average* of Φ^2. But according to Eq. (5–5), we have chosen to express Φ as a *complex* number, that is,

$$\Phi = \psi e^{-i2\pi vt}, \tag{5-5'}$$

in which the *real* part of the time-dependent exponential function periodically changes sign, as shown graphically in Fig. 5–6. Although it is often convenient and sometimes necessary to use complex numbers in derivations and calculations, we must always require that the final results which we are to compare to observed data be *real*, since experimental results are always real. Thus, in order to always obtain a *real* measure for the time-average displacement, regardless of whether we use a sine, cosine, or complex exponential function for time dependence, Eq. (5–10) must be rewritten more generally as

$$\overline{\Phi^*\Phi} = \overline{|\Phi|^2} = \frac{1}{\tau} \int_0^\tau \Phi^*\Phi \, dt, \tag{5-11}$$

where Φ^* is the complex conjugate of the displacement function, and $|\Phi|$ is the absolute value of the displacement function. Then, using the complex exponential forms for time dependence,

$$\Phi = \psi e^{-i2\pi vt} \quad \text{and} \quad \Phi^* = \psi^* e^{+i2\pi vt},$$

we may write Eq. (5–11) as

$$\overline{\Phi^*\Phi} = \overline{|\Phi|^2} = \frac{\psi^*\psi}{\tau} \int_0^\tau e^{i2\pi vt} e^{-i2\pi vt} \, dt, \tag{5-12}$$

where ψ and its complex conjugate ψ^* are both time-independent. Continuing, we have

$$\overline{|\Phi|^2} = \frac{\psi^*\psi}{\tau} \int_0^\tau e^0 \, dt = \frac{\psi^*\psi}{\tau} \int_0^\tau dt,$$

or

$$\overline{|\Phi|^2} = \psi^*\psi = |\psi|^2. \tag{5-13}$$

Thus at any point in three-dimensional space, $\overline{|\Phi|^2}$, the *time-average* value of the square of the absolute displacement function, is equal to $|\psi|^2$, the square of the absolute value of the *time-independent* portion of the complete displacement function. The quantity ψ is called the *space-dependent amplitude function;* it may or may not be complex, depending on the particular application. Basically then, $\psi^*\psi$ or $|\psi|^2$ is a measure of the amount of activity occurring at a particular point in space in a stationary wave. If a uniform three-dimensional wave is *unconfined* and is thus *not* a standing wave, $|\Phi|^2$ is no longer equal or proportional to $|\psi|^2$ but is in fact identical for all points in space, just as we infer from Fig. 5–3 that the time average of the square of the displacement in an unconfined *one*-dimensional traveling wave is identical at all points along the x-axis of propagation.

When, however, a three-dimensional wave is confined, or when boundary conditions are imposed, nodes and antinodes result, and $|\psi|^2$ varies with space position in such a way that its value directly indicates the time average of the square of the absolute displacement at that point. In stationary waves the function $|\psi|^2$ locates the *regions of maximum physical activity and energy density*, the *antinodes*, and the *areas of no activity or energy density*, the *nodes*.

5–6 THE TIME-INDEPENDENT CLASSICAL WAVE EQUATION

We can derive a time-independent wave equation from which the amplitude ψ can be evaluated for stationary waves by inserting Eq. (5–5),

$$\Phi = \psi e^{-i2\pi vt}, \tag{5–5'}$$

into Eq. (5–4), the general differential wave equation,

$$\nabla^2 \Phi = \frac{1}{u^2}\left(\frac{\partial^2 \Phi}{\partial t^2}\right), \tag{5–4'}$$

where u is constant, remembering that

$$\nabla^2 = \frac{\partial^2}{\partial x^2} + \frac{\partial^2}{\partial y^2} + \frac{\partial^2}{\partial z^2}.$$

Thus, assuming that v is not a function of x, y, z, or t, that is, that v is restricted to constant values in stationary-wave solutions of the general wave equation, and differentiating Eq. (5–5), we obtain

$$\frac{\partial \Phi}{\partial x} = \frac{\partial \psi}{\partial x} e^{-i2\pi vt},$$

$$\frac{\partial^2 \Phi}{\partial x^2} = \frac{\partial^2 \psi}{\partial x^2} e^{-i2\pi vt}. \tag{5–14}$$

Similarly,

$$\frac{\partial^2 \Phi}{\partial y^2} = \frac{\partial^2 \psi}{\partial y^2} e^{-i2\pi vt}, \tag{5–15}$$

and

$$\frac{\partial^2 \Phi}{\partial z^2} = \frac{\partial^2 \psi}{\partial z^2} e^{-i2\pi vt}. \tag{5–16}$$

Also

$$\frac{\partial \Phi}{\partial t} = \psi(-i2\pi v)e^{-i2\pi vt},$$

$$\frac{\partial^2 \Phi}{\partial t^2} = \psi(i2\pi v)^2 e^{-i2\pi vt} = -\psi 4\pi^2 v^2 e^{-i2\pi vt}. \tag{5–17}$$

Substitution of Eqs. (5–14), (5–15), (5–16), and (5–17) into Eq. (5–4) yields

$$\frac{\partial^2 \psi}{\partial x^2} + \frac{\partial^2 \psi}{\partial y^2} + \frac{\partial^2 \psi}{\partial z^2} = \frac{-4\pi^2 v^2 \psi}{u^2}, \tag{5–18}$$

and substitution of the Laplacian operator symbol yields

$$\nabla^2 \psi = \frac{-4\pi^2 v^2}{u^2}\, \psi, \tag{5–19}$$

which may be rearranged to

$$\left(\frac{-u^2 \nabla^2}{4\pi^2}\right)\psi = v^2 \psi. \tag{5–20}$$

Eq. (5–20) is valid for three-dimensional systems in which stationary waves result from boundary conditions. We find, just as we found in one-dimensional problems, that solutions exist and may thus be obtained only for a privileged set of discrete values for v^2, which correspond to the squares of allowed frequencies for the standing or stationary waves. The privileged v^2-values are the eigenvalues of the operator $-u^2 \nabla^2/4\pi^2$, and the corresponding ψ-functions which satisfy Eq. (5–20) are the eigenfunctions of the same operator. Thus, we'll usually find that a *series* of eigenfunctions satisfies the conditions of a particular problem, and that a *series* of eigenvalues results. Recall, as a comparative example, the series of displacement equations representing the allowed modes of vibration which were generated as solutions for the one-dimensional violin string.

Furthermore, once we have evaluated the eigenfunctions, we may immediately write ψ^* for each of the solutions by substituting $(-i)$ for i wherever it appears so that we may write

$$\psi^* \psi = |\psi|^2 = \overline{|\Phi|^2}. \tag{5–21}$$

from which we may determine where the energy density is localized for each of the modes.

It is also important to recall that the wave function which we have conveniently separated as the product of a space-dependent part and a time-dependent part, that is,

$$\Phi(x, y, z, t) = \psi(x, y, z)e^{-i2\pi v t}, \tag{5–5'}$$

is satisfactory *only* when v is constant, that is, only when the system is in a stationary state. Thus Eq. (5–21) is a valid relationship only for *stationary states* in which v is restricted to a constant. The determination of $\overline{|\Phi|^2}$ when v is a function of t is considerably more difficult.

Finally, we recall that $u = \lambda v$ and write Eq. (5–19) in a somewhat simpler form,

$$\nabla^2 \psi = -4\pi^2 \psi/\lambda^2, \tag{5–22}$$

to which we'll again refer in Chapter 6.

We'll resist the temptation to develop solutions of Eq. (5–20) for classical boundary-value problems. Rather, let's once again return in analogy to the intriguing pilot waves of Louis de Broglie.

We are trying always to feel our way into something new and unexperienced. We take into it what we have, which is our own experience, in this case of the physical world, and we seek a relevant pattern of form and order. . . . You have, in entering novelty, to use what you know. You would not be able to make meaningful mistakes without analogy. You would not be able to try things out, the failure of which was interesting. You start thinking by the use of analogy. Analogy is not the criterion of truth; it is an instrument of creation, and the sign of the effort of human minds to cope with something novel, something fresh, something unexpected.

—J. Robert Oppenheimer.*

PROBLEMS

5–1 Plot each of the following complex numbers in the complex plane: $1 + i, 1 - i$, $4 - 3i, -4 + 3i, i\pi$.

5–2 Write the complex conjugates and determine the absolute values for each of the complex numbers given in Problem 5–1.

5–3 Maclaurin's series for the expansion of $f(x)$ is given in Problem 3–8. Prove Euler's formula by expanding each of the terms in Eq. (5–9) in a Maclaurin series.

5–4 Assume that we had arbitrarily chosen to use $\sin 2\pi v t$ as the periodic time-dependent function in Eq. (5–5) rather than $e^{-i2\pi v t}$, that is, that $\Phi = \psi \sin 2\pi v t$.

  a) Show that $|\psi|^2$ would still be directly proportional to $\overline{|\Phi|^2}$, or that $\overline{|\Phi|^2} = k|\psi|^2$, where k is a constant. What is the value of k?

  b) What would k be if we had used $\cos 2\pi v t$ as the time-dependent function?

5–5 Show whether or not each of the following displacement functions is an eigenfunction of the Laplacian operator. If so, write the eigenvalue in each case (a, b, and c are constants):

  a) $a(\sin bx)(\cos cy)$,

  b) $a(\sin bx + \cos cy)$.

---

* From "Electron Theory," *Physics Today*, **10**, No. 7, 12–20 (1957).

In our study of classical waves in the last two chapters we have been concerned mostly with the expression and evaluation of displacement as a function of position and time. The total displacement function $\Phi(x, y, z, t)$ for a three-dimensional wave must be determined by finding a suitable function which satisfies both the general differential wave equation *and* any boundary conditions imposed in a specific situation. We also found it very useful to be able to separate $\Phi(x, y, z, t)$ into the product of a time-independent amplitude function $\psi(x, y, z)$, and an appropriate time-dependent function $e^{-i2\pi\nu t}$ for stationary states in which ν was restricted to constant values. In particular, when boundary conditions were imposed, we found the expression for the space-dependent function, $\psi(x, y, z)$, to be quite helpful in allowing us to determine the distribution of energy density in space.

For all of the classical waves which we have studied, the satisfactory displacement function in effect has been a complete source of all information concerning the wave. For example, the displacement function $y(x, t)$ for the standing wave in the violin string (Eq. 4–51) intrinsically contains all measurable information concerning the wave. It allows, through proper operation, the calculation of displacement at any point in space or time. It allows the calculation of the frequencies of the infinite number of allowed modes of vibration, from which all possible motions of the string may be synthesized by simple superposition. In a sense, the satisfactory displacement equation for a wave represents the clearest and most concise definition of the wave which one can construct and represents the essence of all information concerning the wave motion.

6-1 DISPLACEMENT FUNCTIONS FOR PARTICLE WAVES

If, as de Broglie postulated, all particles have symbolic pilot waves associated with them whose frequencies are given by $\nu = E/h$ and whose wavelengths are given by $\lambda = h/mv$, then each individual particle must have associated with it some property which is analogous to the classical displacement. In fact, we might expect to find an appropriate displacement-analog function $\Psi(x, y, z, t)$ which completely describes the particle. We might also expect, *in cases involving stationary states*, to be able to separate the general function $\Psi(x, y, z, t)$ into a space-dependent amplitude function $\psi(x, y, z)$ and a time-dependent function $f(t)$, so that

$$\Psi(x, y, z, t) = \psi(x, y, z)f(t). \qquad (6-1)$$

Such an expectation is the basis for the first postulate of quantum mechanics:

POSTULATE I. *The state of a single particle is described as fully as possible by an appropriate state function,* $\Psi(x, y, z, t)$, *which may be expressed for stationary energy states as the product of a time-independent amplitude function,* $\psi(x, y, z)$ *and a time-dependent function* $f(t)$. *Both functions,* Ψ *and* ψ, *must be single-valued, continuous, and finite for all values of their coordinates, and must be smoothly varying within the boundaries at which they go to zero.*

A function which is single-valued, continuous, finite, and smoothly varying is said to be *well-behaved*. That is, if it is single-valued and continuous, it has only one value at any given point in coordinate space and there are no gaps or discontinuities in the function. Smooth variation of the function is assured by requiring that its first derivative be continuous so that there will be no sharp angular changes between the boundaries at which the function goes to zero. The requirement that the function be finite* will be understood more clearly when we discuss probability relationships later in this chapter. Some examples of *poorly behaved* functions are shown in Fig. 6–1.

The argument for the first postulate, as well as arguments for other postulates which are to follow, will have to lie in the success with which resultant equations predict the experimental measurement of physical properties. Of course, at this point, we know neither the form nor the significance of the state function Ψ for a particle such as an electron. But we have shown that the *classical* displacement function $\Phi(x, y, z, t)$ is separable into the product of a space-dependent function, $\psi(x, y, z)$, and a time-dependent function, $e^{-i2\pi\nu t}$, *in those specific cases in which stationary waves exist* and in which the values of ν are *restricted to discrete constant values*. Because of the direct relationship between the total energy of a particle and the frequency of its associated pilot wave, that is, $E = h\nu$, we might intuitively expect an analogous relationship to exist for particle waves. In fact, we propose that the state function $\Psi(x, y, z, t)$ for a particle is separable into the product of a space-dependent function $\psi(x, y, z)$ and a time-dependent function $f(t)$ *in those specific cases in which stationary energy states exist* and in which E is *restricted to discrete constant values*. This intuitive proposal will be justified through later mathematical treatments and results.

---

* Actually, we'll show that it is necessary that the integral of $\psi^*\psi$ over all space be finite, rather than that ψ itself be finite, for all coordinate values.

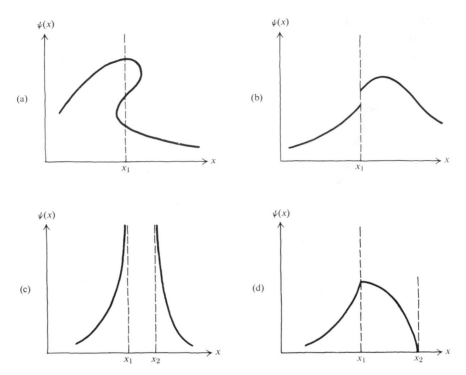

FIG. 6–1   Four examples of poorly behaved functions. (a) Function is not single valued. ψ has three values at x_1. (b) Function is not continuous at x_1. (c) Function is not finite for all values of x, but goes to infinity between x_1 and x_2. (d) Function is not smoothly varying; that is, $d\psi/dx$ is not continuous at x_1. However, the discontinuity in slope at x_2 is allowed, since this is one of the boundaries at which ψ goes to zero.

Recall also that for the classical stationary wave, the time average value of the square of the absolute displacement, $\overline{|\Phi|^2}$, which was given by $\psi^*\psi$ (Eq. 5–13), is a measure of the energy density at a given point in space. *We might also intuitively expect that $\psi^*\psi$ will have an equally important meaning for stationary particle waves.*

6–2   STATIONARY ENERGY STATES

Let's then assume a stationary energy state with E constant in which $\Psi(x, y, z, t)$ may be expressed as the product of a time-independent function, $\psi(x, y, z)$, multiplied by a time-dependent function $f(t)$, and apply the time-independent classical stationary-state equation, Eq. (5–22), to the behavior of the pilot wave associated with a single particle. Substitution of the wavelength of the pilot wave, $\lambda = h/mv$, directly into Eq. (5–22) yields

$$\nabla^2\psi = -4\pi^2 m^2 v^2 \psi/h^2, \tag{6–2}$$

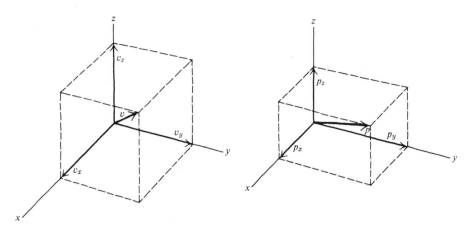

FIG. 6–2   Velocity and momentum of a single particle in terms of component velocities and momenta in the coordinate directions.

where m is the mass of the particle and v is its velocity. Equation (6–2) is the quantum-mechanical analog of Eq. (5–22), the time-independent classical stationary-state equation.

We may rearrange Eq. (6–2) to

$$\nabla^2\psi = \frac{-8\pi^2 m}{h^2}\left(\frac{mv^2}{2}\right)\psi. \qquad (6\text{–}3)$$

But the kinetic energy T of the particle is $mv^2/2$, so that Eq. (6–3) may be written as

$$\left(-\frac{h^2}{8\pi^2 m}\nabla^2\right)\psi = T\psi.$$

The particular collection of constants $h/2\pi$ appears often enough in quantum mechanics that we find it convenient to define it by the new symbol $\hbar$ ("h bar"), so that the above equation may be restated as

$$\left(\frac{-\hbar^2\nabla^2}{2m}\right)\psi = T\psi. \qquad (6\text{–}4)$$

We may also abbreviate Eq. (6–4) further as

$$\hat{T}\psi = T\psi, \qquad (6\text{–}5)$$

where

$$\hat{T} = -\hbar^2\nabla^2/2m. \qquad (6\text{–}6)$$

$\hat{T}$ is called the *kinetic energy operator*. Now the problem we face in solving Eq. (6–4) is similar to that we have encountered in solving classical wave equations. Here, we assume a stationary *energy* state in which E is constant.

Although T is not *generally* constant in a stationary energy state, it might be constant under certain physical conditions where potential energy V is constant. Under these specific conditions, the kinetic energy T of the particle might be restricted by boundary conditions to a single constant value or to a set or spectrum of discrete constant values, similar to the set of discrete frequencies for the violin string. If T is constant, Eq. (6–4) is obviously an eigenvalue equation from which, along with known boundary conditions, we may determine by trial the eigenfunction(s) ψ describing the state(s) of the particle, as well as the exact kinetic-energy eigenvalue in each state. Conversely, if the ψ function representing a particular state of the particle is known by some other means, we may determine by simple substitution into Eq. (6–4) whether or not ψ is an eigenfunction of $\hat{T}$. If it *is* an eigenfunction, then its associated eigenvalue is the *exact* kinetic energy T of the particle in the state ψ. On the other hand, if it turns out that ψ *is not* an eigenfunction of $\hat{T}$, there is no associated eigenvalue kinetic energy, which implies that the kinetic energy T is not restricted to a constant value, which further implies that T cannot be measured experimentally as an exact value in the state ψ.

In order to develop other operators, we may write Eq. (6–6) as

$$\hat{T} = \frac{1}{2m}\left[-\hbar^2\left(\frac{\partial^2}{\partial x^2} \right) - \hbar^2\left(\frac{\partial^2}{\partial y^2} \right) - \hbar^2\left(\frac{\partial^2}{\partial z^2} \right) \right], \qquad (6–7)$$

and we may write the kinetic energy T for the single particle in terms of the component velocities (Fig. 6–2) as

$$T = \frac{mv^2}{2} = \frac{m}{2}(v_x^2 + v_y^2 + v_z^2), \qquad (6–8)$$

or

$$T = \frac{1}{2m}(m^2v_x^2 + m^2v_y^2 + m^2v_z^2). \qquad (6–9)$$

But the linear momentum of the particle is given as $p = mv$, so that

$$T = \frac{1}{2m}(p_x^2 + p_y^2 + p_z^2), \qquad (6–10)$$

where p_x, p_y, and p_z are the component linear momenta in the x-, y-, and z-directions, respectively (Fig. 6–2). Substitution of Eqs. (6–7) and (6–10) into Eq. (6–5) yields

$$\left[-\hbar^2\left(\frac{\partial^2}{\partial x^2} \right) - \hbar^2\left(\frac{\partial^2}{\partial y^2} \right) - \hbar^2\left(\frac{\partial^2}{\partial z^2} \right) \right]\psi = (p_x^2 + p_y^2 + p_z^2)\psi. \qquad (6–11)$$

Since x, y, and z are independent variables, each of the operator terms for a given coordinate direction on the left-hand side of Eq. (6–11) corresponds to the linear momentum in that same direction in the expression on the right-hand side. We might then write, for example,

$$\left[-\hbar^2 \left(\frac{\partial^2}{\partial x^2} \right) \right] \psi = p_x^2 \psi. \tag{6-12}$$

We may consider the operator in brackets as consisting of two identical and repetitive sub-operators, so that Eq. (6–12) may be written as

$$\left[\frac{\hbar}{i} \left(\frac{\partial}{\partial x} \right) \right] \left[\frac{\hbar}{i} \left(\frac{\partial}{\partial x} \right) \right] \psi = p_x p_x \psi, \tag{6-13}*$$

where $i^2 = -1$. Observe that each of the suboperations independently yields p_x; that is,

$$\frac{\hbar}{i} \left(\frac{\partial}{\partial x} \right) \psi = p_x \psi \tag{6-14}$$

or

$$\hat{p}_x \psi = p_x \psi,$$

where

$$\hat{p}_x = \frac{\hbar}{i} \left(\frac{\partial}{\partial x} \right). \tag{6-15}$$

The operator $\hat{p}_x$ is the operator for the x component of linear momentum. Here again, Eq. (6–14) can be solved as an eigenvalue equation only for those specific ψ states in which p_x is constant, which is not *generally* the case. However, if p_x is constant, the state function(s) ψ and associated exact p_x eigenvalue(s) may be determined from Eq. (6–14). The values of p_x so calculated are the *only* exact values which may be observed in experiment. Conversely, if ψ for a given state, as determined by some other means, is *not* an eigenfunction of $\hat{p}_x$, we infer that p_x is not constant in that state and cannot be measured exactly in an experiment. Similarly, the corresponding equations and component linear-momentum operators for the y and z directions are:

$$\frac{\hbar}{i} \left(\frac{\partial}{\partial y} \right) \psi = p_y \psi, \tag{6-16}$$

or

$$\hat{p}_y \psi = p_y \psi, \tag{6-17}$$

---

* The choice of positive roots rather than negative roots for $\hat{p}_x$ in Eq. (6–13) is arbitrary. It is important, however, that we remain consistent in related expressions.

where

$$\hat{p}_y = \frac{\hbar}{i} \left(\frac{\partial}{\partial y} \right) \tag{6–18}$$

and

$$\frac{\hbar}{i} \left(\frac{\partial}{\partial z} \right) \psi = p_z \psi, \tag{6–19}$$

or

$$\hat{p}_z \psi = p_z \psi, \tag{6–20}$$

where

$$\hat{p}_z = \frac{\hbar}{i} \left(\frac{\partial}{\partial z} \right). \tag{6–21}$$

6-3   THE HAMILTONIAN OPERATOR FOR TOTAL ENERGY

Recall that we are trying to determine the form of the time-independent state function ψ for stationary states in which E is constant. If we ignore the rest-mass energy term in Eq. (3–38), which cancels out whenever energy differences are calculated, the total *nonrelativistic* energy, E, of a single particle is given as the sum of the kinetic energy T and the potential energy V:

$$T + V = E. \tag{6–22}$$

In general, the potential energy V may be a function of x, y, z, and t. However, unless light or other radiation happens to be shining on the system, most problems involving atoms and molecules deal with particles moving in force fields in which V is dependent on *position only* and is independent of time. Such a field in which V depends only on position and is independent of time is called a *conservative force field*. For example, in our treatment of the Bohr hydrogen model in Chapter 3, the electron moved in a conservative force field in which $V = -e^2/r$ (Eq. 3–9). In our treatment of the mass or a spring in Chapter 4, the potential energy of the mass was given as $V = \kappa y^2/2$ (Eq. 4–20) and was not dependent on t. Thus, the force was conservative. An example of a nonconservative force field is given by a charge e in a light wave polarized in the x-direction, for which the potential energy is a function of both x and t. In a conservative force field, the system under consideration is not subjected to *external* forces nor does it dissipate energy through internal losses. In effect, its *total energy is conserved*, so that E *remains constant*, even though there may be interconversion of potential and kinetic energies. Thus, particles achieve stationary energy states in conservative force fields.

Multiplication of Eq. (6–22) by ψ yields

$$T\psi + V\psi = E\psi. \tag{6–23}$$

Combination with Eq. (6–5) yields

$$\hat{T}\psi + V\psi = E\psi, \tag{6-24}$$

or

$$(\hat{T} + V)\psi = E\psi, \tag{6-25}$$

which may also be written *in a conservative force field*, where V does not depend on t, as

$$\left[\frac{-\hbar^2}{2m} \nabla^2 + V(x, y, z) \right]\psi = E\psi. \tag{6-26}$$

Since E is always constant in a conservative system, Eq. (6–26) is an extremely important eigenvalue equation because it provides, when used in conjunction with known boundary conditions, the correct state function(s) ψ for a particle in a conservative force field as well as its exact and constant-energy eigenvalue(s). The eigenvalue energies are the only energies available to the particle and no other energies may be measured experimentally. Eq. (6–26) is called the *time independent Schrödinger equation** or the *Schrödinger amplitude equation*. The operator in brackets is called the *Hamiltonian operator* and is given the symbol $\hat{H}$.

$$\hat{H}\psi = E\psi, \tag{6-27}$$

where

$$\hat{H} = \left[-\frac{\hbar^2}{2m} \nabla^2 + V(x, y, z) \right]. \tag{6-28}$$

In order to define $\hat{H}$, we must know the potential energy of the particle as a function of position in the conservative force field which applies in a particular problem. Once the form for $V(x, y, z)$ is known, we may write $\hat{H}$ and then try to discover suitable amplitude functions ψ which satisfy both Eq. (6–26) *and* any boundary conditions which we might also impose on the particle. For most of the problems which we'll study, we'll discover that a *series* of acceptable eigenfunctions will be generated, each individual eigenfunction representing an eigenstate corresponding to a discrete eigenvalue energy.

6–4 SOME PROPERTIES OF LINEAR OPERATORS

Before we introduce time dependence into the Schrödinger equation, it is necessary that we consider a very important property of operators, that is, *linearity*. An operator $\hat{A}$ is said to be linear if it obeys the relationship

$$\hat{A}(\Psi_1 + \Psi_2 + \Psi_3 + \cdots) = \hat{A}\Psi_1 + \hat{A}\Psi_2 + \hat{A}\Psi_3 + \cdots \tag{6-29}$$

---

* E. Schrödinger, *Ann. Physik*, **79**, 361, 489; **80**, 437 (1926).

For example, the operators c (a constant), x, d/dx, and d^2/dx^2 are linear, whereas the operator $\sqrt{}$ is nonlinear, since it does not obey Eq. (6–29). Now consider a linear-operator equation of the form

$$\hat{A}\Psi + \hat{B}\Psi + \cdots = 0, \tag{6–30}$$

where $\hat{A}, \hat{B}, \ldots$ are *different* linear operators. When the constant on the right-hand side of such a linear-operator equation is zero, that is, when Ψ appears to the same degree in each finite term, the equation is said to be *homogeneous*. We'll now state and prove an important relationship which applies to *linear homogeneous equations:*

If $\Psi_1, \Psi_2, \ldots$ individually are satisfactory solutions for a given linear homogeneous equation, then any linear combination of $\Psi_1, \Psi_2, \ldots$ is also a satisfactory solution.

To prove the above relationship, consider two functions, Ψ_1 and Ψ_2, which individually satisfy the same linear homogeneous equation:

$$\hat{A}\Psi_1 + \hat{B}\Psi_1 = 0, \tag{6–31}$$

$$\hat{A}\Psi_2 + \hat{B}\Psi_2 = 0. \tag{6–32}$$

A general form of linear combination of the two functions is $a\Psi_1 + b\Psi_2$, where a and b are constants. Let's now show that the linear-combination function, $a\Psi_1 + b\Psi_2$, also satisfies the same linear homogeneous equation:

$$\hat{A}(a\Psi_1 + b\Psi_2) + \hat{B}(a\Psi_1 + b\Psi_2) = 0, \tag{6–33}$$

$$a\hat{A}\Psi_1 + b\hat{A}\Psi_2 + a\hat{B}\Psi_1 + b\hat{B}\Psi_2 = 0,$$

$$a(\hat{A}\Psi_1 + \hat{B}\Psi_1) + b(\hat{A}\Psi_2 + \hat{B}\Psi_2) = 0. \tag{6–34}$$

Substitution of Eqs. (6–31) and (6–32) into Eq. (6–34) yields

$$a(0) + b(0) = 0,$$

and the relationship is proved. In order to ensure that time-dependent wave functions for particles may be superposed mathematically in simple linear combinations, *we'll insist that all quantum-mechanical operators be linear.*

6–5 DEVELOPMENT OF THE GENERAL TIME-DEPENDENT WAVE EQUATION

The argument for (not derivation of!) the Schrödinger amplitude equation (Eq. 6–26), *which applies only to conservative systems involving standing de Broglie waves*, was based on the assumption that the general state function,

$\Psi(x, y, z, t)$, may be expressed as the product of two separate functions, one dependent only on position, and the other dependent only on time; that is,

$$\Psi(x, y, z, t) = \psi(x, y, z)f(t). \qquad (6\text{-}1')$$

Unfortunately, such a separation of time and position variables is not possible in many of the more complex and chemically interesting problems of quantum mechanics. The lack of general ability to achieve such a separation is perhaps the major barrier against rigorous application of quantum mechanics to chemistry. For those cases in which time dependence cannot be separated or in which time-dependent solutions are specifically sought, we must develop a more general time-dependent equation similar in form to Eq. (6–26). Let's now try to deduce a satisfactory form for such an equation.

First of all, *whatever the form of the general time-dependent wave equation, it must reduce to the form of Eq. (6–26) in the special case of a conservative system,* in which E is a constant. Furthermore, in the *general* case, where V is a function of x, y, z, *and* t, the total energy E is *not* a constant but rather is a function of t, *which means that E itself must not appear as a constant in the general time-dependent equation.* In addition, we also expect that the general time-dependent quantum-mechanical wave equation will be similar in form to the corresponding classical time-dependent displacement equation, as given by Eq. (5–2). In particular, we expect that the general quantum-mechanical wave equation will contain a *time-dependent* operator. Finally, since E is constant in conservative force fields, *it is necessary that the time-dependent operator yield allowed constant values of E as eigenvalues in the special case of a conservative system.*

Let's then propose the following form for the time-dependent differential wave equation for a single particle of mass m:

$$\left[\frac{-\hbar^2}{2m} \nabla^2 + V(x, y, z, t) \right] \Psi = \hat{E}\Psi, \qquad (6\text{-}35)$$

where $\hat{E}$ is a *time-dependent energy operator* such that, in *conservative* systems, where E is constant,

$$\hat{E}\Psi = E\Psi. \qquad (6\text{-}36)$$

Thus, in conservative force fields, where $V(x, y, z, t)$ becomes $V(x, y, z)$, we may substitute Eq. (6–36) into Eq. (6–35) so that

$$\left[\frac{-\hbar^2}{2m} \nabla^2 + V(x, y, z) \right] \Psi = E\Psi.$$

Substitution of Eq. (6–1) in the above equation yields

$$\left[\frac{-\hbar^2}{2m} \nabla^2 + V(x, y, z) \right] [\psi f(t)] = E[\psi f(t)].$$

But since the operator is time independent, $f(t)$ can be factored from each side so that

$$\left[\frac{-\hbar^2}{2m}\nabla^2 + V(x,y,z)\right]\psi = E\psi$$

or

$$\hat{H}\psi = E\psi,$$

and it is thus shown that the general equation reduces to the time-independent Schrödinger amplitude equation (Eq. 6–26). That is, we may meet the requirement of reducibility from the general case to the conservative system case if we postulate the relationship given by Eq. (6–36).

We must now determine the form of the time-dependent energy operator $\hat{E}$. Recall that many classical phenomena, such as the diffraction of light, are explained by assuming that the total wave displacement at any given coordinate position may be considered to be a linear combination of the displacements of several constituent waves (the principle of superposition). By analogy, in order to provide a mathematical scheme capable of describing interference phenomena observed for particle waves, for example, electron diffraction, *we must insist that $\hat{E}$ be a linear operator*, so that Eq. (6–35) then becomes a linear homogeneous equation,

$$\hat{H}\Psi - \hat{E}\Psi = 0,$$

whose properties we have discussed in the previous section. That is, if Ψ_1 and Ψ_2 are two functions which individually satisfy the time-dependent differential wave equation (Eq. 6–35), then a linear combination such as $\Psi_1 + \Psi_2$ must also satisfy the *time-dependent* differential wave equation. Promising linear operators which we might consider are c, $c(\partial/\partial t)$, $c(\partial^2/\partial t^2)$, and similar differential operators containing higher-order derivatives, where c is a constant. Recall that such operators are linear because

$$c(\Psi_1 + \Psi_2) = c\Psi_1 + c\Psi_2, \tag{6–37}$$

$$c\frac{\partial}{\partial t}(\Psi_1 + \Psi_2) = c\frac{\partial\Psi_1}{\partial t} + c\frac{\partial\Psi_2}{\partial t}, \tag{6–38}$$

$$c\frac{\partial^2}{\partial t^2}(\Psi_1 + \Psi_2) = c\frac{\partial^2\Psi_1}{\partial t^2} + c\frac{\partial^2\Psi_2}{\partial t^2}. \tag{6–39}$$

Operators such as $c(\ )^{1/2}$ or $c(\ )^2$ are *not* satisfactory operators, since they are not linear. That is,

$$c(\Psi_1 + \Psi_2)^{1/2} \neq c\Psi_1^{1/2} + c\Psi_2^{1/2}, \tag{6–40}$$

$$c(\Psi_1 + \Psi_2)^2 \neq c\Psi_1^2 + c\Psi_2^2. \tag{6–41}$$

We'll then expect that the operator $\hat{E}$ will be of the form c, $c(\partial/\partial t)$, $c(\partial^2/\partial t^2)$, or that it might contain some higher-order time derivative.

Let's first look at the operator c and immediately rule it out because $\hat{E}$ cannot *generally* be a constant since in such a case:

$$c\Psi = E\Psi,$$

which would mean that

$$c = \hat{E} = E,$$

and Eq. (6–35) would under *all* conditions reduce to the conservative system form of Eq. (6–26), which we cannot allow. Thus, $\hat{E}$ *will have to involve a time derivative* and might, for example, be either

$$c\left(\frac{\partial}{\partial t}\right) \quad \text{or} \quad c\left(\frac{\partial^2}{\partial t^2}\right). \tag{6–42}$$

But how does Ψ depend on t? Recall that in order to develop the *time-independent* wave equation, we have had to assume that in a conservative system

$$\Psi(x, y, z, t) = \psi(x, y, z)f(t). \tag{6–1'}$$

Let's substitute Eq. (6–1) into Eq. (6–36), which is an eigenvalue equation in conservative systems:

$$\hat{E}\psi(x, y, z)f(t) = E\psi(x, y, z)f(t). \tag{6–43}$$

Since $\hat{E}$ involves a time derivative and not a space derivative, $\psi(x, y, z)$ may be factored from each side, and Eq. (6–43) may then be simplified to

$$\hat{E}f(t) = Ef(t). \tag{6–44}$$

We must now select a suitable form for $f(t)$. Recall that for classical waves, time dependence may be expressed conveniently in terms of four basic periodic functions:

$$\sin 2\pi\nu t,\ \cos 2\pi\nu t,\ e^{+i2\pi\nu t},\ \text{or}\ e^{-i2\pi\nu t}.$$

In developing the general time dependent quantum mechanical wave equation, we'll insist on the *general* validity of de Broglie's postulates, which assign a frequency $\nu = E/h$ and a wavelength $\lambda = h/mv$ to the pilot waves associated with a particle, so that $f(t)$ for periodic time dependence in the *pilot wave* will be given by

$$\sin\left(\frac{2\pi Et}{h}\right),\ \cos\left(\frac{2\pi Et}{h}\right),\ e^{+i2\pi Et/h},\ \text{or}\ e^{-i2\pi Et/h}$$

that is,

$$\sin\left(\frac{Et}{\hbar}\right),\ \cos\left(\frac{Et}{\hbar}\right),\ e^{+iEt/\hbar},\ \text{or}\ e^{-iEt/\hbar} \tag{6–45}$$

or some linear combination of these terms. Although a strict interpretation of de Broglie's postulates, which are consistent with the theory of relativity, requires that E be the total *relativistic* energy, we'll assume, as did Schrödinger, that for a nonrelativistic treatment we may ignore the rest-mass-energy term, m_0c^2, in Eq. (3–39) and *define E as the classical total energy*, $T + V$, so that it has the same meaning as in Eq. (6–26). Of course, such a redefinition of E involves a corresponding change in the calculated value of v, the frequency of the hypothetical associated pilot wave, but this is really no bother in the Schrödinger theory since v is not experimentally observable anyway. Recall, in comparison, that we are able to measure *wavelengths* for particle waves through diffraction measurements.

Let's now substitute the two promising forms for $\hat{E}$ given by Eq. (6–42) and the four possible basic forms of $f(t)$ given by Eq. (6–45) into Eq. (6–44) in order to determine appropriate forms for both $\hat{E}$ in general and $f(t)$ in conservative systems. We'll first try $c(\partial^2/\partial t^2)$ as the operator with the sine form of $f(t)$:

$$c \frac{\partial^2}{\partial t^2} \left[\sin\left(\frac{Et}{\hbar} \right) \right] = E \sin\left(\frac{Et}{\hbar} \right).$$

Double differentiation yields

$$- c\left(\frac{E^2}{\hbar^2} \right)\sin\left(\frac{Et}{\hbar} \right) = E \sin\left(\frac{Et}{\hbar} \right),$$

or

$$c = \frac{-\hbar^2}{E} \quad \text{(tentative)},$$

which would mean that $\hat{E}$ would have to be:

$$\hat{E} = c\left(\frac{\partial^2}{\partial t^2} \right) = \frac{-\hbar^2}{E} \left(\frac{\partial^2}{\partial t^2} \right) \text{(tentative)}.$$

But such a form for $\hat{E}$ cannot be allowed since it introduces E as a constant into the general time-dependent form of the wave equation (Eq. 6–35). Recall that E as such must not appear in the correct form for the time-dependent wave equation; E may appear only in special solutions! A similar trial of $c(\partial^2/\partial t^2)$ as the operator with either $\cos(Et/\hbar)$ or with $e^{\pm iEt/\hbar}$ also leads to a result in which c contains the term E. Furthermore, all higher-order time derivatives are also unsatisfactory, since they all introduce E to some higher power into the constant c and thus into the general wave equation. Since any other acceptable periodic time function would have to be some linear combination of the three forms given in (6–45), it is apparent that $\hat{E}$ cannot be of the form $c(\partial^2/\partial t^2)$, nor can it include a time derivative higher in order than the first derivative.

The *one* remaining trial form for $\hat{E}$ is $c(\partial/\partial t)$. But neither $\sin(Et/\hbar)$ nor $\cos(Et/\hbar)$ is an eigenfunction of $c(\partial/\partial t)$, so that the sine and cosine functions are eliminated, which leaves *only* the functions $e^{\pm iEt/\hbar}$. Let's then substitute the *one remaining satisfactory form of $\hat{E}$*, that is $c(\partial/\partial t)$, along with the *two remaining satisfactory forms for $f(t)$*, that is $e^{\pm iEt/\hbar}$, into Eq. (6–44) in order to evaluate c:

$$c\left(\frac{\partial e^{\pm iEt/\hbar}}{\partial t} \right) = E e^{\pm iEt/\hbar},$$

$$c\left(\pm \frac{iE}{\hbar} \right) e^{\pm iEt/\hbar} = E e^{\pm iEt/\hbar},$$

or

$$c = \pm \frac{\hbar}{i},$$

and therefore

$$\hat{E} = c\left(\frac{\partial}{\partial t} \right) = \pm \frac{\hbar}{i}\left(\frac{\partial}{\partial t} \right). \tag{6–46}$$

Either the positive or the negative version of Eq. (6–46) *is a satisfactory time-dependent energy operator since neither contains E as a constant.* We'll *arbitrarily* select the negative solution which corresponds to the negative exponential form for $f(t)$ in conservative systems; that is,

$$f(t) = e^{-iEt/\hbar}, \tag{6–47}$$

so that we may now write, by substitution of Eq. (6–47) into Eq. (6–1),

$$\Psi(x, y, z, t) = \psi(x, y, z)e^{-iEt/\hbar} \tag{6–48}$$

as the *time-dependent wave function* in a conservative system.

Substitution of $\hat{E} = -(\hbar/i)(\partial/\partial t)$ into Eq. (6–35) yields

$$\left[\frac{-\hbar^2}{2m} \nabla^2 + V(x, y, z, t) \right]\Psi = \frac{-\hbar}{i}\left(\frac{\partial \Psi}{\partial t} \right), \tag{6–49}$$

which is the *general time-dependent Schrödinger equation.** Note that the complex conjugate of Eq. (6–49) is

$$\left[\frac{-\hbar}{2m} \nabla^2 + V(x, y, z, t) \right]\Psi^* = \frac{\hbar}{i}\left(\frac{\partial \Psi^*}{\partial t} \right), \tag{6–50}$$

which corresponds to the choice of the positive solution for $\hat{E}$ in Eq. (6–46). The time-dependent Schrödinger equation is often written in shorter operator form as

$$\hat{H}\Psi = \hat{E}\Psi. \tag{6–51}$$

---

* E. Schrödinger, *Ann. Physik*, **81**, 109 (1926).

It is important to recognize that the general time-dependent *quantum-mechanical* differential wave equation which we have developed (Eq. 6–49) differs from the *classical* time-dependent differential wave equation (Eq. 5–2) in two significant ways:

1. The time derivative in the quantum-mechanical wave equation is a *first derivative*, whereas the time derivative in the classical wave equation is a second derivative.

2. For conservative systems the quantum-mechanical wave function (state function) $\Psi(x, y, z, t)$ *must be complex*, whereas the classical wave function $\Phi(x, y, z, t)$ may always be real.

The above two important differences between the quantum-mechanical and the classical wave equations result from an insistence that we adhere in general to de Broglie's two basic postulates, $\nu = E/h$ and $\lambda = h/mv$, with the exception that we redefine E as the total classical energy rather than the total relativistic energy, and in addition that we require all quantum-mechanical operators to be linear.

On the basis of the expected validities of Eqs. (6–49) and (6–26), let's now write the second postulate of quantum mechanics:

POSTULATE II. a) *The possible state functions* $\Psi(x, y, z, t)$ *for a single particle are given by the solution of the time-dependent Schrödinger wave equation:*

$$\left[\frac{-\hbar^2}{2m} \nabla^2 + V(x, y, z, t) \right] \Psi = \left[\frac{-\hbar}{i} \left(\frac{\partial}{\partial t} \right) \right] \Psi, \qquad (6\text{–}49')$$

b) *In the special case of conservative systems, where V is not dependent on t, and* $\Psi(x, y, z, t) = \psi(x, y, z)e^{-iEt/\hbar}$, *the possible time-independent amplitude functions* $\psi(x, y, z)$ *for a single particle are given by solution of the time-independent Schrödinger equation,*

$$\left[\frac{-\hbar^2}{2m} \nabla^2 + V(x, y, z) \right] \psi = E\psi, \qquad (6\text{–}26')$$

6-6 GENERAL QUANTUM-MECHANICAL OPERATORS

Thus far in this chapter we have developed several quantum-mechanical operators, each of which is related to some specific quantity, or *dynamical variable*, which is used to describe the system. Some of the operators, such as $\hat{p}_x$, $\hat{p}_y$, $\hat{p}_z$, $\hat{T}$, and $\hat{H}$ (in conservative systems), are time-independent, that is, neither t nor a derivative of t appears in the operator expression. The operations of such time-independent operators on the *time-independent* amplitude function $\psi(x, y, z)$ in conservative systems were shown to yield the only allowed values for the corresponding dynamical variables under those conditions in which the dynamical variable is restricted to stationary states. For example,

$$\hat{T}\psi = T\psi, \tag{6–5'}$$

$$\hat{p}_x\psi = p_x\psi, \tag{6–19'}$$

$$\hat{H}\psi = E\psi. \tag{6–27'}$$

Furthermore, only when a dynamical variable is constant in a stationary state can it be determined exactly. Thus, *the only possible exact values which may be observed for the dynamical variables, T, p_x, and E are those given by the eigenvalues of the respective eigenvalue equations,* Eqs. (6–5'), (6–19'), and (6–27'). Dynamical variables which can be measured experimentally are called *observables*.

We'll now show, for a single particle in a *conservative* force field, that the operation of time-independent operators on the *general* function $\Psi(x, y, z, t)$ yields the same eigenvalues as those given by the solutions of the above three equations. Let's, for example, substitute Eq. (6–48) into Eq. (6–19'):

$$\hat{p}_x\left[\frac{\Psi(x, y, z, t)}{\exp(-iEt/\hbar)}\right] = p_x\left[\frac{\Psi(x, y, z, t)}{\exp(-iEt/\hbar)}\right].$$

Cancellation of the exponential terms yields

$$\hat{p}_x\Psi = p_x\Psi. \tag{6–52}$$

Furthermore, recall that it is necessary that time-dependent operators, such as $\hat{E}$, operate on $\Psi(x, y, z, t)$ rather than on $\psi(x, y, z)$. For example,

$$\hat{E}\Psi = E\Psi. \tag{6–36'}$$

In writing Eq. (6–52) we have argued from the specific case of a conservative system to the general case involving time dependence. Let's now postulate that relationships of the form of Eq. (6–52) are *generally* valid:

POSTULATE III.   a)   *For every dynamical variable there is a corresponding linear operator $\hat{A}$. If the dynamical variable is capable of exact experimental determination, its only possible exact values A are those given by the eigenvalues of the equation*

$$\hat{A}\Psi = A\Psi. \tag{6–53}$$

b)   *In the special case of conservative systems and for time-independent operators, the only possible exact values of the dynamical variable are also given by*

$$\hat{A}\psi = A\psi. \tag{6–54}$$

TABLE 6–1

Important Quantum Mechanical Operators

Dynamical variable	Operator symbol	Operation
Position	$\hat{x}$ $\hat{y}$ $\hat{z}$	x (multiplication by x) y (multiplication by y) z (multiplication by z)
Time	$\hat{t}$	t (multiplication by t)
Linear momentum		
x-component	$\hat{p}_x$	$\dfrac{\hbar}{i}\left(\dfrac{\partial}{\partial x}\right)$
y-component	$\hat{p}_y$	$\dfrac{\hbar}{i}\left(\dfrac{\partial}{\partial y}\right)$
z-component	$\hat{p}_z$	$\dfrac{\hbar}{i}\left(\dfrac{\partial}{\partial z}\right)$
Kinetic energy T	$\hat{T}$	$\dfrac{-\hbar^2}{2m}\nabla^2$
Potential energy V	$\hat{V}$	$V(x, y, z, t)$ (general)
		$V(x, y, z)$ (conservative system) (multiplication by V)
Total energy $T + V$ (Hamiltonian)	$\hat{H} = \hat{T} + \hat{V}$ $= \hat{T} + V$	$\dfrac{-\hbar^2}{2m}\nabla^2 + V(x, y, z, t)$ (general)
		$\dfrac{-\hbar^2}{2m}\nabla^2 + V(x, y, z)$ (conservative system)
Total energy $T + V$ (time operator)	$\hat{E}$	$\dfrac{-\hbar}{i}\left(\dfrac{\partial}{\partial t}\right)$

As we have already observed, many of the important quantum-mechanical operators are time-independent and force fields of interest in atomic and molecular systems are often conservative, so that the time-independent relationship given by Eq. (6–54) is very important.

A summary of the quantum-mechanical operators we have developed is given in Table 6–1. We have added position operators $\hat{x}$, $\hat{y}$, and $\hat{z}$ and the time operator $\hat{t}$, which may appear to be somewhat trivial since they involve mere multiplication by x, y, z or t. However, such operators are important in forming new operators for other dynamical variables. The general prescription for constructing a new operator is:

1. Write the classical expression, in terms of coordinates, time, and component linear momenta, for the dynamical variable corresponding to the desired operator.

2. Substitute position operators and the time operator for coordinate positions and time wherever they appear in the classical expression. (This really involves leaving the coordinates and time just as they are!)

3. Substitute component linear-momentum operators for component linear momenta wherever they appear in the classical expression.

As an example, let's show how we may construct the operator for angular momentum about the z-axis through proper combination of fundamental position and linear-momentum operators. The classical expression for angular momentum about the z-axis is given as

$$L_z = xp_y - yp_x. \tag{6–55}$$

Substituting linear-momentum operators according to our prescription yields the operator for angular momentum about the z-axis, $\hat{L}_z$, as

$$\hat{L}_z = x\hat{p}_y - y\hat{p}_x, \tag{6–56}$$

or

$$\hat{L}_z = \frac{\hbar}{i}\left(x\frac{\partial}{\partial y} - y\frac{\partial}{\partial x}\right). \tag{6–57}$$

It is certainly now apparent that $\Psi(x, y, z, t)$ and $\psi(x, y, z)$ are powerful functions. By suitable operation on Ψ or ψ, using appropriate operators, we may in theory generate all the allowed exact values for all of the observables pertaining to a given particle. The amplitude function ψ is particularly important since many of the systems of interest in atomic and molecular studies are conservative, so that time-independent operators may be used. It is now time that we interpret more clearly the meaning of ψ in conservative systems.

6–7 THE SIGNIFICANCE OF ψ IN CONSERVATIVE SYSTEMS

In our treatment of classical waves in Chapter 5, we showed that $\psi(x, y, z)$ was a time-*independent* function such that for stationary classical waves $|\psi|^2$ or $\psi^*\psi$ was a measure of the relative energy density at a selected coordinate position. Thus, $\psi^*\psi$ was shown to be a maximum at antinode or loop positions and zero at node positions.

In addition, we have allowed for the possibility that ψ may be a complex function involving i. But *the observables which we experimentally measure for any particle or system of particles are real.* Thus, *useful* quantum-mechanical expressions, although they may involve intermediate complex numbers, *must yield real results.* The Schrödinger amplitude equation (Eq. 6–27), which is an eigenvalue equation in conservative force fields, may be written as

$$\hat{T}\psi + V\psi = E\psi. \tag{6–58}$$

Equation (6–58), which may involve complex numbers (ψ may be complex) may be converted to a form involving only real numbers by multiplication by ψ^*. Thus

$$\psi^*\hat{T}\psi + \psi^*V\psi = \psi^*E\psi, \tag{6–59}$$

or

$$\psi^*\hat{T}\psi + V\psi^*\psi = E\psi^*\psi. \tag{6–60}$$

Since $\psi^*\psi = |\psi|^2$, a real number,

$$\psi^*\hat{T}\psi + V|\psi|^2 = E|\psi|^2. \tag{6–61}$$

Since the second and third terms in Eq. (6–61) are real, the first term, $\psi^*\hat{T}\psi$, must also be real. Thus, Eq. (6–61) contains only real quantities. In order to investigate the real quantity $\psi^*\hat{T}\psi$ more fully, we must temporarily divert our attention to a method for expressing average quantities.

6-8 AN EXPRESSION FOR AVERAGE QUANTITIES

Let's assume that we wish to calculate the average grade $\overline{G}$ on an examination which has been given to a class of n students. Let n_i be the number of students who have earned the grade G_i. Then

$$\overline{G} = \frac{\sum G_i n_i}{\sum n_i} = \frac{\sum G_i n_i}{n},$$

where the summations are taken over all possible grades. In a large group of students, the fraction n_i/n of students who earn the grade G_i is also equal to the probability p_i that a randomly selected individual student receives the grade G_i. Thus

$$\overline{G} = \sum G_i(n_i/n) = \sum G_i p_i. \tag{6–62}$$

For example, consider the following specific distribution for which the average grade $\overline{G}$ is 73.0:

G_i	n_i	$p_i = n_i/n$	$G_i p_i$
90	2	0.10	9.0
80	6	0.30	24.0
70	8	0.40	28.0
60	4	0.20	12.0
	$n = \sum n_i = 20$		$\sum G_i p_i = 73.0$

In order to extend Eq. (6–62) to an infinite number of possible grades, each involving a differential probability, we write

$$\bar{G} = \int_{\substack{\text{all possible} \\ \text{grades}}} G \, dp. \tag{6–63}$$

The form of Eq. (6–63) is applicable to the measurement of other average quantities. For example, the average potential energy $\bar{V}$ for a single particle in three dimensions in a conservative force field, considering all space available to the particle, is given by

$$\bar{V} = \int_{\text{all space}} V(x, y, z) \, dp, \tag{6–64}$$

where dp is the differential probability that the particle has a potential energy between V and $V + dV$. This is the same as the probability that the particle is within the volume element $dx \cdot dy \cdot dz$ at the position x, y, z, since the potential energy is directly related to the position. We may express the probability in *finite* terms as

$$P = dp / (dx \, dy \, dz) \tag{6–65}$$

where P is the probability *per unit volume* that the particle is within the differential volume element $dx \, dy \, dz$. P is also referred to as the *probability density*. For convenience, we designate the volume element as $d\tau$, so that

$$dp = P \, dx \, dy \, dz = P \, d\tau. \tag{6–66}$$

Then Eq. (6–64) may be written

$$\bar{V} = \int_{-\infty}^{\infty} \int_{-\infty}^{\infty} \int_{-\infty}^{\infty} VP \, dx \, dy \, dz = \int_{\substack{\text{all} \\ \text{space}}} VP \, d\tau \tag{6–67}$$

6–9 BACK TO THE MEANING OF ψ

We may now multiply both sides of Eq. (6–61) by the differential space element $d\tau$,

$$\psi^* \hat{T} \psi \, d\tau + V \, |\psi|^2 \, d\tau = E \, |\psi|^2 \, d\tau, \tag{6–68}$$

and integrate over all space available to the particle, recalling that E is constant in a conservative force field,

$$\int_{-\infty}^{+\infty} \psi^* \hat{T} \psi \, d\tau + \int_{-\infty}^{+\infty} V \, |\psi|^2 \, d\tau = E \int_{-\infty}^{+\infty} |\psi|^2 \, d\tau. \tag{6–69}$$

A comparison of Eq. (6–67) with the second term on the left-hand side of Eq. (6–69) reveals that $\int_{-\infty}^{+\infty} V \, |\psi|^2 \, d\tau$ would be the average potential energy $\bar{V}$ if

$|\psi|^2$ *were equal to P, the probability density.* This is exactly what Born* originally postulated in 1926. Although we have argued in terms of the meaning of ψ in a conservative force field, we'll state the fourth postulate of quantum mechanics in terms of the more general time-dependent function $\Psi(x, y, z, t)$.

POSTULATE IV. a) *The term* $|\Psi|^2\, d\tau$ *or* $\Psi^*\Psi\, d\tau$ *is the time-dependent probability that a single particle exists at a given time t in the space element* $d\tau$, *that is, between x, y, z and* $(x + dx)$, $(y + dy)$, *and* $(z + dz)$.

b) *For the special case of a conservative system, in which the single particle is restricted to an energy eigenstate, the time-independent probability that the particle exists in the space element* $d\tau$ *is* $|\psi|^2\, d\tau$ *or* $\psi^*\psi\, d\tau$.

We may show the relationship between (a) and (b) in Postulate IV by noting that $\Psi = \psi \exp(-iEt/\hbar)$ in conservative force fields. Then, if the energy of the particle is restricted to the single value E,

$$\Psi^*\Psi = \psi^* \exp(iEt/\hbar)\psi \exp(-iEt/\hbar) = \psi^*\psi. \qquad (6\text{--}70)$$

In a conservative force field the probability per unit volume at position x, y, z, or *probability density*, may be expressed as the probability that the particle exists in the differential element at position x, y, z divided by the volume of the element or

$$P = \frac{|\psi|^2\, d\tau}{d\tau} = |\psi|^2 = \psi^*\psi. \qquad (6\text{--}71)$$

Because of its immediate association with the probability of finding a particle, ψ is often called the *probability amplitude*. The fourth postulate of quantum mechanics seems intuitively acceptable, since we have already shown in classical mechanics that $|\psi|^2$ represents any of several quantities comparable to the probability of finding a particle. That is, for elastic waves on springs and in strings, $|\psi|^2$ was a direct measure of the energy density at a given point in space. In addition, the square of either the magnetic amplitude or the electrical amplitude at any point in an electromagnetic wave is directly proportional to the *energy density* at that point, and according to the theory of relativity, $m = E/c^2$, so that for particle waves we might expect that $|\psi|^2$ would be a direct measure of *mass density*. That is, in an electromagnetic wave, the energy density may be thought of as a measure of concentration of photons at a given position or as a measure of the probability per unit volume of finding an individual photon at that same position. In the same manner, for an electron wave in a conservative system, the energy density, as given by $|\psi|^2$, is considered a measure of the probability per unit volume of finding the electron at a given position.

---

* M. Born, Z. *Physik,* **37**, 863; **38**, 803 (1926).

Based on Postulate IV, it now follows that, in conservative force fields,

$$\int_{-\infty}^{+\infty} V \, |\psi|^2 \, d\tau = \int_{-\infty}^{+\infty} VP \, d\tau = \bar{V}. \qquad (6\text{–}72)$$

Before we return to Eq. (6–72) we shall concern ourselves with the total probability of finding the particle in all of space.

6–10 NORMALIZED WAVE FUNCTIONS

Since the single particle under consideration must be found somewhere in space, it is necessary that the total probability of finding the particle in all space, $\int_{-\infty}^{+\infty} |\psi|^2 \, d\tau$, be a finite constant. It is mathematically convenient to define the total probability of finding the particle in all space as unity. According to this definition, when the probability of an event is unity, the event is certain to occur. When the probability of finding a particle in all of space is unity, we are certain to find the particle *somewhere* in space. We thus desire that the amplitude function, ψ, obey the condition

$$\int_{-\infty}^{+\infty} |\psi|^2 \, d\tau = \text{total probability of finding the particle} = 1, \qquad (6\text{–}73)$$

which is called the *normalization condition*. A wave function which obeys Eq. (6–73) is said to be *normalized*.

6–11 AVERAGE VALUES FOR DYNAMICAL VARIABLES

Substitution of Eqs. (6–72) and (6–73) into Eq. (6–69) yields, for *normalized amplitude functions*,

$$\int_{-\infty}^{+\infty} \psi^* \hat{T} \psi \, d\tau + \bar{V} = E. \qquad (6\text{–}74)$$

But in a conservative force field the total energy E is constant and

$$\bar{T} + \bar{V} = E, \qquad (6\text{–}75)$$

where $\bar{T}$ is the average or mean kinetic energy of the particle taken over all *positions in space available to the particle*. In a system of many particles, $\bar{T}$ is the average kinetic energy for all of the particles assuming that all of the space is available to each of the particles. Comparison of Eq. (6–74) and Eq. (6–75) yields, for normalized amplitude functions in a conservative system,

$$\bar{T} = \int_{-\infty}^{+\infty} \psi^* \hat{T} \psi \, d\tau. \qquad (6\text{–}76)$$

Note that it is *not* a requirement of Eq. (6–76) that ψ or ψ^* be an eigenfunction of $\hat{T}$, a condition previously required for solution of Eqs. (6–4) and (6–5).

We may now also rewrite Eq. (6–72) for comparison:

$$\bar{V} = \int_{-\infty}^{+\infty} V|\psi|^2 \, d\tau = \int_{-\infty}^{+\infty} \psi^* \hat{V} \psi \, d\tau. \tag{6–77}$$

We recall that in developing the definition for $\bar{V}$ given by Eq. (6–72) we used *normal* probabilities, so that the amplitude functions ψ^* and ψ in Eq. (6–77) are really normalized functions. Since $\hat{V}$ is the quantum-mechanical operator for V, Eqs. (6–76) and (6–77) are identical in form. When the amplitude functions are *not normalized*, it is easy to show by rearrangement of Eq. (6–69) that $\bar{T}$ and $\bar{V}$, in conservative force fields, are given as

$$\bar{T} = \frac{\int_{-\infty}^{+\infty} \psi^* \hat{T} \psi \, d\tau}{\int_{-\infty}^{+\infty} \psi^* \psi \, d\tau}, \tag{6–78}$$

and

$$\bar{V} = \frac{\int_{-\infty}^{+\infty} \psi^* \hat{V} \psi \, d\tau}{\int_{-\infty}^{+\infty} \psi^* \psi \, d\tau}. \tag{6–79}$$

The similarity of Eqs. (6–78) and (6–79) now leads to another general postulate. Before stating the postulate, however, we note that the operators $\hat{T}$ and $\hat{V}$ in Eqs. (6–78) and (6–79) are time-independent ($\hat{V}$ is time-independent by virtue of the restriction to conservative force fields). We have previously shown that time-*dependent* operators must operate on $\Psi(x, y, z, t)$, the more general state function, rather than on $\psi(x, y, z)$, the time-independent function. Recognizing the greater generality of Ψ over ψ, we then write the fifth postulate of quantum mechanics as follows.

POSTULATE V. a) *The expected mean value $\bar{A}$ of a series of measurements of an observable A made over a large number of particles each in the state represented by* Ψ *is*

$$\bar{A} = \frac{\int_{-\infty}^{+\infty} \Psi^* \hat{A} \Psi \, d\tau}{\int_{-\infty}^{+\infty} \Psi^* \Psi \, d\tau}, \tag{6–80}$$

where $\hat{A}$ is the quantum-mechanical operator for the observable.
b) *For the special case of conservative systems in which the operator $\hat{A}$ does not depend explicitly on time,*

$$\bar{A} = \frac{\int_{-\infty}^{+\infty} \psi^* \hat{A} \psi \, d\tau}{\int_{-\infty}^{+\infty} \psi^* \psi \, d\tau}. \tag{6–81}$$

c) *For the special case of conservative systems and time-independent operators and where the wave functions are normalized,*

$$\bar{A} = \int_{-\infty}^{+\infty} \psi^* \hat{A} \psi \, d\tau. \tag{6–82}$$

Equation (6–82) is, of course, identical in form to Eqs. (6–76) and (6–77), since the latter two equations involve time-independent operators and normalized wave functions. Since $\bar{A}$ represents the *probable result of an experimental measurement of A*, it is also called the *expectation value*. Postulate V, called the *mean-value postulate*, allows the calculation of *average* or *mean* values for observables for which stationary-state values cannot be measured exactly. Note that if ψ *happens to be* an eigenfunction of $\hat{A}$, then $\hat{A}\psi = A\psi$, and Eq. (6–81) reduces to

$$\bar{A} = \frac{\int_{-\infty}^{+\infty} \psi^* A\psi \, d\tau}{\int_{-\infty}^{+\infty} \psi^* \psi \, d\tau} = A.$$

That is, *every* measurement of the observable for a particle in the eigenstate represented by ψ is the eigenvalue A for that state, so that the average value is also the eigenvalue.

6–12 SUMMARY

In this chapter we have developed a number of basic relationships and postulates of quantum mechanics, each of which will be very useful in later chapters in which we begin to solve some concrete problems. At this point the student may feel somewhat as though he were drowning in a sea of operating principles and guidelines which he has not yet had a chance to apply. We are now, however, prepared to look immediately at several simple applications of quantum mechanics.

Perhaps the single most important equation we have presented in this chapter is the time-independent Schrödinger equation, $\hat{H}\psi = E\psi$, which will allow us to calculate allowed energy levels and corresponding spatial probabilities for particles in conservative systems. In addition, it will allow us to determine the form of the function ψ, which is the *operand* for the operators corresponding to *all* observables in conservative force fields.

. . . in order to account for certain facts, one had to tolerate the rude intrusion (groben Eingriff) of quite new and incomprehensible postulates, which were called quantum conditions and quantum postulates. These were gross dissonances in the symphony of classical mechanics—and yet they were curiously chiming in with it, as if they were being played on the same instrument . . . . If the old mechanics had failed entirely, that would have been tolerable, for thus the ground would have been cleared for a new theory. But as it was, we were faced with the difficult problem of saving its soul, *whose breath could be palpably detected in this microcosm, and at the same time persuading it, so to speak, not to consider the quantum conditions "rude intruders" but something arising out of the inner nature of the situation itself.*

— ERWIN SCHRÖDINGER, Address on receiving the Nobel Prize, Stockholm, December 12, 1933.*

---

* From the translation in E. Schrödinger, *Science and the Human Temperament*, W. W. Norton & Co., 1935. Used by permission of George Allen & Unwin, Ltd., London.

PROBLEMS

6–1 Which of the following operators are linear operators?

a) x b) d^2/dx^2 c) $(\ )^3$ d) $(d/dx)^2$ e) $(x^2 - 6)$

6–2 Indicate whether or not each of the following equations involving $\Psi(x, t)$ is a linear homogeneous equation (a and b are constants):

a) $\partial^2\Psi/\partial x^2 = a$ b) $a\Psi = b\Psi$

c) $a(\Psi)^{1/2} = b\,\partial\Psi/\partial x$ d) $a(\partial\Psi/\partial t)^2 = b\Psi$

6–3 Write the correct quantum mechanical operators for each of the following:

a) yp_x b) $xy - p_x p_y$ c) $\frac{1}{2}mv_x^2$

6–4 Write the quantum-mechanical operators associated with each of the following dynamical variables:

a) The spherical coordinate r

b) Total energy $T + V$ in a conservative system

c) Angular momentum L_x about the x-axis ($L_x = yp_z - zp_y$)

6–5 The equation for the total nonrelativistic energy of a single particle may be written as $p^2/2m + V = E$, where $p^2 = p_x^2 + p_y^2 + p_z^2$. Show how appropriate substitution of the operators $\hat{p}_x, \hat{p}_y, \hat{p}_z, \hat{V}$, and $\hat{E}$ leads to the time dependent Schrödinger wave equation.

6–6 Complete the following integration over the dimensions of a rectangular box of dimensions a, b, and c in the x-, y-, and z-directions, respectively:

$$\int xy^2z^3\,d\tau.$$

6–7 Following is a distribution of potential energies in a system of particles:

V, arbitrary units	Number of particles having energy V
10	6
20	17
30	30
40	42
50	34
60	21
70	10
80	5

a) What is the probability that a randomly selected particle will have a potential energy of 20 units? 80 units?

b) Calculate the average potential energy.

c) What is the most probable potential energy?

6–8 Consider the two time-independent wave functions ψ_1 and ψ_2, each of which is an eigenfunction of the Hamiltonian operator $\hat{H}$; that is, each function satisfies the time-independent Schrödinger equation. The corresponding eigenvalues are E_1 and E_2.

a) Show that the linear combination $\psi_1 + \psi_2$ is *not* an eigenfunction of $\hat{H}$ if E_1 and E_2 have *different* values.

b) Under what special condition is $\psi_1 + \psi_2$ an eigenfunction of $\hat{H}$? Are all linear combinations of ψ_1 and ψ_2 eigenfunctions under this same condition?

6–9 What must be the value of the constant a for the function $\psi = iax$ to be normalized between $x = 0$ and $x = L$?

6–10 What must be the value of the constant α in order that the function $\Psi = i\alpha xy^2 z^3$ is normalized in the interval $0 \leqslant x \leqslant a, 0 \leqslant y \leqslant b, 0 \leqslant z \leqslant c$.

Many of the problems concerned with atomic and molecular structure require
the determination of allowed energy levels. We have previously seen that the
transition of an electron from one allowed Bohr orbit to another results in a
discrete spectral line in the visible or ultraviolet region. Similarly, transitions
from one allowed vibrational energy state to another or from one allowed
rotational energy state to another are also related to discrete spectral frequencies
in the infrared and microwave regions. In addition, in the construction of models
for atoms and molecules, we are very much concerned with the relative locations
of electrons and nuclei and with the probabilities of finding electrons as functions
of position.

Furthermore, in most atomic and molecular systems, we deal with conservative force fields and time-independent operators, so that the Schrödinger
amplitude equation is the single most important equation for the evaluation of
allowed energy states and probabilities of location.

We'll again write the Schrödinger amplitude equation as

$$\hat{H}\psi = E\psi,$$

or

$$\left[\frac{-\hbar^2}{2m} \nabla^2 + V(x, y, z) \right]\psi = E\psi. \tag{7-1}$$

Let's now outline the general steps in the solution of Eq. (7-1) for a typical
system:

1. Write V as a function of x, y, and z and insert in Eq. (7-1). [The Hamiltonian operator $\hat{H}$ cannot be specifically expressed until $V(x, y, z)$ is known.]

2. Solve the resultant eigenvalue equation (Eq. 7-1) to obtain a *general* solution
   for ψ. More than one general form of solution may be possible, and
   constants are usually introduced.

3. Adjust ψ to obey each boundary condition which may be imposed. For
   bound particles this procedure usually introduces integers in the ψ function
   and defines one or more of the constants.

104

4. Normalize ψ so that $\int \psi^*\psi \, d\tau = 1$ within the range of the coordinate system. This procedure usually leads to evaluation of a final constant.

5. Use the final equation for ψ, which now satisfies the Schrödinger amplitude equation, obeys the boundary conditions, and is normalized, to evaluate the allowed steady-state values A for any *exact* observable quantity according to *Postulate III*:

$$\hat{A}\psi = A\psi \qquad (7\text{--}2)$$

if ψ is an eigenfunction of $\hat{A}$.

6. If ψ should turn out *not* to be an eigenfunction of $\hat{A}$ for a desired dynamical variable, compute the average value of the desired observable from *Postulate V* as

$$\bar{A} = \int_{-\infty}^{+\infty} \psi^*\hat{A}\psi \, d\tau. \qquad (7\text{--}3)$$

7. Write ψ^* by substituting $(-i)$ for i wherever it occurs in ψ. Then write $\psi^*\psi$ as a function of position, which is the probability distribution, where

$$\psi^*\psi = |\psi|^2 = probability \ density.$$

We are now prepared to consider the application of quantum mechanics to some very simple systems.

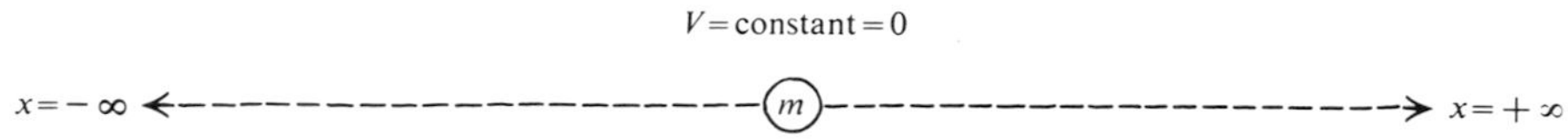

$$V = \text{constant} = 0$$

$$x = -\infty \ \longleftarrow \text{-------------------}(m)\text{-------------------} \longrightarrow \ x = +\infty$$

FIG. 7–1   A free particle moving in the x direction only.

7–1 FREE PARTICLE IN ONE DIMENSION

Consider a single particle of mass m which, in the absence of a force field, is free to move anywhere in the x-direction (Fig. 7–1). That is, there are no external forces operating on the particle and the potential energy is a constant which for convenience we'll assume to be *zero*. (If the particle were to move in a *force field*, potential energy would depend on position.) We write the Schrödinger amplitude equation in one dimension as

$$\frac{-\hbar^2}{2m}\left(\frac{d^2\psi}{dx^2}\right) = (E - V)\psi = E\psi, \qquad (7\text{--}4)$$

where ψ is a function of x and $E = T + V = T$. Thus the total energy E is given by the kinetic energy T, and is constant in the conservative force field. Equation (7–4) may be rearranged to

$$\frac{d^2\psi}{dx^2} = \frac{-2mE}{\hbar}\psi, \qquad (7\text{--}5)$$

which is an eigenvalue equation whose form we have often encountered in previous treatments of classical waves. The amplitude function $\psi(x)$, which represents the *complete description of the particle*, must be found by solving Eq. (7–5). It can be shown readily by substitution that either of the following two functions is a possible solution:

$$\psi_1 = A \exp\left[\frac{i}{\hbar}(2mE)^{1/2}x\right],\tag{7–6}$$

or

$$\psi_2 = B \exp\left[\frac{-i}{\hbar}(2mE)^{1/2}x\right],\tag{7–7}$$

where A and B are constants.

For the free particle, the only reasonable boundary condition seems to be that ψ remain finite rather than diverge as $x \to \pm\infty$. This requires that the quantity $(2mE)^{1/2}$ be *real* in order to preserve the periodic and hence non-divergent form $\exp(\pm i\alpha x)$, which in turn *requires that E be positive*. To understand this, observe what happens to ψ_1 when E is negative:

$$\psi_1 = A \exp\left(\frac{i}{\hbar}(-2m|E|)^{1/2}x\right) = A \exp\left(\frac{i^2}{\hbar}(2m|E|)^{1/2}x\right)$$

$$= A \exp\left(-(2m|E|)^{1/2}x/\hbar\right),$$

which means that $\psi_1 \to \infty$ and is thus divergent as $x \to -\infty$. Similarly, for negative vaues of E,

$$\psi_2 = B \exp\left(+(2m|E|)^{1/2}x/\hbar\right),$$

which means that $\psi_2 \to \infty$ as $x \to +\infty$. In each case the negative value of E violates the limit condition. The restriction that $E \geqslant 0$ is the *only* restriction on E. A continuous spectrum of all possible E values from 0 to $+\infty$ is permissible, and energies are not quantized.

If we next attempt to normalize ψ_1 or ψ_2 through the usual normalization requirement, that is,

$$\int_{-\infty}^{+\infty}\psi^*\psi\,dx = 1,$$

we immediately discover that the integral is divergent, because $\psi^*\psi$ remains finite as $x \to \pm\infty$. For example, for the ψ_1 function,

$$\int_{-\infty}^{+\infty}\psi^*\psi\,dx = \int_{\infty}^{\infty}A^* \exp\left(\frac{-i}{\hbar}(2mE)^{1/2}x\right)A \exp\left(\frac{i}{\hbar}(2mE)^{1/2}x\right)dx$$

$$= A^*A\int_{-\infty}^{+\infty}dx,$$

which is divergent. This means that there is no finite value assignable to A which can reduce the normalization integral to unity, or to any other constant value. Thus, the amplitude function for the free particle is not normalizable in the usual sense. This result is expected on physical grounds, since there is no reason to expect the probability of existence of the free particle to vanish at any point in its coordinate space. Otherwise, the particle wouldn't truly be free. The lack of ability to normalize the wave function, as in our first example, is really unusual, but the truly free particle represents an unreal situation. That is, no single particle can exist without interaction with any other particle in the universe.

Let's now use the operator postulate (Postulate IIb)

$$\hat{p}_x \psi = p_x \psi$$

in order to determine the allowed exact values for the momentum. The allowed eigenvalues of the observable, p_x, are obtained by the operation of $\hat{p}_x$ on either ψ_1 or ψ_2. Operation first on ψ_1 yields

$$\hat{p}_x \psi_1 = \frac{\hbar}{i} \frac{d}{dx} A \exp\left[\frac{i}{\hbar} (2mE)^{1/2} x \right]$$

$$= \frac{\hbar}{i} \frac{i}{\hbar} (2mE)^{1/2} A \exp\left[\frac{i}{\hbar} (2mE)^{1/2} x \right].$$

Cancellation of terms and reinsertion of ψ_1 gives

$$\hat{p}_x \psi_1 = +(2mE)^{1/2} \psi_1,$$

from which

$$p_x = +(2mE)^{1/2}. \tag{7–8}$$

Since E may have any positive value and is not quantized, then also p_x may have any *positive* value and is not quantized. Similarly, operation of $\hat{p}_x$ on ψ_2 yields

$$\hat{p}_x \psi_2 = -(2mE)^{1/2} \psi_2,$$

from which

$$p_x = -(2mE)^{1/2}. \tag{7–9}$$

Apparently then, if ψ_2 is chosen as the wave function for the free particle, the momentum, which is a vector quantity, must be negative. Note, in comparison, that for the classical particle for which $V = 0$,

$$E = T = \frac{mv_x^2}{2} = \frac{m^2 v_x^2}{2m} = \frac{p_x^2}{2m}, \tag{7–10}$$

so that

$$p_x = \pm (2mE)^{1/2}, \tag{7–11}$$

where $(2mE)^{1/2}$ is the *magnitude* of the linear momentum in the x direction.

Equation (7–11) is consistent with the forms for momentum given by Eqs. (7–8) and (7–9). Regardless of whether we choose ψ_1 or ψ_2, the energy of the free particle is not quantized. Correspondingly, neither is the momentum, as given by Eq. (7–11), quantized. However, if ψ_1 is chosen, the continuous range of momentum values available to the particle is restricted to *positive* values, which means that motion of the particle is restricted to the $+x$ direction. On the other hand, if ψ_2 is chosen to represent the free particle, the particle is allowed any *negative* value for the momentum and the motion of the particle is thus restricted to the $-x$ direction.

Let's now calculate the relative probability density $\psi^*\psi$ (ψ is not normalized) as a function of x, using ψ_1 for the particle moving in the $+x$ direction:

$$\psi_1^*\psi_1 = A^* \exp\left[-\frac{i}{\hbar}(2mE)^{1/2}x\right]A \exp\left[\frac{i}{\hbar}(2mE)^{1/2}x\right]$$

$$= A^*A = |A|^2 = \text{constant.} \tag{7–12}$$

Note that the probability density is independent of x; that is, there is an equal probability of finding the particle at any point along its path from $x = -\infty$ to $x = +\infty$. The particle is totally *nonlocalized*. The results are analogous to those which we obtained for the classical free traveling wave in a string of infinite length (Fig. 5–3), in which we observed no nodes or antinodes to indicate localization of action.

Similarly, the use of ψ_2 for the particle moving in the $-x$ direction yields

$$\psi_2^*\psi_2 = B^*B = |B|^2 = \text{constant.}$$

Apparently then, if ψ_1 and ψ_2 are both *equally representative* solutions, that is, if the particle has an equal probability of moving in either direction (which is not necessary),

$$|A|^2 = |B|^2 \tag{7–13}$$

A more general wave function which allows the free particle to move in *either* direction may be written as the linear combination (in this case, addition) of ψ_1 and ψ_2. Thus

$$\psi = \psi_1 + \psi_2. \tag{7–14}$$

Note that ψ is also a solution of the time-independent Schrödinger equation, since ψ_1 and ψ_2, each of which has the *same eigenvalue energy* E, obey the same linear homogeneous equation,

$$\hat{H}\psi - E\psi = 0,$$

and therefore any linear combination of ψ_1 and ψ_2 is also a solution. That is,

$$\hat{H}\psi_1 - E\psi_1 = 0, \qquad \hat{H}\psi_2 - E\psi_2 = 0;$$

therefore,

$$\hat{H}(\psi_1 + \psi_2) - E(\psi_1 + \psi_2) = 0.$$

Wave functions representing states having the same energy are said to be *degenerate*. It is important, however, to realize that in the *general* case, if the eigenvalue energies E_1 and E_2 corresponding to the eigenfunctions ψ_1 and ψ_2 are *not equal*, that is, if the eigenfunctions are *nondegenerate*, each of the functions obeys a *different* linear homogeneous equation:

$$\hat{H}\psi_1 - E_1\psi_1 = 0, \qquad \hat{H}\psi_2 - E_2\psi_2 = 0,$$

in which case $(\psi_1 + \psi_2)$ is *not* an eigenfunction of $\hat{H}$. (See Problem 6–8.)

Substitution of Eqs. (7–6) and (7–7) for ψ_1 and ψ_2 in Eq. (7–14) yields

$$\psi = A \exp\left[\frac{i}{\hbar}(2mE)^{1/2}x\right] + B \exp\left[\frac{-i}{\hbar}(2mE)^{1/2}x\right]. \qquad (7\text{–}15)$$

If the particle has an equal probability of moving in either direction, $|A|^2$ must be equal to $|B|^2$ according to Eq. (7–13). On the other hand, if the particle is restricted to movement in the $(+x)$-direction *only*, B must equal zero, in which case the *second* term in Eq. (7–15) vanishes. Conversely, if the particle is restricted to motion in the $(-x)$-direction *only*, A must equal zero, in which case the *first* term of Eq. (7–15) vanishes.

Finally, it is important to emphasize once more that the energy of a free particle is not restricted to discrete values. The ionized electron is a good example of a free particle in three-dimensional space. Once the electron escapes from the coulombic force field created by the nucleus (once it ionizes) its position is in no way localized and it may assume any exact energy level. In fact, the continuum we have previously observed in the Balmer series spectrum (Fig. 3–2) for hydrogen is produced when free ionized electrons drop from a continuous spectrum of nondiscrete energy states to second allowed energy levels within individual hydrogen atoms.

7–2 SINGLE PARTICLE IN A ONE-DIMENSIONAL BOX

Just as we have observed a resultant localization of action (nodes and anti-nodes) whenever we have confined classical waves, we'll discover that the physical confinement of de Broglie waves also gives rise to a localization of probability for finding the associated particle and leads to quantization of energy levels. Let's, for example, confine the free particle we have just considered in such a way that it can move in only the x-direction within a one-dimensional potential energy well of infinite depth, which we'll refer to simply as a "box." In order to ensure that the particle remains in the box we'll assume that $V = 0$ everywhere in the box and $V = \infty$ everywhere outside the box. Such a box, of length L, is illustrated in Fig. 7–2. The Schrödinger amplitude

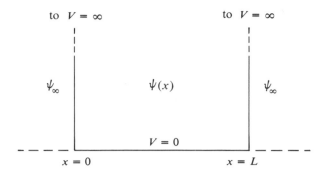

FIG. 7–2.   Potential energy as a function of x position for the particle in a one-dimensional box.

equation for the regions outside the box, where $V = \infty$ and the wave function is $\psi_\infty(x)$, is

$$\frac{-\hbar}{2m} \frac{d^2\psi_\infty}{dx^2} = (E - \infty)\psi_\infty.$$

Neglecting E compared to ∞,

$$\frac{d^2\psi_\infty}{dx^2} = \infty\psi_\infty, \quad \psi_\infty = \frac{1}{\infty} \frac{d^2\psi_\infty}{dx^2} = 0,$$

which means that $\psi_\infty{}^*\psi_\infty = 0$, so that the particle has zero probability of existence outside the box.

Within the box, where $V = 0$ and where the wave function is $\psi(x)$, the Schrödinger equation is

$$\frac{-\hbar^2}{2m}\left(\frac{d^2\psi}{dx^2}\right) = E\psi, \tag{7–16}$$

where $E = T + V = T$. Equation (7–16) is identical to Eq. (7–4) for the free particle, so that Eqs. (7–6) and (7–7) are valid solutions, at least as far as the Schrödinger equation is concerned.

But in order that the probability of finding the particle outside the box be zero, $\psi^*\psi$ must be zero outside the box. Furthermore, if the wave function is to be continuous, $\psi^*\psi$ must also be zero at the walls of the box and ψ itself must be zero at the walls. We then write the two *boundary conditions* as

$$\psi = 0 \quad \text{when} \quad x = 0, \tag{7–17}$$

$$\psi = 0 \quad \text{when} \quad x = L. \tag{7–18}$$

In addition to satisfying the Schrödinger amplitude equation, an acceptable amplitude function must *also* satisfy each of the imposed boundary conditions.

Let's then apply the first boundary condition to ψ_1 as given by

$$\psi_1 = A \exp \left[\frac{i}{\hbar} (2mE)^{1/2} x \right]. \tag{7-6'}$$

When $x = 0$, then $\psi_1 = 0$, so that

$$0 = A \exp (0) = A.$$

But A cannot be zero, because if it were, ψ_1 as given by Eq. (7–6) would *always* be equal to zero, and $\psi_1^* \psi_1$ would also always be zero at *any* value of x, so that the particle would have a zero probability of existing anywhere! In other words, if $A = 0$, the particle cannot exist in the box! A similar attempt to apply the first boundary condition to ψ_2 as given by Eq. (7–7) leads to the equally impossible conclusion that $B = 0$. Thus, although ψ_1 or ψ_2 is each individually a satisfactory solution to the Schrödinger amplitude equation, neither is able to satisfactorily incorporate the first boundary condition.

Fortunately, we may solve the dilemma by constructing a new solution through linear combination. We have shown in the previous section that if ψ_1 and ψ_2 is each a satisfactory solution to the Schrödinger amplitude equation and if each of the functions corresponds to the same eigenvalue energy, then *any* linear combination of ψ_1 and ψ_2 is also a satisfactory solution. We may arbitrarily write ψ, a proposed new solution (the simplest linear combination), as

$$\psi = \psi_1 + \psi_2, \tag{7-19}$$

or

$$\psi = A \exp \left[\frac{i}{\hbar} (2mE)^{1/2} x \right] + B \exp \left[-\frac{i}{\hbar} (2mE)^{1/2} x \right]. \tag{7-20}$$

Equation (7–20) is identical to Eq. (7–15), which we have applied to the motion in *either* direction of a free particle. The physical significance of Eq. (7–20) is this: *we must allow the particle to move in either direction if we are to keep it in the box*. We have already seen that if *either* A or B is zero, so that motion of the particle is restricted to one direction only, the probability of finding the particle in the box is zero. Now let's apply the first boundary condition to Eq. (7–20), that is, $\psi = 0$ when $x = 0$. Substitution of $x = 0$ and $\psi = 0$ in Eq. (7–20) yields

$$0 = A + B,$$

or

$$B = -A. \tag{7-21}$$

Then Eq. (7–20) becomes

$$\psi = A \left\{ \exp \left[\frac{i}{\hbar} (2mE)^{1/2} x \right] - \exp \left[-\frac{i}{\hbar} (2mE)^{1/2} x \right] \right\}, \tag{7-22}$$

a function which now satisfies both the Schrödinger amplitude equation *and* the first boundary condition. If we let $y = (2mE)^{1/2}x/\hbar$, we may write Eq. (7-22) more simply as

$$\psi = A(e^{iy} - e^{-iy}). \tag{7-23}$$

But, according to Euler's formula (Eq. 5-9),

$$e^{iy} = \cos y + i \sin y \qquad \text{and} \qquad e^{-iy} = \cos y - i \sin y,$$

so that

$$e^{iy} - e^{-iy} = 2i \sin y, \tag{7-24}$$

and Eq. (7-23) becomes, after substitution for y,

$$\psi = i2A \sin\left[(2mE)^{1/2}x/\hbar\right] = C \sin\left[(2mE)^{1/2}x/\hbar\right], \tag{7-25}$$

where $C = i2A$ (an imaginary number only if A is real).

We are now ready to impose the second boundary condition, that is, $\psi = 0$ when $x = L$. Substitution of $\psi = 0$ and $x = L$ into Eq. (7-25) yields

$$0 = C \sin\left[(2mE)^{1/2}L/\hbar\right]. \tag{7-26}$$

Since $C \neq 0$, $\sin[(2mE)^{1/2}L/\hbar]$ must be equal to zero, which is possible whenever the argument of the sine is an integral multiple of π, that is, when

$$(2mE)^{1/2}L/\hbar = \pm n\pi, \tag{7-27}$$

where $n = 0, 1, 2, 3, \ldots, \infty$.

From Eq. (7-27), then,

$$(2mE)^{1/2}/\hbar = \pm n\pi/L \tag{7-28}$$

Substitution of Eq. (7-28) into Eq. (7-25) yields

$$\psi = C \sin\left(\pm \frac{n\pi x}{L}\right) = \pm C \sin\left(\frac{n\pi x}{L}\right). \tag{7-29}$$

The amplitude functions ψ given by Eq. (7-29) now satisfy the Schrödinger amplitude equation and both boundary conditions. That is, when x is 0 or L, $\psi = 0$. The $\pm$ sign in Eq. (7-29) indicates that there are two satisfactory series of functions, either the positive series or the negative series, each containing one function for each value of n.

Our final task is to evaluate C by applying the normalization condition, that is,

$$\int_0^L \psi^*\psi \, dx = 1. \tag{7-30}$$

The normalization condition ensures that each wave function is such that the total probability for finding the particle between $x = 0$ and $x = L$ (in the "box") is unity. From Eq. (7-29) we may write ψ^* as

$$\psi^* = \pm C^* \sin\left(\frac{n\pi x}{L}\right). \tag{7-31}$$

Substitution of *either* the positive or negative versions of Eqs. (7–29) and (7–31) into Eq. (7–30) yields

$$C^*C \int_0^L \sin^2 \left(\frac{n\pi x}{L}\right) dx = 1. \qquad (7\text{–}32)$$

Integration of Eq. (7–32) yields

$$C^*C \left[\frac{x}{2} - \frac{\sin(2\pi nx/L)}{4\pi n/L}\right]_0^L = 1,$$

or

$$C^*C \left(\frac{L}{2}\right) = 1 \quad \text{or} \quad |C|^2 = \frac{2}{L}.$$

When C is expressed as a real number, Eq. (7–29) becomes

$$\psi = \pm \left(\frac{2}{L}\right)^{1/2} \sin\left(\frac{n\pi x}{L}\right). \qquad (7\text{–}33)$$

However, because probability is given by $|\psi|^2$ rather than by ψ, it is immaterial whether C is chosen as the positive or negative square root. That is, it is not really necessary to carry both a positive *and* a negative function for each value of n, since *either* the positive or the negative function alone can be used to calculate probabilities and to generate exact and expectation values for desired observables, including energy levels, for any given value of n. Thus, following custom, we discard the set of negative functions and adopt the positive series of normalized functions for the single particle in the one-dimensional box as

$$\psi = \left(\frac{2}{L}\right)^{1/2} \sin\left(\frac{n\pi x}{L}\right) \qquad (7\text{–}34)$$

where $n = 1, 2, 3, \ldots, \infty$. Note that we have eliminated the solution for which $n = 0$, since when $n = 0$, then $\psi = 0$ and $\psi^*\psi = 0$ everywhere. That is, when $n = 0$, *the particle does not exist.* According to Eq. (7–34), there is an *infinite number* of satisfactory discrete functions, each involving a different integral value for n, the *quantum number.*

It is enlightening now to compare Eq. (7–34) to Eq. (4–52) for the stretched string fixed at both ends. Note that the forms are identical. That is, confinement of a de Broglie wave within a "box" leads to the same results we obtain when we confine an elastic wave in a string by clamping the ends so that the wave is not allowed beyond a finite length.

7-3 ALLOWED ENERGY LEVELS AND PROBABILITY DISTRIBUTIONS FOR A PARTICLE IN A ONE-DIMENSIONAL BOX

The allowed values for the energy of the particle in the one-dimensional box are given by squaring and rearranging Eq. (7–28), which gives

$$E = n^2 h^2 / 8mL^2, \qquad (7\text{–}35)$$

where $n = 1, 2, 3, \ldots, \infty$. Equation (7–35) expresses the stationary-state energies available to the particle in the box. *No other exact energies are allowed nor can other exact energies be experimentally observed.* The state of the particle characterized by a particular value for one of the allowed energy levels is called a *quantum state*. For example, in the first *quantum state*, or *ground state*, $n = 1$, and according to Eq. (7–35) the energy is,

$$E_1 = h^2/8mL^2.$$

It is important to note that the allowed energy states given by Eq. (7–35) could also have been easily generated by the operation

$$\hat{T}\psi = T\psi, \tag{7–36}$$

according to the operator postulate (Postulate II), since $E = T$. In using Eq. (7–36), $\hat{T} = (-\hbar^2/2m)(d^2/dx^2)$, and the ψ functions are given by Eq. (7–34).

In order to determine where the particle might be localized within the one-dimensional box we'll now develop the expression for $\psi^*\psi$, the probability density, as a function of x. The complex conjugate of ψ is equal to ψ, since i does not appear in the equation for ψ. Thus, for any value of n,

$$\psi^*\psi = |\psi|^2 = \psi^2 = \frac{2}{L}\sin^2\left(\frac{n\pi x}{L}\right), \tag{7–37}$$

where $n = 1, 2, 3, \ldots, \infty$. According to Eq. (7–37), the probability density is a function of both x and n.

The allowed amplitude functions for various values of the quantum number n are plotted in Fig. 7–3. Corresponding energy levels and probability distributions are also shown. As for the vibrating string fixed at both ends, n is equal to the number of antinodes or loops. Note that the energy of a quantum state increases as the square of n. In the case of the fixed vibrating string we also noted that the energy in the string was higher for higher values of n. Since in a system of fixed total energy higher-energy levels are statistically less probable at thermal equilibrium, we would expect that in a system consisting of a large number of particles at equilibrium, distributed over all of the allowed quantum states, most of the particles would be in the ground state, for which $n = 1$.

The magnitude of the energy levels, and thus the magnitude of the separation between energy levels, is inversely proportional to mL^2. Thus, for large masses or large values of L (large "boxes"), the allowed energy levels are very close together and appear to be continuous, to the best of our ability to measure such changes. In other words, *large masses in large spaces behave classically.* For such systems, quantum mechanics reduces to classical mechanics. When m and L are small, however, the energy in the second allowed energy state may be appreciably above the energy in the ground state and the energy difference

| Quantum number | Energy | Amplitude function $\psi = \left(\frac{2}{L}\right)^{1/2} \sin\left(\frac{n\pi x}{L}\right)$ | Probability density $\psi^2 = \frac{2}{L}\sin^2\left(\frac{n\pi x}{L}\right)$ |

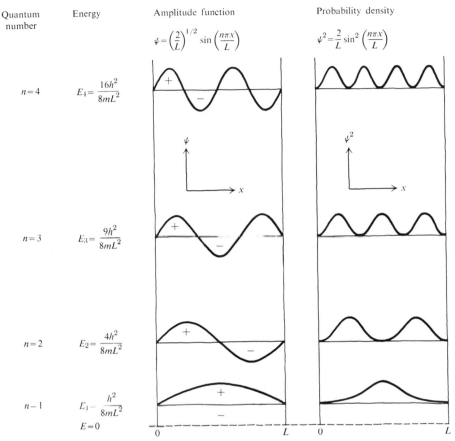

FIG. 7–3   Energy levels, amplitude functions, and probability densities for the single particle in the one-dimensional box.

may become experimentally detectable. Two specific examples will immediately clarify the difference between "classical" and "quantum" systems.

Let's first consider a 10-g marble in a 100-cm one-dimensional box. According to Eq. (7–35), the lowest energy the marble may have is that energy for which $n = 1$ (ground-state energy), which is

$$E_1 = \frac{n^2 h^2}{8mL^2} = \frac{(1)^2(6.63 \times 10^{-27})^2}{8(10)(100)^2}$$

$$= 5.48 \times 10^{-59} \text{ erg.}$$

The next highest allowed energy level is

$$E_2 = (2)^2 E_1 = 21.9 \times 10^{-59} \text{ erg.}$$

If the marble were to drop in energy from the first excited state E_2 to the ground state E_1, the energy lost would be $\epsilon = E_2 - E_1 = 16.4 \times 10^{-59}$ erg. In turn, if this minute amount of energy were to be emitted as a single photon, the wavelength of the emitted radiation, as calculated from the Planck-Einstein relationship

$$\epsilon = h\nu = hc/\lambda, \tag{7-38}$$

would be

$$\lambda = \frac{hc}{\epsilon} = \frac{(6.63 \times 10^{-27})(3 \times 10^{10})}{16.4 \times 10^{-59}} = \sim 10^{42} \text{ cm},$$

and the corresponding period of the photon would be

$$\tau = \frac{1}{\nu} = \frac{h}{\epsilon} = \frac{6.63 \times 10^{-27}}{16.4 \times 10^{-59}} = 4 \times 10^{31} \text{ sec}$$

$$= \sim 10^{24} \text{ years},$$

which is a long time indeed to wait for a classical measurement! Because of the closeness in spacing between adjacent energy levels of the marble and the resulting limitations in our ability to measure the corresponding energy changes, the spectrum of energy states available to the marble appears experimentally as if it were continuous.

As a second example, let's calculate the first two allowed energy levels for a mass the size of an electron ($m = 9.1 \times 10^{-28}$ g) confined in a one-dimensional box about the size of an atom ($L = 3 \times 10^{-8}$ cm). The ground-state energy is

$$E_1 = \frac{n^2 h^2}{8mL^2} = \frac{(1)^2 (6.63 \times 10^{-27})^2}{8(9.1 \times 10^{-28})(3 \times 10^{-8})^2} = 6.7 \times 10^{-12} \text{ erg},$$

and the energy in the second energy level is

$$E_2 = (2)^2 E_1 = 26.8 \times 10^{-12} \text{ erg}.$$

The energy evolved when the electron undergoes a transition from E_2 to E_1 is given as

$$\epsilon = E_2 - E_1 = 20.1 \times 10^{-12} \text{ erg}.$$

If this amount of energy is emitted as a single photon, the wavelength of the emitted radiation is

$$\lambda = \frac{hc}{\epsilon} = \frac{(6.63 \times 10^{-27})(3 \times 10^{10})}{20.1 \times 10^{-12}} = 9.9 \times 10^{-6} \text{ cm} = 990 \text{ Å}.$$

The radiation emitted is a high-energy photon in the ultraviolet region, *which can be detected quite easily experimentally.* In general, we are able to measure readily most differences in allowed electron energies within atoms.

7-4 INTERPRETATION OF PROBABILITY AND PROBABILITY DENSITY

The probability distributions shown in Fig. 7–3 are quite surprising to one unaccustomed to treating quantum systems. In each state, the probability of finding the particle at the wall is zero. This is not unexpected, since we insisted on this behavior when we defined the boundary conditions. However, when n is greater than one, there are points of zero probability for finding the electron at values for x other than 0 or L. For example, for the second allowed state, in which $n = 2$, the probability of finding the electron at $x = L/2$ is zero and we calculate maximum probability densities at $x = L/4$ and $x = 3L/4$. Let's consider the $n = 2$ state in more detail to be certain that we properly understand the meaning of the terms "probability" and "probability density." The curve for $|\psi_2|^2$ versus x is again shown in Fig. 7–4. The probability that the particle exists between any two positions x_1 and x_2, is given as

$$\int_{x_1}^{x_2} |\psi_2|^2 \, dx = \int_{x_1}^{x_2} \psi_2^2 \, dx, \qquad (7\text{–}39)$$

where $|\psi_2|^2$ is the probability density, or probability *per unit length of* x. Graphically, the probability is simply the area under the curve between x_1 and x_2 (the shaded area in Fig. 7–4). Normalization of ψ ensures that the *total* area under the curve from $x = 0$ to $x = L$ is unity, so that the total probability is unity, and we are thus certain to find the particle somewhere in the box.

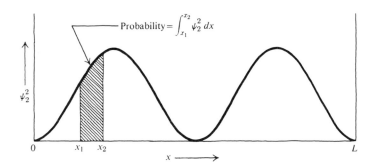

FIG. 7–4   A graphical representation of the probability that a single particle exists between x_1 and x_2 in a one-dimensional box. ψ_2^2 is the probability density for the second allowed quantum state ($n = 2$).

Let's now pose a very interesting question. If the single particle has a finite probability of existing between $x = 0$ and $x = L/2$ and also has a finite probability of existing between $x = L/2$ and $x = L$, but has zero probability of existing at the position $x = L/2$, how does it get from one side of the box to the other? Apparent paradoxes such as this often arise when we invoke classical macroscopic experience in an attempt to understand microscopic behavior. However, note that the problem* doesn't even arise in the more refined *relativistic* treatment of Dirac,† which is mathematically much more complicated than the nonrelativistic Schrödinger treatment which we have used. In brief, when the problem of the particle in the box is solved by Dirac's relativistic approach, which is a more rigorous and correct approach, one of the important consequences is the *disappearance* of the nodes of zero probability which we have shown, through a nonrelativistic treatment, to arise within the box. In place of the nodes there result *approximate nodes* at which the probability density, although very small, is nevertheless *finite*. In a similar way, even though our later nonrelativistic treatment of the electron in the hydrogen atom will lead to nodes, or surfaces in space at which the probability density for finding the electron is zero, we must realize that, in terms of the more correct relativistic treatment, the "Schrödinger nodes" are not true nodes at all, but rather are approximate nodes which represent regions of very small but finite probability density.

7–5 THE HEISENBERG UNCERTAINTY PRINCIPLE

According to Eq. (7–12), the classical momentum p_x of the particle in the box is related to the total energy E by

$$p_x^2 = 2mE, \tag{7–40}$$

and since the energy E of the particle is quantized according to

$$E = \frac{n^2 h^2}{8mL^2}, \tag{7–35'}$$

it follows by combination of Eqs. (7–40) and (7–35') that

$$p_x^2 = \frac{n^2 h^2}{4L^2}$$

which implies that the linear momentum is restricted to either of two discrete values

$$p_x = +\frac{nh}{2L} \quad \text{or} \quad p_x = -\frac{nh}{2L} \tag{7–41}$$

This, in turn, implies that for a given quantum state (for a given value of n) the allowed *magnitude* of the momentum for the particle is inversely proportional to

---

* A good recent review of the node problem is given by R. E. Powell, *J. Chem. Ed.*, **45**, 558–563 (1968).

† P. A. M. Dirac, *Proc. Roy. Soc.* (London), **A117**, 610; **A118**, 351 (1928).

the size of the box. Furthermore, the momentum may be either positive or negative. That is, the particle may be moving in either direction. In fact quantum mechanics implies that *it cannot be known in which direction the particle is moving*. In addition, although we can express the probabilities associated with different positions within the box, we cannot know *exactly* where within the box the particle is located. We can know only that it is somewhere in the box.

All of this is not surprising when we consider the particle to be governed by wave motion. Consider, for example, the standing wave in a stretched string. In which direction is the wave traveling? The answer, of course, is that the wave is traveling in both directions simultaneously. In fact, the interference of waves traveling simultaneously in both directions is precisely what produces the standing-wave pattern. Secondly, we might ask where the wave is located. The most precise answer we may give is that it is located between the fixed ends. By analogy to the classical wave, we may consider the particle in the box to be traveling in either direction with a linear momentum $+nh/2L$ or $-nh/2L$ (we don't know which). Furthermore, we may not designate its exact position any more precisely than to say that it is in the box between $x = 0$ and $x = L$. Operations on the wave function for the particle will not give us any more detailed information. Thus, according to quantum mechanics, a basic *theoretical uncertainty* exists in the accuracy of measurement of either position or momentum or both. Since we *cannot know* whether the momentum is $+nh/2L$ or $-nh/2L$ the uncertainty in momentum, Δp_x, may be expressed as

$$\Delta p_x = \frac{+nh}{2L} - \left(\frac{-nh}{2L}\right) = \frac{nh}{L}. \tag{7–42}$$

Furthermore, since we cannot hope to know the location of the particle any more accurately than to say it is in the box of length L, we may state the uncertainty in position as

$$\Delta x = L. \tag{7–43}$$

Combination of Eqs. (7–42) and (7–43) yields

$$\Delta x \, \Delta p_x = nh, \tag{7-44}$$

which is one form of *Heisenberg's uncertainty principle*. According to Eq. (7–44), when $n = 1$ (in the ground state), the product of uncertainty in position times uncertainty in momentum is a minimum and is equal to Planck's constant. As the separation between the walls of the box approaches infinity, the particle behavior approaches that of an unconfined free particle. That is, the uncertainty in position approaches infinity and the uncertainty in momentum approaches zero. Quantum mechanics gives no indication of the exact location of a free particle, any more than one might indicate the exact location of a free traveling wave. If it is not confined, it may be anywhere. On the other hand, as we confine a particle to smaller and smaller dimensions in order to define its

x-position more closely, we increase the uncertainty in the statement of momentum because we increase the magnitude of the minimum momentum, $nh/2L$, required for existence of the particle. In fact, if we attempt to locate a particle *exactly*, that is, if $L = \Delta x = 0$, the minimum momentum required for existence is either $(+)$ or $(-)$ infinity and the uncertainty in linear momentum, Δp_x, is infinite.

It is also apparent from Eq. (7–44) that higher values of n lead to greater uncertainties. That is, if the particle is in the second quantum state ($n = 2$),

$$\Delta x \, \Delta p_x = 2h.$$

If it is in the third quantum state ($n = 3$),

$$\Delta x \, \Delta p_x = 3h,$$

and so forth. But we have mentioned in Chapter 2 that at thermal equilibrium low quantum states are statistically much more probable than higher states and that the ground state is the most probable of all. We might then write the Heisenberg uncertainty principle more generally as

$$\Delta x \, \Delta p_x \geq h,$$

or, in order to emphasize the statistical importance of the ground state, as

$$\Delta x \, \Delta p_x \approx h. \tag{7–45}$$

That is, for any particle, the minimum magnitude of the product of uncertainty in position and uncertainty in linear momentum in the same direction is *about equal* to Planck's constant.

7–6 WAVE PACKETS AND THE UNCERTAINTY PRINCIPLE

There is another, perhaps more fundamental, way to appreciate the uncertainty principle. Let's return to a consideration of the free particle moving in one dimension in the absence of a force field, which we have discussed at the beginning of this chapter. We have shown that the time-independent wave function for the particle moving in the $(+x)$-direction is

$$\psi(x) = A \, \exp \left[\frac{i}{\hbar} (2mE)^{1/2} x \right], \tag{7–6'}$$

and that the linear momentum is

$$p_x = (2mE)^{1/2}, \tag{7–10'}$$

so that we may write $\psi(x)$ in terms of p_x as

$$\psi(x) = A \, \exp \left(\frac{ip_x x}{\hbar} \right). \tag{7–46}$$

It then follows, according to Eq. (6–48), that the complete time-dependent wave function $\Psi(x, t)$ for the free particle is

$$\Psi(x, t) = A \exp\left(\frac{ip_x x}{\hbar}\right) \exp\left(\frac{-iEt}{\hbar}\right), \qquad (7\text{--}47)$$

where A is a constant. But in deriving Eq. (7–47), we have used the de Broglie postulates $E = h\nu$ and $\lambda = h/p_x$ which, upon resubstitution into Eq. (7–47), yield the fundamental equation for displacement in a one-dimensional de Broglie wave traveling in the $(+x)$-direction:

$$\Psi(x, t) = A \exp\left[i2\pi\left(\frac{x}{\lambda} - \nu t\right)\right]. \qquad (7\text{--}48)$$

Substitution of the *wave number* $\bar{\nu}$, which is defined as $1/\lambda$ and is equal to the number of waves per unit length, into the above equation gives

$$\Psi(x, t) = A e^{i2\pi(\bar{\nu}x - \nu t)}, \qquad (7\text{--}49)$$

where $\bar{\nu}$ and ν are the wave number and the frequency of the de Broglie pilot wave associated with a free particle whose exact momentum is p_x. At any given moment, the probability density $\Psi^*\Psi$ as a function of position x is given from Eq. (7–49) as the *constant* $|A|^2$. Thus, it once again follows that the free particle of exact momentum p_x has an equal probability of being found anywhere along the x-coordinate from $-\infty$ to $+\infty$. Although we can define the momentum of the particle exactly, its position is totally uncertain.

But let's now try to construct a wave function which describes the particle existing at a given time within some finite region between two selected values of x. Such a wave function will have to be nonzero and finite between the two values of x and zero everywhere else. In order to devise such a function, it is necessary that we first review the classical concept of wave groups. It is well known from classical wave theory that one may mathematically construct a *wave group* or *wave packet*, such as that shown in Fig. 7–5, by superposing an infinite number of simple monochromatic periodic waves of infinitesimally differing wave numbers and frequencies, each having its own velocity, called the *phase velocity*. For example, two such superpositions, each involving only *two* contributing monochromatic waves, are shown in Fig. 7–6. When only *two* contributing waves are used, an infinite number of successive wave packets is produced. However, if an *infinite* number of contributing monochromatic waves is used, it is possible to superpose them in such a way that a *single* packet is formed. In order to do this, we require that the contributing waves cancel each other everywhere except in one localized region corresponding to the wave group, in the center of which the crests of all the waves coincide. At a given moment in time at this single point, the positive displacements of all the contributing periodic waves add up to a maximum total displacement. At no other point at this same time do the crests of all the waves similarly coincide. In fact, by

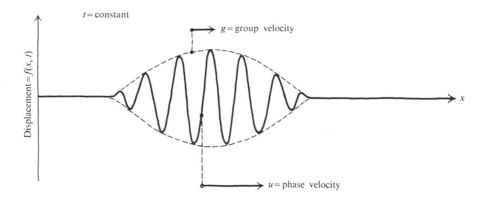

FIG. 7–5   A qualitative representation of the wave packet formed by the superposition of the real parts of an infinite number of monochromatic periodic waves of infinitesimally differing wave numbers and frequencies. The packet may be constructed in such a way that the group velocity g is less than the phase velocity u.

properly selecting the range of wave numbers from which the simple periodic waves are chosen, we can construct the total wave group so that, at distances more than a few wavelengths removed from the point of coincidence of the wave crests, the displacements are so random that we are almost equally likely to encounter negative displacements as we are to encounter positive displacements in the real parts of the contributing periodic waves, so that the total displacement approaches zero.   Furthermore, if the phase velocities of the contributing periodic waves are not exactly the same, the wave group can be made to move in the x-direction with a *group velocity* g, which is less than the phase velocities u of the contributing periodic waves.

Since u is larger than g, the individual pilot waves may be imagined to constantly move through the wave packet from rear to front while the packet moves forward at a slower speed.   Let's now determine how the group velocity of such a packet depends on the choice of wave numbers and frequencies of the contributing monochromatic waves.   For simplicity, we'll consider the superposition of just *two* contributing monochromatic waves, having equal amplitudes A but differing infinitesimally with respect to wave numbers and frequencies.   Using Eq. (7–49), we may write the individual displacements for the two contributing waves as

$$\Psi_1(x, t) = A \exp\left[i2\pi(\bar{\nu}x - \nu t)\right],$$

$$\Psi_2(x, t) = A \exp\left\{i2\pi[(\bar{\nu} + d\bar{\nu})x - (\nu + d\nu)t]\right\}.$$

According to the principle of superposition, the total displacement for the combined wave is

$$\Psi(x, t) = \Psi_1(x, t) + \Psi_2(x, t), \qquad (7\text{–}50)$$

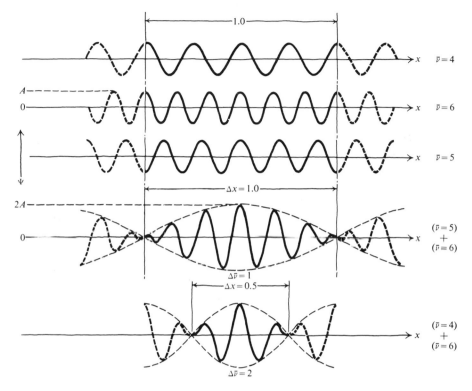

FIG. 7-6 Graphical superpositions of the real parts of monochromatic periodic waves of equal amplitudes. Whenever two such waves are superposed, $\Delta \bar{\nu} \, \Delta x = 1$ (t is constant in each of the above plots).

or

$$\Psi(x, t) = A \exp \left[i2\pi (\bar{\nu} x - \nu t) \right]$$
$$+ A \exp \left[i2\pi (\bar{\nu} x - \nu t) \right] \exp \left\{ i2\pi [(d\bar{\nu}) x - (d\nu) t] \right\}$$
$$= (1 + \exp \left\{ i2\pi [(d\bar{\nu}) x - (d\nu) t] \right\})(A \exp \left[i2\pi (\bar{\nu} x - \nu t) \right]). \quad (7\text{–}51)$$

An examination of the above equation reveals that the second term in boldface parentheses is identical to Eq. (7–49) for a monochromatic traveling wave of amplitude A. However, the real value of the *first* term in bold parentheses varies periodically from $(1 - 1)$ to $(1 + 1)$, so that the first term serves to modulate the amplitude of the traveling wave defined by the second term in such a way that the real part of the amplitude periodically varies from 0 to $2A$, thus defining the ends and centers, respectively, of successive wave packets. The net effect of modulation of amplitude is clearly evident in the superposed waves shown in Fig. 7–6. The phase velocities u of the individual monochromatic waves can be evaluated by considering the second term of Eq. (7–51), but the velocity

g of the groups which result from the amplitude modulation must be evaluated from the first term. That is, we first write the phase velocity as

$$u = \lambda v = v/\bar{v}$$

and by comparing the form of the first term to that of the second term, we write the group velocity as

$$g = dv/d\bar{v}. \tag{7–52}$$

It can be shown that Eq. (7–52) is general, that is, that it also applies to the group velocity of a packet formed from an *infinite* number of contributing monochromatic waves in which the rate of change of frequency with wave number is given by $dv/d\bar{v}$.

Now let's use the classical concept of the wave packet in order to describe a free particle momentarily existing at some value of t within a defined region, Δx. If we properly superpose an infinite number of simple de Broglie waves, we may construct such a wave packet, and because the probability of finding the particle at a given position of x is related to the square of the absolute displacement for the total wave, which is zero everywhere outside the packet, it is necessary that the particle be located within the dimensions of the packet. Furthermore, for de Broglie waves

$$v = E/h \quad \text{and} \quad \bar{v} = 1/\lambda = p_x/h,$$

so that

$$dv = dE/h \quad \text{and} \quad d\bar{v} = dp_x/h.$$

Then, according to Eq. (7–52), the group velocity of the packet is given as

$$g = dE/dp_x, \tag{7–53}$$

where E is the relativistic energy and p_x is the relativistic momentum in the x-direction. But, according to Eq. (3–39),

$$E^2 = c^2 p_x^2 + m_0^2 c^4,$$

so that

$$2E \, dE = 2c^2 p_x \, dp_x$$

and

$$\frac{dE}{dp_x} = \frac{c^2 p_x}{E}.$$

Therefore, Eq. (7–53) becomes

$$g = \frac{c^2 p_x}{E}.$$

But $E = mc^2$ and $p_x = mv_x$, so that

$$g = c^2 \left(\frac{mv_x}{mc^2} \right) = v_x.$$

That is, *the group velocity of the packet of de Broglie waves is exactly equal to the velocity of the particle whose motion it governs.*

Note, however, that in associating the position and velocity of a free particle with the position and velocity of a packet of de Broglie waves, we introduce an important new consideration. That is, *the wave packet is spread over a region of space, whereas in classical physics the center of mass of a particle is treated as a point.* The relations which we have developed in this section are relativistically correct, and it is apparent that if the phase velocities of all of the contributing waves are identical, the packet must move with the same velocity as the waves, that is, $v_x = u$, so that, according to Eq. (3–46), $v_x = c$, which implies that the packet describes the behavior of a photon traveling in a vacuum. Furthermore, the shape of such a photon's wave packet is preserved as it moves in the *x*-direction with time. However, if the phase velocities of all of the contributing waves are *not* the same, *a condition which is required for all particles having velocities less than the speed of light in a vacuum,* it can be shown that the packet becomes lower and wider, or spreads out, as it moves in the *x*-direction with increasing time. For example, immediately after we momentarily observe the position of an electron in such a way that its wave packet may be defined, the packet begins to dissipate or disperse and continues to disperse as long as we don't try to observe the particle. However, as soon as we once more locate the particle as a corpuscle, the pilot waves immediately condense again into a small wave packet whose size depends on the accuracy of observation of position, and the process of dispersion then begins anew.

An important question remains to be answered: how is the size of the wave packet, which is a measure of uncertainty in the location of the particle, related to the distribution of wave numbers in the contributing monochromatic pilot waves? Once more consider the graphical superpositions of the real parts of two different sets of simple periodic waves shown in Fig. 7–6. Each of the contributing waves has the same amplitude. For the first superposition, $\Delta \bar{\nu} = 1$, and for the second, $\Delta \bar{\nu} = 2$. It is apparent that the distance Δx, which separates the nearest two points of total destructive interference, is dependent on $\Delta \bar{\nu}$. For the particular examples given in Fig. 7–6, when $\Delta \bar{\nu} = 1$,

$$\Delta x = 1.0,$$

and when $\Delta \bar{\nu} = 2$,

$$\Delta x = 0.5.$$

That is, the greater the difference in wave numbers, the nearer to the center of the packet are the positions at which total destructive interference occurs. It is clearly evident, in addition, that when $\Delta \bar{\nu} = 0$,

$$\Delta x = \infty,$$

since identical waves which are in phase at any one point are in phase at all points. In general, we may write

$$\Delta\bar{\nu}\,\Delta x = 1.$$

It can also be shown that, if we combine an *infinite* number of simple mono-chromatic waves within the wave number range $\Delta\bar{\nu}$ in order to construct a single packet, then

$$\Delta\bar{\nu}\,\Delta x = \text{constant} \approx 1/2\pi. \tag{7–54}$$

In mathematically formulating such a packet, it is usually customary to select different amplitudes, or weighting factors, for each of the component waves, in order to construct a packet whose shape leads to a distribution for probability of location of the particle which is consistent with a statistical probability distribution. Thus, the exact value of the constant in Eq. (7–54) depends on the amplitudes assigned, but is approximately $1/2\pi$. Referring to Eq. (7–54), we note that if the range of wave numbers of the contributing waves is large, we may construct a relatively narrow packet. In fact, if we choose to combine simple periodic waves whose range of wave numbers is infinite, we may in the limit construct a collapsed packet which approaches a simple vertical line, for which $\Delta x \rightarrow 0$. But if we are to associate the wave with a particle, it is important to note that each wave number $\bar{\nu}$ must be directly related to an exact value of an x-directional linear momentum, p_x. That is

$$\bar{\nu} = 1/\lambda = p_x/h,$$

which implies that the range $\Delta\bar{\nu}$ of wave numbers of monochromatic pilot waves used in constructing the wave packet is related to the range Δp_x of corresponding momentum values of the associated particle through the relationship

$$\Delta\bar{\nu} = \Delta p_x/h. \tag{7–55}$$

Substitution of Eq. (7–55) into Eq. (7–54) yields

$$\Delta x\,\Delta p_x \approx \hbar. \tag{7–56}$$

The meaning of Eq. (7–56) is this: we may write a wave function which momentarily locates a free particle within a given region Δx of one-dimensional space, but *only* if, in order to do it, we are willing to use an infinite number of pilot waves, corresponding to a spread Δp_x in momentum values. The greater the uncertainty in momentum, the less the uncertainty in position, and vice versa. Some illustrative wave functions are shown in Fig. 7–7.

Finally, we may extend Eq. (7–56) to the other coordinate directions, recognizing at the same time that the uncertainty product given by Eq. (7–56) is a *minimum* uncertainty. That is, uncertainty products greater than $\hbar$ can be experimentally realized. Thus

$$\Delta x\,\Delta p_x \gtrsim \hbar, \qquad \Delta y\,\Delta p_y \gtrsim \hbar, \qquad \Delta z\,\Delta p_z \gtrsim \hbar.$$

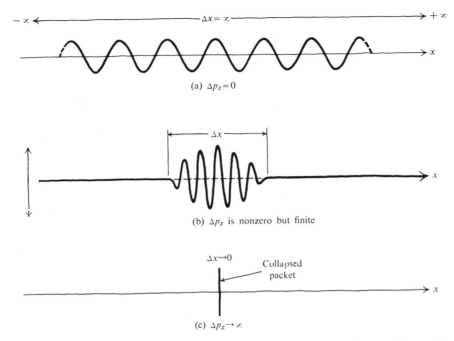

(a) $\Delta p_x = 0$

(b) Δp_x is nonzero but finite

$\Delta x \rightarrow 0$ Collapsed packet

(c) $\Delta p_x \rightarrow \infty$

FIG. 7-7   The real parts of several wave functions which may be used, according to the uncertainty principle, in order to describe a particle in one dimension at some specific time when (a) the momentum is known exactly (a limiting case) so that $\Delta p_x = 0$, (b) the uncertainty in momentum, Δp_x, lies within the nonzero finite range, Δp_x, and (c) the uncertainty in momentum approaches infinity, that is, $\Delta p_x \rightarrow \infty$ (a limiting case).

Note, however, that the position in a given coordinate direction and the momentum in another *different* coordinate direction, for example, x and p_y, can be measured precisely at the same time. That is, it is theoretically possible for

$$\Delta x \, \Delta p_y = 0, \quad \text{or} \quad \Delta y \, \Delta p_z = 0.$$

Theoretical uncertainty in measurement arises only when one attempts to determine simultaneously position and momentum in the *same* coordinate direction.

Although we observe that the Heisenberg uncertainty principle arises from the Schrödinger wave approach, it should be mentioned that the uncertainty principle was the *beginning* of the quantum-mechanical formalism based on matrix mechanics which was developed by Werner Heisenberg* in 1925.

The nature of the physical problem which arises when we attempt to measure simultaneously the exact position and the exact momentum of a particle may be

---

* W. K. Heisenberg, *Z. Physik*, **33**, 879 (1925); M. Born, W. K. Heisenberg, and P. Jordan, *Z. Physik*, **35**, 557 (1926).

made somewhat clearer if we consider a specific example. Let's assume that we desire to determine the exact position of a very small particle. We might use a microscope and a suitable source of illumination. To see a particle, it is necessary that *at least* one photon of light be reflected back from the particle into the microscope. The resolving power of an optical system, that is, the power to closely define the position of a particle, depends on the wavelength of the light used for illumination and increases as the wavelength gets smaller. Thus, we obtain maximum definition of position for the particle when we use illumination of minimum wavelength. But the shorter the wavelength for a photon, the higher is its energy and the more will it disturb the momentum of the particle whose position it is measuring. To measure position *exactly* we must use a photon whose wavelength approaches zero or whose frequency and energy approach infinity. Such a photon, in the process of measuring the position of the particle, will so upset the particle that its momentum will be totally uncertain. In short, *in order to make any physical measurement, we must disturb the system whose properties we are measuring.* In order to measure the position of a particle in a given coordinate direction we must disturb its momentum in that same direction. Such disturbances are usually important only for very small particles.

We conclude this chapter with a note concerning a gratuitous but sometimes useful interpretation of the meaning of ψ in conservative systems. Even though it is philosophically inconsistent with the presently accepted practice of associating $\psi^*\psi$ with the *probability* of locating a *corpuscular* particle, we sometimes find it mathematically or conceptually convenient to think of a particle as being simultaneously distributed or smeared throughout the entire region of its confinement.

We may thus imagine a particle in confinement, such as an electron in an atom, as a cloudlike distribution of the electron which exists to a partial extent everywhere in the region of confinement. The term $\psi^*\psi$ may then be interpreted as the density of the negative charge cloud, or as the fraction of the electron's charge which occurs within a unit space element at a particular point in space. Such an interpretation was first made by Schrödinger in his original radiation theory but has since been generally replaced by Born's probability interpretation. Nevertheless, in constructing models for atoms and molecules, we'll often find the charge-cloud concept to be both useful and comfortable compared with a model in which the electron constantly flits from one point to another.

Something very queer about it! A player put a ball on the table and hit it with a cue. Watching the rolling ball, Mr. Tompkins noticed to his great surprise that the ball began to "spread out." This was the only expression he could find for the strange behavior of the ball, which, moving across the green field, seemed to become more and more washed out, losing its sharp contours.

"... the uncertainty of motion," said the Professor. "The owner of the billiard room has collected here several objects which suffer, if I may so express myself, from quantum elephantism. *Actually all bodies in nature are subject to quantum laws, but the so-called quantum constant is very very small ... For these balls here, however, this constant is much larger—about unity—and you may easily see with your own eyes phenomena which science succeeded in discovering only by using very sensitive and sophisticated methods of observation."*

—GEORGE GAMOW, *Mr. Tompkins in Wonderland.**

PROBLEMS

7–1 Show by substitution that Eqs. (7–6) and (7–7) are both satisfactory solutions of Eq. (7–5).

7–2 Using ψ_1 as given by Eq. (7 6) in the mean-value postulate (Postulate V), calculate the expectation value $\bar{p}_x$ for the momentum of a free particle restricted to the x-direction. Compare your answer to Eq. (7–8) and comment.

7–3 Consider the free single particle in the usual one-dimensional box in which the potential energy is some constant value V_c rather than zero:

a) Write the Schrödinger equation (time independent).
b) Write a wave function ψ in *sine* form which satisfies the above equation.
c) Write the boundary conditions if the box width is L.
d) Prove that quantization of energy must result if the function from (b) is to meet the boundary conditions, and write the expression for quantized energy levels.

7–4 a) Write $\psi(x)$ for the free particle in a *sine* form which satisfies the Schrödinger amplitude equation [Eq. (7–5)].
b) Show, by operation of first $\hat{p}_x$ and then $\widehat{p_x^2}$ on $\psi(x)$, that the sine form of ψ accommodates movement of the free particle in both the positive and negative x-direction.

7–5 Using the fact that the half-wavelength of a single particle in a one-dimensional box must fit between the walls an integral number of times, and the de Broglie relationship $\lambda = h/p$, derive Eq. (7–35).

7–6 Prove that $\psi = A \cos\left[\dfrac{2\pi}{h} (2mE)^{1/2} x \right]$ satisfies the Schrödinger equation for the traditional particle in the one dimensional box of length L. Is ψ as given above an acceptable state function for $0 \leqslant x \leqslant L$? Why or why not?

7–7 What is the minimum energy a baseball (weight = 9 oz) must have in order to exist in a one-dimensional box 60 ft long?

7–8 Small nuclei have diameters in the neighborhood of 10^{-13} cm. What is the lowest energy level which a proton may have in a one-dimensional box whose length is 10^{-13} cm? Comment on your answer.

7–9 Calculate the first four energy levels for a particle of mass 9.109×10^{-28} g (an electron) confined in a one-dimensional box whose length is 10 Å. What is the

---

* From *Mr. Tompkins in Wonderland*, Cambridge University Press, London, 1960. Reprinted through permission of the publisher.

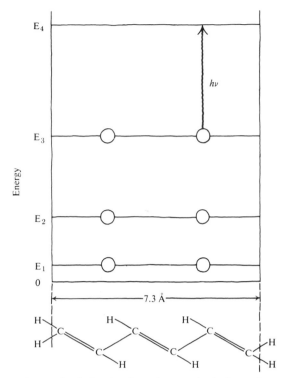

FIG. 7–8   Energy levels for the six π electrons in the hexatriene molecule.

wavelength of the light emitted when the particle drops from the first excited state to the ground state?

7–10 Because of delocalization, each of the six π electrons in the hexatriene molecule behaves approximately as if it were contained in a one-dimensional box whose length extends about one-half a bond length beyond each of the terminal carbons and is given as 7.3 Å (see Fig. 7–8). (a) Calculate the first four energy levels for a π electron in hexatriene. (b) We'll show later that, according to the Pauli exclusion principle, no more than two electrons may simultaneously occupy a single energy level, so that the six electrons completely fill the first three energy levels. What is the wavelength of the photon absorbed when a single electron undergoes a transition from the third to the fourth energy level? Compare with the experimentally observed absorption band at 2580 Å.

7–11 Using Postulate V, calculate the average position for a single particle in a one-dimensional box.

7–12 Use Postulate V to calculate the expectation value for v_x^2 for a single particle of mass m in its first excited state in a one-dimensional box of length a.

7–13 Show that when the kinetic-energy operator $\hat{T}$ is used according to Postulate III with ψ as given by Eq. (7–34) for the single particle in a one-dimensional box, the resultant stationary-state values for the kinetic energy of the particle are given by Eq. (7–35).

7–14 The diameter of an atom is about 10^{-8} cm. In order to locate an electron in an atom with a microscope, we must use radiation whose wavelength is *at least* as small as 10^{-8} cm.

   a) What is the momentum of a photon having this wavelength?

   b) When the photon strikes the electron, it may change the electron's momentum by an amount varying from the total momentum of the photon down to an infinitesimal amount, depending on the nature of the collision. This creates an uncertainty in the final momentum of the electron. Calculate this uncertainty, Δp.

   c) If we assume that the uncertainty in position Δx is given by the diameter of the atom, calculate $\Delta x \, \Delta p$, the product of uncertainty in position and uncertainty in momentum. Comment on your answer.

7–15 An electron is confined in a one-dimensional box of length 1.0 Å. What is the minimum uncertainty in its velocity? Estimate the minimum energy of the electron from the uncertainty in velocity.

7–16 The energy of a single mass moving only in the x direction on a spring fixed at the other end is

$$E = \left(mv_x^2 + \kappa x^2 \right)/2$$

where κ is the force constant.

   a) Is the system conservative? How do you know this?

   b) Write the Schrödinger equation for the system.

Before we extend the one-dimensional treatment of the single particle to three-dimensional space, it is convenient to introduce and discuss in some detail the relationships among three important properties: (1) the orthogonality of wave functions, (2) the Hermitian property of quantum-mechanical operators, and (3) the reality of physical observables generated through quantum mechanics.

8-1 ORTHOGONALITY OF WAVE FUNCTIONS

Recall that for the single particle in the one-dimensional box there is a series of acceptable wave functions generated through Postulate II (Eq. 7–16), each of which is an eigenfunction of the Hamiltonian operator $\hat{H}$ and each of which conforms to appropriate boundary conditions. These are given by Eq. (7–34). For example, each of the following is an acceptable wave function:

$$\psi_1 = \left(\frac{2}{L}\right)^{1/2} \sin \frac{\pi x}{L}, \qquad \psi_2 = \left(\frac{2}{L}\right)^{1/2} \sin \frac{2\pi x}{L}, \qquad \psi_3 = \left(\frac{2}{L}\right)^{1/2} \sin \frac{3\pi x}{L},$$

and so forth.

We'll now show that each of the generated eigenfunctions is, in a special way, *independent* of any other, and that this independence is expressed in a property called *orthogonality*. By definition, two wave functions, ψ_m and ψ_n, where m and n are two different quantum numbers, are *orthogonal* if

$$\int_{\text{all space}} \psi_m^* \psi_n \, d\tau = 0. \tag{8-1}$$

For the single particle in the one-dimensional box, Eq. (8–1) may be written as

$$\int_0^L \psi_m^* \psi_n \, dx = 0, \tag{8-2}$$

where

$$\psi_m^* = \psi_m = \left(\frac{2}{L}\right)^{1/2} \sin \frac{m\pi x}{L} , \qquad (8\text{–}3)$$

$$\psi_n = \left(\frac{2}{L}\right)^{1/2} \sin \frac{n\pi x}{L} . \qquad (8\text{–}4)$$

We may prove easily that any two of the generated wave functions for the particle in the one-dimensional box are orthogonal by substitution of Eqs. (8–3) and (8–4) into Eq. (8–2):

$$\int_0^L \psi_m^* \psi_n \, dx = \frac{2}{L} \int_0^L \sin \frac{m\pi x}{L} \sin \frac{n\pi x}{L} \, dx. \qquad (8\text{–}5)$$

But one of the fundamental trigonometric identities is

$$\sin \alpha \cdot \sin \beta = \tfrac{1}{2} \cos (\alpha - \beta) - \tfrac{1}{2} \cos (\alpha + \beta).$$

Therefore Eq. (8–5) may be written as

$$\int_0^L \psi_m^* \psi_n \, dx = \frac{1}{L} \int_0^L \left\{ \cos\left[(m - n)\pi \frac{x}{L} \right] - \cos\left[(m + n)\pi \frac{x}{L} \right] \right\} dx$$

which, if $m \neq n$, may be integrated to

$$\frac{1}{L} \left\{ \frac{L}{(m - n)\pi} \cdot \sin\left[(m - n)\pi \frac{x}{L} \right] - \frac{L}{(m + n)\pi} \cdot \sin\left[(m + n)\pi \frac{x}{L} \right] \right\}_0^L = 0,$$

because $m - n$ and $m + n$ are integers.

Since any one of the wave functions for the particle in the one-dimensional box is orthogonal to any other of the wave functions, the entire set of solutions is referred to as an *orthogonal set*. Furthermore, since each wave function in the set is not only orthogonal to all others, but is also normalized, the set of solutions given by Eq. (7–34) is called an *orthonormal* set (*orthogonal and normalized*).

Later on, in order to describe certain physical systems, we'll find it convenient to construct new wave functions through linear combination of orthonormal eigenfunctions. It is important to recognize that such linear-combination functions are not *generally* orthogonal. In fact, in order to ensure orthogonality whenever the new linear-combination functions contain orthonormal eigenfunctions *in common*, we must carefully choose proper coefficients for the terms in the linear combination. For example, consider the two wave functions ψ_I and ψ_{II} generated by arbitrary linear combinations of the orthonormal eigenfunctions ψ_m and ψ_n for the particle in the one-dimensional box,

$$\psi_I = a\psi_m + b\psi_n, \qquad \psi_{II} = c\psi_m + d\psi_n,$$

where a, b, c, and d are constants to be determined.

Although either of the two linear-combination functions ψ_I or ψ_{II} may under certain circumstances represent a *physically satisfactory description* of a system, note that neither ψ_I nor ψ_{II} is an eigenfunction of the Hamiltonian operator for the particle in the one-dimensional box unless the eigenvalue energies E_m and E_n are identical (see Problem 6-8 and Section 7-1). In order for ψ_I and ψ_{II} to be orthogonal, it is necessary that

$$\int_0^L \psi_I^* \psi_{II} \, dx = 0.$$

But, since all of the eigenfunctions for the particle in the one-dimensional box are real, $\psi_I^* = \psi_I$,

$$\int_0^L \psi_I \psi_{II} \, dx = 0, \qquad \text{or} \qquad \int_0^L (a\psi_m + b\psi_n)(c\psi_m + d\psi_n) \, dx = 0.$$

Multiplication of terms, followed by rearrangement, yields:

$$ac\int_0^L \psi_m\psi_m \, dx + ad\int_0^L \psi_m\psi_n \, dx + bc\int_0^L \psi_n\psi_m \, dx + bd\int_0^L \psi_n\psi_n \, dx = 0.$$

$$(8\text{-}6)$$

Since ψ_m and ψ_n are orthogonal, the integrals in the second and third terms in Eq. (8-6) are zero, and since ψ_m and ψ_n are each normalized, the integrals in the first and fourth terms are unity, so that

$$ac + bd = 0,$$

which is a requirement if ψ_I and ψ_{II} are to be orthogonal. It is easily shown, for example, that if $a = b$ and $c = -d$ the linear combination functions

$$\psi_I = a(\psi_m + \psi_n), \qquad \psi_{II} = c(\psi_m - \psi_n),$$

where a or c may assume any constant value, are orthogonal in the interval from $x = 0$ to $x = L$ (see Problem 8-2). If, in addition, ψ_I and ψ_{II} are also to be *normalized*, it may be shown (see Problem 8-2) that a and c each must be equal to $1/\sqrt{2}$, so that

$$\psi_I = \frac{1}{\sqrt{2}}(\psi_m + \psi_n) \qquad \text{and} \qquad \psi_{II} = \frac{1}{\sqrt{2}}(\psi_m - \psi_n) \qquad (8\text{-}7)$$

are now *orthonormalized*.

It is easy to show that linear-combination functions which do *not* contain any orthonormal eigenfunction in common are *always* orthogonal (see Problem 8-3), and that linear-combination functions which contain *only one* orthonormal eigenfunction in common can *never* be made orthogonal (see Problem 8-4).

8-2 HERMITIAN OPERATORS:
THE ASSURANCE OF REAL OBSERVABLES

Thus far, we have shown in the specific case of the single particle in the one-dimensional box that:

1. *Each of the eigenvalue energies is a real quantity.*

2. *The eigenfunctions of the Hamiltonian operator, each of which corresponds to a different eigenvalue energy, form an orthogonal set.*

These two very important conditions arise as a mathematical result of a property, called the *Hermitian property, which is possessed by all quantum-mechanical operators.* An operator $\hat{A}$ is said to be *Hermitian* if it obeys the relationship

$$\int \varphi_1{}^* \hat{A} \varphi_2 \, d\tau = \int \varphi_2 \hat{A}^* \varphi_1{}^* \, d\tau \tag{8-8}$$

where φ_1 and φ_2 are any two *well-behaved* functions whatever and where the integration is performed over the entire range of configurational space in which the two functions exist. The definition given by Eq. (8–8) also includes the specific case in which $\varphi_1 = \varphi_2$.

As a sample proof of the Hermitian nature of the quantum-mechanical operators given in Table 6–1 let's show that the linear momentum operator $\hat{p}_x$ is Hermitian over some finite range of one-dimensional space between the limits $x = a$ and $x = b$. That is, let's show that

$$\int_a^b \varphi_1^* \hat{p}_x \varphi_2 \, dx = \int_a^b \varphi_2 \hat{p}_x^* \varphi_1^* \, dx, \tag{8-9}$$

where φ_1 and φ_2 are functions of x.

Substitution of $(\hbar/i)(d/dx)$ for $\hat{p}_x$ and $(-\hbar/i)(d/dx)$ for $\hat{p}_x^*$ into Eq. (8–9) yields

$$\frac{\hbar}{i} \int_a^b \varphi_1^* \frac{d\varphi_2}{dx} \, dx = -\frac{\hbar}{i} \int_a^b \varphi_2 \frac{d\varphi_1^*}{dx} \, dx \tag{8-10}$$

which may be simplified and rearranged to

$$\int_{x=a}^{x=b} \varphi_1^* \, d\varphi_2 + \int_{x=a}^{x=b} \varphi_2 \, d\varphi_1^* = 0. \tag{8-11}$$

But recall, for the general variables u and v, that

$$\int u \, dv + \int v \, du = \int d(uv)$$

so that Eq. (8–11) may be written as

$$\int_{x=a}^{x=b} d(\varphi_1^* \varphi_2) = \varphi_1^* \varphi_2 \Big]_{x=a}^{x=b} = 0. \tag{8-12}$$

In evaluating the integral in Eq. (8–12) we must use the values of $\varphi_1^*\varphi_2$ only at the end points of the range. But we have insisted that all well-behaved functions have integrable squares, that is, that $\int \varphi_1^*\varphi_1\,dx$ and $\int \varphi_2^*\varphi_2\,dx$ each be finite, which means that φ_1 and φ_2 must each vanish at the end points of the range (for example, at 0 and L for the particle in the one-dimensional box). Thus

$$\varphi_1^*\varphi_2 \Big]_{x=a}^{x=b} = 0$$

or

$$0 = 0,$$

and $\hat{p}_x$ is thus proved to be Hermitian.

Now let's show that *Hermitian operators always yield observables which are real numbers.* Even though i may appear as part of an operator or in the wave function itself, it is important that i should not appear in the allowed value for an observable if quantum mechanics is to have physical significance. We have shown in Chapter 6 that observables appear in quantum mechanics either as *eigenvalues A* according to the operator postulate (Postulate III),

$$\hat{A}\psi = A\psi, \tag{6–54'}$$

or as expectation or *average values* $\bar{A}$ according to the mean-value postulate (Postulate V),

$$\bar{A} = \int \psi^*\hat{A}\psi\,d\tau / \int \psi^*\psi\,d\tau, \tag{6–81'}$$

where the integrals are taken over all configurational space. For convenience, we have chosen to write Eqs. (6–54') and (6–81') for *conservative* systems. That is, we have used ψ rather than Ψ for the wave function.

First, let's show that Hermitian operators always yield real eigenvalues for exact observables, as given by Eq. (6–54'). The eigenvalue A_m generated as a result of the operation of $\hat{A}$ on an acceptable eigenfunction ψ_m is defined by

$$\hat{A}\psi_m = A_m\psi_m. \tag{8–13}$$

We may write the complex conjugate of Eq. (8–13) as

$$\hat{A}^*\psi_m^* = A_m^*\psi_m^*. \tag{8–14}$$

Multiplication of Eq. (8–13) by ψ_m^* followed by integration over all available space yields

$$\int \psi_m^*\hat{A}\psi_m\,d\tau = A_m \int \psi_m^*\psi_m\,d\tau. \tag{8–15}$$

In similar fashion, multiplication of Eq. (8–14) by ψ_m followed by integration over all available space yields

$$\int \psi_m\hat{A}^*\psi_m^*\,d\tau = A_m^* \int \psi_m^*\psi_m\,d\tau. \tag{8–16}$$

If $\hat{A}$ is a Hermitian operator, as defined by Eq. (8–8), the left-hand sides of Eqs. (8–15) and (8–16) are equal and combination of the two equations yields

$$A_m = A_m^*,$$

which is possible only when A_m is a real number. Thus, the physical significance of eigenvalues is assured by requiring that quantum-mechanical operators be Hermitian.

Now, let's also show that Hermitian operators always yield real expectation values for observables, as given by Eq. (6–81'). Consider the evaluation of a certain observable A under conditions such that exact eigenvalues are not observed. The average value or expectation value $\bar{A}_n$ for the observable in the state characterized by the wave function ψ_n is given by

$$\bar{A}_n = \int \psi_n^* \hat{A} \psi_n \, d\tau \Big/ \int \psi_n^* \psi_n \, d\tau, \tag{8–17}$$

where the integrations are performed over all available space. The complex conjugate of Eq. (8–17) is

$$\bar{A}_n^* = \int \psi_n \hat{A}^* \psi_n^* \, d\tau \Big/ \int \psi_n \psi_n^* \, d\tau. \tag{8–18}$$

But, if $\hat{A}$ is Hermitian,

$$\int \psi_n^* \hat{A} \psi_n \, d\tau = \int \psi_n \hat{A}^* \psi_n^* \, d\tau,$$

and the combination of Eqs. (8–17) and (8–18) then leads to

$$\bar{A}_n = \bar{A}_n^*,$$

which can be true only if $\bar{A}_n$ is a real quantity. Thus, the physical significance of the average value or expectation value of an observable is assured by requiring that all quantum-mechanical operators be Hermitian.

We have shown in the last section that the eigenfunctions for the single particle in the one-dimensional box form an orthogonal set. Now let's show under what specific conditions the Hermitian nature of the Hamiltonian operator $\hat{H}$ guarantees the orthogonality of wave functions. For convenience, we'll again restrict the argument to conservative systems. Recall that in any conservative system, the acceptable state functions for a particle are given, according to Postulate II, by the eigenfunctions of the time-independent Schrödinger equation:

$$\hat{H}\psi = E\psi.$$

Thus, for the two eigenfunctions ψ_m and ψ_n corresponding to the eigenvalue energies E_m and E_n we may write

$$\hat{H}\psi_m = E_m\psi_m, \tag{8–19}$$

and

$$\hat{H}\psi_n = E_n\psi_n. \tag{8–20}$$

The complex conjugate of Eq. (8–19) is

$$\hat{H}^*\psi_m^* = E_m^*\psi_m^*. \qquad (8\text{–}21)$$

But since $\hat{H}$ is Hermitian, E_m^* is equal to E_m because E_m is real, and

$$\hat{H}^*\psi_m^* = E_m\psi_m^*. \qquad (8\text{–}22)$$

Multiplication of Eq. (8–20) by ψ_m^* followed by integration over all coordinate space yields

$$\int \psi_m^* \hat{H}\psi_n \, d\tau = E_n \int \psi_m^*\psi_n \, d\tau, \qquad (8\text{–}23)$$

and multiplication of Eq. (8–22) by ψ_n followed by integration over all coordinate space yields

$$\int \psi_n \hat{H}^*\psi_m^* \, d\tau = E_m \int \psi_m^*\psi_n \, d\tau. \qquad (8\text{–}24)$$

Since $\hat{H}$ is a Hermitian operator, as defined by Eq. (8–8), the left-hand sides of Eqs. (8–23) and (8–24) are equal, so that the combination of the two equations leads to

$$(E_m - E_n)\int \psi_m^*\psi_n \, d\tau = 0. \qquad (8\text{–}25)$$

It then follows that for two *different* eigenfunctions, for which E_m *is not equal to* E_n,

$$\int \psi_m^*\psi_n \, d\tau = 0 \qquad (m \neq n),$$

which is the condition for orthogonality. Thus, *eigenfunctions of the Hamiltonian operator (which are the state functions of the particle) are guaranteed to be orthogonal provided they are associated with different energy eigenvalues, i.e., with nondegenerate states.* The converse, however, is not true. That is, the fact that two different eigenfunctions of $\hat{H}$ are orthogonal does not mean that they must *necessarily* have different eigenvalue energies. We'll now show that even when the eigenvalue energies of two wave functions are *the same*, that is, even when the states are degenerate, it is still possible to construct a suitable set of orthogonal functions by simple superposition. For example, consider the two *normalized* eigenfunctions ψ_m and ψ_n which have the common eigenvalue energy E:

$$\hat{H}\psi_m = E\psi_m,$$

and

$$\hat{H}\psi_n = E\psi_n.$$

If the two degenerate states are not orthogonal,

$$\int \psi_m^*\psi_n \, d\tau = b \neq 0, \qquad (8\text{–}26)$$

where b is a finite constant.   Let's, through appropriate superposition, construct the new function ψ'_n, where

$$\psi'_n = \psi_n - b\psi_m.$$

We may now show that ψ_m and ψ'_n are orthogonal; that is,

$$\int \psi_m^* \psi'_n \, d\tau = \int \psi_m^* \psi_n \, d\tau - b \int \psi_m^* \psi_m \, d\tau = b - b = 0.$$

In addition,

$$\hat{H}\psi'_n = \hat{H}(\psi_n - b\psi_m) = \hat{H}\psi_n - b\hat{H}\psi_m = E\psi_n - bE\psi_m = E\psi'_n$$

so that ψ'_n is also shown to be an eigenfunction of $\hat{H}$. We'll use a similar super-position in our later treatment of degenerate hybrid orbitals for the hydrogen atom.

8–3 A SINGLE PARTICLE IN A THREE-DIMENSIONAL BOX

Let's now consider a single particle confined in a three-dimensional box of dimensions a, b, and c in the x-, y-, and z-directions, respectively. In order to ensure that the particle remains in the box, we'll require that the potential energy V be zero everywhere within the box and infinite everywhere outside the box. Then within the box, where $V = 0$, the Schrödinger amplitude equation may be written as

$$\frac{\partial^2\psi(x, y, z)}{\partial x^2} + \frac{\partial^2\psi(x, y, z)}{\partial y^2} + \frac{\partial^2\psi(x, y, z)}{\partial z^2} = \frac{-2mE}{\hbar^2}\psi(x, y, z). \quad (8\text{–}27)$$

Equation (8–27) cannot be solved as it is, because of its dependence on three variables. We have, however, been successful before in separating the variables in similar linear differential equations by assuming that the state function may be expressed as a *product* of several functions, each function dependent on one of the variables. Let's assume then that

$$\psi(x, y, z) = \varphi(x)\varphi(y)\varphi(z), \quad (8\text{–}28)$$

where $\varphi(x)$ depends only on x, $\varphi(y)$ depends only on y, and $\varphi(z)$ depends only on z. In order to determine the satisfactory forms for $\varphi(x)$, $\varphi(y)$, and $\varphi(z)$, we'll now substitute Eq. (8–28) into Eq. (8–27):

$$\varphi(y)\varphi(z)\frac{d^2\varphi(x)}{dx^2} + \varphi(x)\varphi(z)\frac{d^2\varphi(y)}{dy^2}$$
$$+ \varphi(x)\varphi(y)\frac{d^2\varphi(z)}{dz^2} = \frac{-2mE}{\hbar^2}\varphi(x)\varphi(y)\varphi(z). \quad (8\text{–}29)$$

Note that the derivatives in Eq. (8–29) are now expressed as *total* derivatives rather than partial derivatives, since each of the φ-functions is dependent on only one of the coordinate variables. Division of Eq. (8–29) by $\varphi(x)\varphi(y)\varphi(z)$

with slight rearrangement yields

$$-\frac{\hbar^2}{2m}\frac{1}{\varphi(x)}\frac{d^2\varphi(x)}{dx^2} - \frac{\hbar^2}{2m}\frac{1}{\varphi(y)}\frac{d^2\varphi(y)}{dy^2} - \frac{\hbar^2}{2m}\frac{1}{\varphi(z)}\frac{d^2\varphi(z)}{dz^2} = E.$$

Since E is constant in a conservative force field ($E = T = $ constant in this specific case), the sum of the three terms on the left-hand side of Eq. (8–30) is constant, *regardless* of particular values of x, y, or z. Furthermore, a change in the value of x cannot change the value of either the second or the third term on the left-hand side of Eq. (8–30), since each is independent of x. Thus the first term itself must be constant. By identical argument the second term and the third term on the left side are also constants. Defining the constants respectively as E_x, E_y, and E_z, where $E_x + E_y + E_z = E$, we write the three independent equations

$$-\frac{\hbar^2}{2m}\frac{1}{\varphi(x)}\frac{d^2\varphi(x)}{dx^2} = E_x, \tag{8–30}$$

$$-\frac{\hbar^2}{2m}\frac{1}{\varphi(y)}\frac{d^2\varphi(y)}{dy^2} = E_y, \quad \text{and} \tag{8–31}$$

$$-\frac{\hbar^2}{2m}\frac{1}{\varphi(z)}\frac{d^2\varphi(z)}{dz^2} = E_z, \tag{8–32}$$

which can be arranged to the more familiar forms

$$\frac{d^2\varphi(x)}{dx^2} = \frac{-2mE_x}{\hbar^2}\varphi(x), \tag{8–33}$$

$$\frac{d^2\varphi(y)}{dy^2} = \frac{-2mE_y}{\hbar^2}\varphi(y), \tag{8–34}$$

$$\frac{d^2\varphi(z)}{dz^2} = \frac{-2mE_z}{\hbar^2}\varphi(z). \tag{8–35}$$

Equations (8–33), (8–34), and (8–35) are one-dimensional eigenvalue equations identical to that for the particle in the one-dimensional box (Eq. 7–16). The individual normalized eigenfunctions are thus identical in form to Eq. (7–34) and may be written as

$$\varphi(x) = \left(\frac{2}{a}\right)^{1/2}\sin\frac{n_x\pi x}{a}, \tag{8–36}$$

$$\varphi(y) = \left(\frac{2}{b}\right)^{1/2}\sin\frac{n_y\pi y}{b}, \tag{8–37}$$

$$\varphi(z) = \left(\frac{2}{c}\right)^{1/2}\sin\frac{n_z\pi z}{c}, \tag{8–38}$$

where n_x, n_y, and n_z are the quantum numbers in each of the coordinate directions such that

$$n_x = 1, 2, 3, \ldots, \infty,$$

$$n_y = 1, 2, 3, \ldots, \infty,$$

$$n_z = 1, 2, 3, \ldots, \infty.$$

Substitution of Eqs. (8–36), (8–37), and (8–38) into Eq. (8–28) yields as the amplitude function for the particle in the three-dimensional box

$$\psi(x, y, z) = \left(\frac{2}{a}\right)^{1/2} \left(\frac{2}{b}\right)^{1/2} \left(\frac{2}{c}\right)^{1/2} \sin \frac{n_x \pi x}{a} \sin \frac{n_y \pi y}{b} \sin \frac{n_z \pi z}{c}, \qquad (8\text{–}39)$$

where n_x, n_y, and n_z may each have independently any integral value other than zero. If any value of n were zero, $\psi(x, y, z)$ would always be zero and the particle could not exist, since $\psi^*\psi$ would *always* be zero. Since $\psi(x)$, $\varphi(y)$, and $\varphi(z)$ are each individually normalized within their respective one-dimensional limits, it follows (see Problem 8–7) that $\psi(x, y, z)$ as given by Eq. (8–39) must also be normalized within the dimensions of the box.

According to Eq. (7–35) it also follows that the allowed energy levels for the single particle in the three-dimensional box are given by

$$E = E_x + E_y + E_z = \frac{h^2 n_x^2}{8ma^2} + \frac{h^2 n_y^2}{8mb^2} + \frac{h^2 n_z^2}{8mc^2}, \qquad (8\text{–}40)$$

or

$$E = \frac{h^2}{8m}\left(\frac{n_x^2}{a^2} + \frac{n_y^2}{b^2} + \frac{n_z^2}{c^2}\right). \qquad (8\text{–}41)$$

E_x, E_y, and E_z are apparently the rectangular components of the total energy E.

8–4 ALLOWED ENERGY LEVELS IN A CUBIC BOX

If the box is cubic and each edge is equal to L, then Eq. (8–41) becomes

$$E = \frac{h^2}{8mL^2}(n_x^2 + n_y^2 + n_z^2). \qquad (8\text{–}42)$$

The lowest allowed energy state, or *ground state*, is that state for which each of the three quantum numbers is equal to one, that is, $n_x = 1$, $n_y = 1$, and $n_z = 1$. We will refer to this state as the (111) state. Then, according to Eq. (8–42),

$$E_{(111)} = 3h^2/8mL^2.$$

The next highest energy state is that state in which any one of the quantum numbers is two but the other two quantum numbers are each one. There are

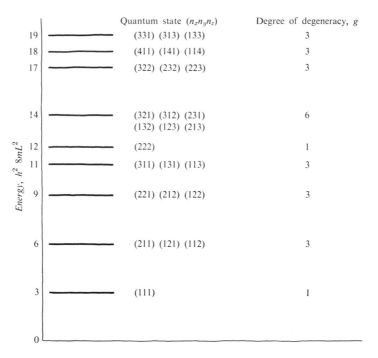

FIG. 8-1   Allowed energy levels for the particle in the cubic box.

three such states:

$$(211) \quad \text{state:} \quad n_x = 2, n_y = 1, n_z = 1,$$

$$(121) \quad \text{state:} \quad n_x = 1, n_y = 2, n_z = 1,$$

$$(112) \quad \text{state:} \quad n_x = 1, n_y = 1, n_z = 2.$$

For each of the above states, the energy calculated from Eq. (8–42) is identical. That is,

$$E_{(211)} = E_{(121)} = E_{(112)} = 6h^2/8mL^2.$$

The existence of several distinct quantum states having the same energy level has been referred to in Section 8–2 as *degeneracy*. Thus, the energy level $6h^2/8mL^2$ is said to be *triply degenerate* or *threefold degenerate* or it is said to have a *multiplicity* of three, or a *degree of degeneracy* of three, or a *quantum weight* of three, or a *statistical weight* of three. In addition, such a state is often called a *triplet* state. Apparently there are enough equivalent terms available to suit any taste. We prefer to use the term *degree of degeneracy* and give this term the symbol g. Thus, for the energy level $6h^2/8mL^2$, three degenerate quantum states, (211), (121), and (112), exist, and $g = 3$. Higher energy levels, corresponding quantum states, and corresponding degrees of degeneracy are given in Fig. 8–1 for the single particle in a cubic box.

It is interesting to note that the degeneracy we observe for the higher quantum states appears by virtue of the symmetry of the cube. If we had assumed $a \neq b \neq c$ in calculating energy levels from Eq. (8–41), such degeneracy would not occur. In fact, if we were to *distort* the cube, for example, by pushing one wall in slightly, levels which were previously degenerate would split into separate energy levels. The splitting of degenerate electronic energy levels into new separate levels may be observed, for example, if we distort atomic symmetry by applying a strong directional electrical or magnetic field to a symmetrical atom. The corresponding loss of degeneracy results in the creation of new lines in the emission spectrum of the atom. The splitting of lines in atomic spectra in the presence of a directional magnetic field is called the *Zeeman effect*, whereas the splitting of lines in the presence of a directional electrical field is known as the *Stark effect*.

8–5 PROBABILITY DENSITIES FOR A PARTICLE IN A CUBIC BOX

For a cubic box of edge L, Eq. (8–39) becomes

$$\psi(x, y, z) = \left(\frac{2}{L}\right)^{3/2} \sin \frac{n_x \pi x}{L} \sin \frac{n_y \pi y}{L} \sin \frac{n_z \pi z}{L}, \qquad (8\text{–}43)$$

and since

$$\psi(x, y, z) = \varphi(x)\varphi(y)\varphi(z), \qquad (8\text{–}28')$$

we may write the contributing one-dimensional functions as

$$\varphi(x) = \left(\frac{2}{L}\right)^{1/2} \sin \frac{n_x \pi x}{L}, \qquad (8\text{–}44)$$

$$\varphi(y) = \left(\frac{2}{L}\right)^{1/2} \sin \frac{n_y \pi y}{L}, \qquad (8\text{–}45)$$

$$\varphi(z) = \left(\frac{2}{L}\right)^{1/2} \sin \frac{n_z \pi z}{L}. \qquad (8\text{–}46)$$

Amplitude functions in each of the coordinate directions are plotted for the (111) state in Fig. 8–2. The corresponding probability densities in each of the coordinate directions are

$$\varphi(x)^*\varphi(x) = \varphi^2(x) = \frac{2}{L} \sin^2 \frac{n_x \pi x}{L}, \qquad (8\text{–}47)$$

$$\varphi(y)^*\varphi(y) = \varphi^2(y) = \frac{2}{L} \sin^2 \frac{n_y \pi y}{L}, \qquad (8\text{–}48)$$

$$\varphi(z)^*\varphi(z) = \varphi^2(z) = \frac{2}{L} \sin^2 \frac{n_z \pi z}{L}. \qquad (8\text{–}49)$$

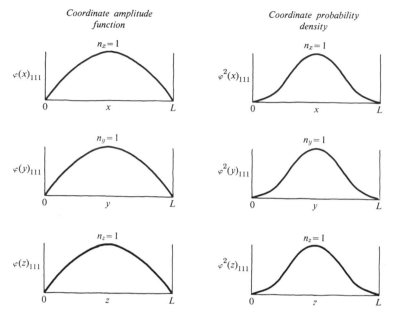

FIG. 8–2   Amplitude functions and probability densities for the x, y, and z coordinate directions for the (111) quantum state of the particle in a cubic box.

Directional probability densities for the (111) quantum state are also plotted in Fig. 8–2. The φ^2-values represent the probability *per unit length* that the particle exists at a given coordinate position.

The probability density for the particle in three dimensions is given as

$$\psi^*\psi = \varphi(x)^*\varphi(x)\varphi(y)^*\varphi(y)\varphi(z)^*\varphi(z),$$

or

$$\psi^2 = \varphi^2(x) \cdot \varphi^2(y) \cdot \varphi^2(z), \qquad (8\text{–}50)$$

which is the probability per unit *volume* that the particle exists at the point x, y, z. Graphical multiplication of the φ- and φ^2-probability functions given for the (111) state in Fig. 8–2 leads to the three-dimensional representations for ψ and ψ^2 shown in Fig. 8–3. In the three-dimensional plots the density of the amplitude cloud or of the probability cloud represents a direct measure of ψ or ψ^2, respectively. Also note from Fig. 8–3 that the particle has the highest probability density at the center of the cube and has zero probability density at the walls. Thus each of the walls is a *nodal plane*.

Figure 8–4 is a contour plot of a cross section of the three-dimensional ψ^2-plot of Fig. 8–3 through the plane $x = L/2$. Since

$$\psi^2 = \frac{8}{L^3} \sin^2\left(\frac{n_x\pi x}{L}\right) \sin^2\left(\frac{n_y\pi y}{L}\right) \sin^2\left(\frac{n_z\pi z}{L}\right),$$

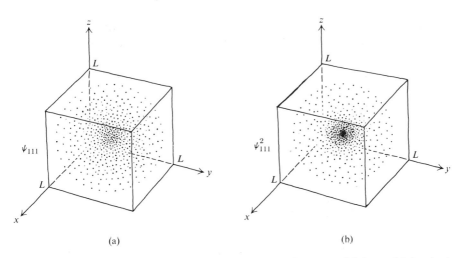

FIG. 8–3   Amplitude function ψ and probability density ψ^2 for the particle in a cubic box in the (111) state. The density of the cloud is proportional to the value of ψ or ψ^2 at a particular coordinate position.

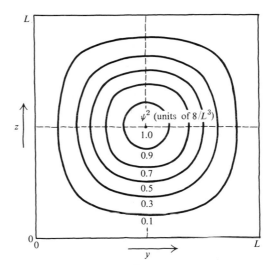

FIG. 8–4   Contour plot of the probability density ψ^2 in the cross-sectional plane of Fig. 8–3(b) for which $x = L/2$. The values given for ψ^2 are in units of $8/L^3$.

the probability density ψ^2 has its maximum value $8/L^3$ when each of the sin^2 terms is equal to 1. In the (111) state this maximum occurs at $x = L/2$, $y = L/2$, and $z = L/2$. Thus for convenience the contour plot in Fig. 8–4 presents ψ^2-values in units of $8/L^3$. The contour lines represent different constant values for ψ^2. In the case of the symmetrical (111) state, contours in the planes for which $y = L/2$ and $z = L/2$ would be identical to those in the $(x = L/2)$-plane.

In pictorializing atomic and molecular orbitals, it is often convenient to represent three-dimensional ψ- and ψ^2-plots by drawing the boundary surface which encloses all points having ψ- or ψ^2-values above some relatively small value such as 0.10 or 0.02 of the maximum. Thus in order to present the general shape of the ψ^2-plot for the (111) state, we may draw the three-dimensional boundary surface which encloses all points for which ψ^2 (in units of $8/L^3$) has a value larger than 0.10. Such boundary surface plots for ψ and for ψ^2 are shown for the (111) state in Fig. 8–5.

Figure 8–6 shows φ-functions and φ^2-functions for the (211) quantum state. In all respects, the one-dimensional φ- and φ^2-plots are identical to those for the second quantum state ($n = 2$) for the particle in the *one-dimensional* box. The corresponding three-dimensional amplitude (ψ) plots and probability density (ψ^2) plots are once more obtained by multiplication of the φ- and φ^2-functions according to Eqs. (8–28) and (8–50) and are shown graphically as boundary surface plots in Fig. 8–7. Note that the sign of the cloud is important for ψ-plots but that all clouds are positive for ψ^2 probability plots. Corresponding plots of ψ and ψ^2 for the degenerate (121) and (112) states are also presented for comparison in Fig. 8–7. For each of the degenerate states (211), (121), and (112), the ψ^2 probability plot is represented as two boundary surface lobes on one of the coordinate axes, and in each case, the probability of finding the particle in the center of the box is zero, whereas the probability of finding the particle in the centers of the lobes is maximum.

Higher quantum states become progressively more complex to show graphically as boundary surface plots. As a few final examples, boundary surfaces for ψ^2 in the (221) state and in the (222) state are shown in Fig. 8–8.

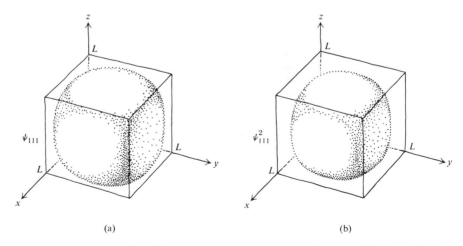

(a)   (b)

FIG. 8–5   Boundary surfaces for ψ and ψ^2 for the (111) state of the particle in a cubic box. Nodal planes occur at each of the walls.

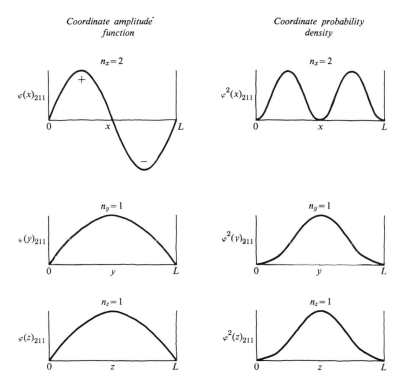

FIG. 8–6 Amplitude functions and probability densities in the x, y, and z coordinate directions for the (211) quantum state of the particle in a cubic box.

It is important that we develop a sound understanding of the ψ- and ψ^2-plots in a relatively simple system such as the particle in the three-dimensional box, since we'll encounter many counterparts of these functions and distributions in atomic and molecular systems.

8–6 FINITE BARRIERS AND THE TUNNEL EFFECT

We have shown in Sections 7–1 and 7–2 that the general initial forms of the wave function for the free particle in one dimension and of the wave function for the particle in a one-dimensional box are identical and are given by slight rearrangement of Eq. (7–20) as

$$\psi = A \exp\left[i\left(2mE/\hbar^2\right)^{1/2}x\right] + B \exp\left[-i\left(2mE/\hbar^2\right)^{1/2}x\right], \quad (8\text{–}51)$$

in which E is constant for conservative systems. If we define

$$\alpha \equiv \left(2mE/\hbar^2\right)^{1/2} \quad (8\text{–}52)$$

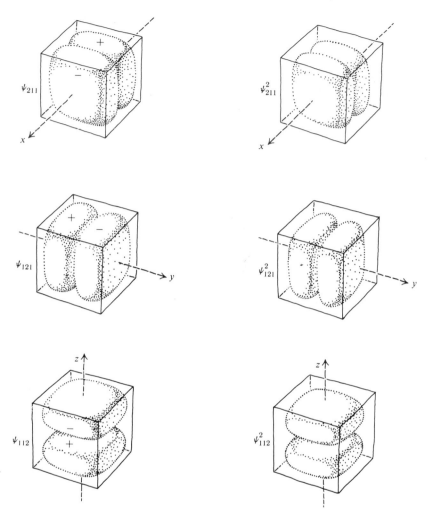

FIG. 8-7   Amplitude and probability density boundary surfaces for the particle in the cubic box in the degenerate quantum states (211), (121), and (112).

we may write Eq. (8–51) in the simpler form:

$$\psi = Ae^{i\alpha x} + Be^{-i\alpha x}. \tag{8-53}$$

In the above equation recall that the first term is the wave function for the particle traveling in the $(+x)$-direction, whereas the second of these terms is the wave function for the particle traveling in the $(-x)$-direction. We have also shown that the probability per unit length, or probability density, for

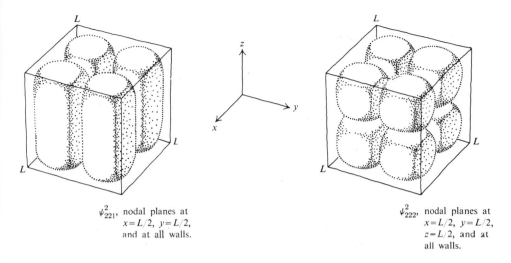

ψ^2_{221}, nodal planes at
$x=L/2$, $y=L/2$,
and at all walls.

ψ^2_{222}, nodal planes at
$x=L/2$, $y=L/2$,
$z=L/2$, and at
all walls.

FIG. 8–8   Boundary surfaces for the probability density ψ^2 for the particle in a cubic box in the (221) and (222) quantum states. Maximum probability density occurs at the center of each of the lobes.

finding the particle traveling in the $(+x)$-direction at a given coordinate position x is given as $|A|^2$ (Eq. 7–8). Similarly, $|B|^2$ is the probability density for the particle traveling in the $(-x)$-direction. For a free particle which is restricted to motion in either the $(+x)$-direction or the $(-x)$-direction *only*, either $|A|^2$ *or* $|B|^2$ must be equal to zero.

We have shown, however (Eq. 7–21), that if we confine a particle within a one-dimensional box (within an infinitely deep potential energy well), it is necessary that $|A|^2$ be equal to $|B|^2$ (Fig. 8–9); otherwise the particle would

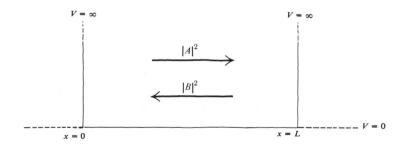

FIG. 8–9   For the single particle confined within a one-dimensional box, it is necessary that the probability density $|A|^2$ for the particle moving in the $(+x)$-direction be equal to the probability density $|B|^2$ for the particle moving in the $(-x)$-direction.

have a finite probability of escaping from the box. The situation is understood more clearly if we imagine the one-dimensional box to contain a *large number* of particles, the behavior of each particle being governed by the wave function given in Eq. (8–53). The *probability* that a *single particle* is moving in the (+ x)-direction is the same as the *fraction* of particles in the *many-particle system* which is moving in the (+ x)-direction. If, for example, $|A|^2$ were larger than $|B|^2$, more particles would be moving from left to right than from right to left, which, by simple particle balance, implies that either additional particles are being introduced through the left wall of the box or that particles are "leaking" out through the right wall of the box! Of course, neither of these solutions is acceptable for a particle *confined* between two *infinite* potential-energy barriers.

Let's now consider a somewhat different and quite interesting variation of the particle in the box problem. Specifically, let's begin by placing a single particle within a one-dimensional box in which the potential-energy barrier is infinite at the left end but *finite* and equal to V_0 at the right. We'll again arbitrarily define the potential energy as zero everywhere within the box. Furthermore, let the finite potential-energy barrier have a *finite thickness a* such that the potential energy is again zero beyond, that is to the right of, the barrier. The system is shown in Fig. 8–10 (a). For convenience, we have set $x = 0$ at the left edge of the barrier.

Consider a conservative system, in which the total energy of the particle is constant and equal to E. It seems obvious from classical mechanics that if E is less than V_0, the particle should not have sufficient energy to escape from the box (region I) and thus should have zero probability of existing in region III to the right of the finite barrier. The results of quantum mechanics, however, are quite different. In fact, we'll now show that *even if E is less than V_0, the single particle still has a finite probability for existence beyond and to the right of the finite barrier.*

Let's first write the time-independent Schrödinger equation for each of the three regions. The Schrödinger equation (Eq. 7–4) may be rearranged to

$$\frac{d^2\psi}{dx^2} = 2\,\frac{m}{\hbar^2}\,(V - E)\psi. \tag{8–54}$$

For region I, where $V = 0$, Eq. (8–54) may be written as

$$\frac{d^2\psi_I}{dx^2} = -\alpha^2\psi_I \quad \text{(region I)},$$

where α is defined according to Eq. (8–52). For region II, where $V = V_0$, the Schrödinger equation may be written as

$$\frac{d^2\psi_{II}}{dx^2} = \beta^2\psi_{II} \quad \text{(region II)},$$

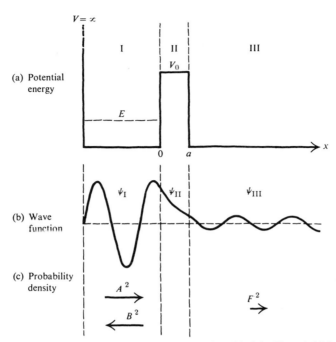

FIG. 8–10 Penetration of a finite potential-energy barrier of height V_0 and thickness a by a single particle whose energy is E. (a) Potential energy as a function of x. (b) A qualitative sketch of the entire wave function in terms of the component functions in each region. (c) Probability densities on each side of the barrier.

where

$$\beta = \left[\frac{2m(V_0 - E)}{\hbar^2} \right]^{1/2}. \tag{8-55}$$

Since $V = 0$ in region III, the Schrödinger equation for region III is identical to that for region 1; that is,

$$\frac{d^2\psi_{III}}{dx^2} = -\alpha^2\psi_{III} \quad \text{(region III)}.$$

Acceptable solutions of the Schrödinger equations in each of the regions may be verified easily by substitution to be

$$\psi_I = Ae^{i\alpha x} + Be^{-i\alpha x}, \tag{8-56}$$

$$\psi_{II} = Ce^{\beta x} + De^{-\beta x}, \tag{8-57}$$

$$\psi_{III} = Fe^{i\alpha x}. \tag{8-58}$$

Note that ψ_I and ψ_{II} each contain terms to allow for particle movement in *either* direction in regions I and II. However, we have not included a negative exponential term, such as $Ge^{-i\alpha x}$, in ψ_{III} because in the system we have described we do not allow for the possibility of particles moving in the $(-x)$-direction in region III. There is no source for such motion.

In a qualitative sense, the entire wave function might look something like that shown in Fig. 8–10 (b). Of course, the exact shape and size of the function depend on the values of E, V_0, a, and m. Furthermore, the function is complex, so that only the real part is shown.

The problem now is to evaluate the constants A, B, C, D, and F in ψ_I, ψ_{II}, and ψ_{III} through the application of appropriate boundary conditions. In order for the entire function to be well-behaved, that is, continuous and smooth, it is necessary that ψ and also $d\psi/dx$ be continuous at the positions $x = 0$ and $x = a$. The *continuity conditions* are then

$$\psi_I = \psi_{II} \quad \text{(at } x = 0\text{)}, \tag{8–59}$$

$$\frac{d\psi_I}{dx} = \frac{d\psi_{II}}{dx} \quad \text{(at } x = 0\text{)}, \tag{8–60}$$

$$\psi_{II} = \psi_{III} \quad \text{(at } x = a\text{)}, \tag{8–61}$$

$$\frac{d\psi_{II}}{dx} = \frac{d\psi_{III}}{dx} \quad \text{(at } x = a\text{)}. \tag{8–62}$$

Substitution of ψ_I, ψ_{II}, and ψ_{III} from Eqs. (8–56), (8–57), and (8–58) into the four continuity equations yields, respectively,

$$A + B = C + D, \tag{8–63}$$

$$i\alpha(A - B) = \beta(C - D), \tag{8–64}$$

$$Ce^{\beta a} + De^{-\beta a} = Fe^{i\alpha a}, \tag{8–65}$$

$$\beta Ce^{\beta a} - \beta De^{-\beta a} = i\alpha Fe^{i\alpha a}. \tag{8–66}$$

The above set of four equations contains five unknowns, so that we cannot directly solve for each of the unknowns. We may, however, solve for any one of the unknowns in terms of any other. For our purpose, it is most instructive to solve for A in terms of F, so that we may then derive an expression for the probability that the particle penetrates, or is transmitted through, the barrier. We define the *transmission coefficient* Γ as:

$$\Gamma = \frac{|F|^2}{|A|^2}. \tag{8–67}$$

Since $|A|^2$ is the probability density for the particle moving from left to right in region I so that it impinges on the left side of the barrier, and $|F|^2$ is the probability density for the particle moving to the right in region III *after emergence* from the barrier, the ratio $|F|^2/|A|^2$ is the probability that a single particle penetrates or is transmitted *through* the barrier. For a system of many particles, the transmission coefficient Γ is the *fraction* of particles which penetrates the barrier and emerges in region III.

To show that penetration of the barrier is possible we must now show that $\Gamma > 0$. Solution of Eqs. (8–63) and (8–64) for A and B in terms of C and D yields

$$A = \frac{C(i\alpha + \beta)}{2i\alpha} + \frac{D(i\alpha - \beta)}{2i\alpha}, \qquad (8\text{–}68)$$

$$B = \frac{C(i\alpha - \beta)}{2i\alpha} + \frac{D(i\alpha + \beta)}{2i\alpha}, \qquad (8\text{–}69)$$

and solution of Eqs. (8–65) and (8–66) for C and D in terms of F gives

$$C = \frac{F(\beta + i\alpha)}{2\beta} e^{(i\alpha - \beta)a}, \qquad (8\text{–}70)$$

$$D = \frac{F(\beta - i\alpha)}{2\beta} e^{(i\alpha + \beta)a}. \qquad (8\text{–}71)$$

Now let's substitute Eqs. (8–70) and (8–71) into Eq. (8–68):

$$A = F\left[\frac{(\beta^2 + 2i\alpha\beta - \alpha^2)}{4i\alpha\beta} e^{(i\alpha - \beta)a} + \frac{(\alpha^2 + 2i\alpha\beta - \beta^2)}{4i\alpha\beta} e^{(i\alpha + \beta)a}\right]$$

$$= \frac{Fe^{i\alpha a}}{4i\alpha\beta}[(\beta^2 + 2i\alpha\beta - \alpha^2)e^{-\beta a} + (\alpha^2 + 2i\alpha\beta - \beta^2)e^{\beta a}]$$

$$= \frac{Fe^{i\alpha a}}{4i\alpha\beta}[2i\alpha\beta(e^{\beta a} + e^{-\beta a}) + (\alpha^2 - \beta^2)(e^{\beta a} - e^{-\beta a})]. \qquad (8\text{–}72)$$

Multiplication of each side of Eq. (8–72) by its complex conjugate yields

$$|A|^2 = |F|^2\left[\frac{(e^{\beta a} + e^{-\beta a})^2}{4} + \frac{(\alpha^2 - \beta^2)^2}{16\alpha^2\beta^2}(e^{\beta a} - e^{-\beta a})^2\right]. \qquad (8\text{–}73)$$

Examination of Eq. (8–73) reveals that Γ, which is given by $|F|^2/|A|^2$, *goes to zero only when βa is infinite*. In order to estimate the magnitude of Γ, we may simplify Eq. (8–73) by noting that

1. If α and β are of similar magnitude, $(\alpha^2 - \beta^2)$ is small, so that the second term in Eq. (8–73) may be neglected relative to the first term.

2. If $\beta a \gg 1$, $e^{-\beta a}$ is small compared to $e^{\beta a}$, so that the first term in the brackets becomes $e^{2\beta a}/4$.

Thus Eq. (8–73) may be written in the *approximate* form

$$\Gamma = \frac{|F|^2}{|A|^2} \cong 4e^{-2\beta a}.$$

Substitution of β from Eq. (8–55) gives

$$\Gamma \cong 4 \exp \left[\frac{-2a}{\hbar} \sqrt{2m(V_0 - E)} \right]. \qquad (8\text{–}74)$$

It is clear that unless V_0, a, or m is infinite, Γ will not be equal to zero and, therefore, *the particle for any value of E will have a finite probability for penetrating the barrier.* Note also that Γ increases as m decreases, or as the thickness a or height V_0 of the barrier decreases.

Experimental evidence seems to indicate that barrier penetration, referred to as *quantum-mechanical tunneling*, actually does occur. Consider, for example, the escape of an α-particle from a radioactive nucleus. An approximate curve for the potential energy of an α-particle as a function of distance r from the center of the nucleus is shown in Fig. 8–11. Note that the curve is *in general* similar to the curve for the simple potential-energy barrier given previously in Fig. 8–10(a). The negative slope at large values of r is due to coulombic repulsion between the positively charged α-particle and the positively charged nucleus. The positive slope at low values of r is due to nuclear attraction. The kinetic energy E of the α-particle within the nucleus (to the left of the barrier) is less than the potential energy needed for escape, as indicated by the height of the barrier. Thus, *according to classical theory, the α-particle should not be emitted.*

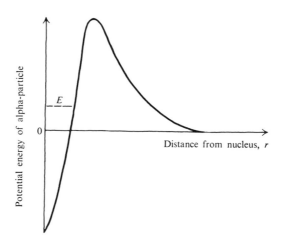

FIG. 8–11   Energy diagram showing the forces which hold an α-particle in the nucleus of a typical radioactive element. The repulsion at large values of r is due to coulombic forces, whereas the strong attraction at low values of r is due to nuclear attraction. The kinetic energy of the α-particle within the nucleus is E.

Nevertheless, the emission of α-particles from radioactive nuclei under classically unfavorable energy conditions *is* experimentally observed. In order to theoretically correlate the observed relationship between nuclear half-life and α-particle kinetic energies Gamow*, and later Gurney and Condon,† assumed α-particle emission to occur through quantum-mechanical tunneling, and derived equations in good agreement with observed data.

Because of the small mass of the electron and the exponential dependence of Γ on m, tunneling appears to be important in many processes in which electrons are transferred, such as electrode reactions, oxidation-reduction reactions in solution, and cold emission of electrons.

The umbrella-like *inversion* of pyramidal molecules such as NH_3, PH_3, and AsH_3 is believed to occur through a process in which the N atom, for example, tunnels through the potential energy barrier associated with the plane of the three H atoms. Experimental spectral observations indicate that the energy of inversion is not sufficient to allow the N atom to pass *over* the barrier. A qualitative representation of the inversion of NH_3 and the associated potential-energy relationship is shown in Fig. 8–12.

In addition, quantum-mechanical tunneling may be important in the biological processes of aging and the occurrence of spontaneous mutations and tumors. Löwdin‡ has recently speculated that such phenomena may be related to the tunneling of hydrogen-bonded protons through potential-energy barriers between specific pairs of nucleotide bases in the DNA (deoxyribonucleic acid) molecule. Such proton tunneling could result in a tautomeric transformation of normal base pairs and thus result in a change in the genetic code. Rough comparisons indicate that the rate of proton tunneling may be of the same magnitude as the rate of spontaneous mutation.

"Now, let us get it in reverse," he suggested, *"and see if the ball can get out of the crater without rolling over the top,"* and he threw the ball back into the hole. *For a while nothing happened, and Mr. Tompkins could hear only the slight rumbling of the ball rolling to and fro in the crater. Then, as if by a miracle, the ball suddenly appeared in the middle of the outer slope and quietly rolled down to the table.*
"What you see here is a very good representation of what happens in radioactive alpha-decay," said the woodcarver, putting the model back into its place, *"only there, instead of the ordinary quantum-oak barrier, you have the barrier of repulsive electric force. But in principle there is no difference whatever. Sometimes these electric barriers are so 'transparent' that the particle escapes in a small fraction of a second; sometimes they are so 'opaque' that it takes many billion years, as for example in the case of the Uranium nucleus."*
"But why aren't all nuclei radioactive?" asked Mr. Tompkins.

---

* G. Gamow, *Z. Physik*, **51**, 204 (1928).
† R. W. Gurney and E. U. Condon, *Phys. Rev.*, **33**, 127 (1929).
‡ P. O. Löwdin, *Advan. Quantum Chem.*, **2**, 216 (1965).

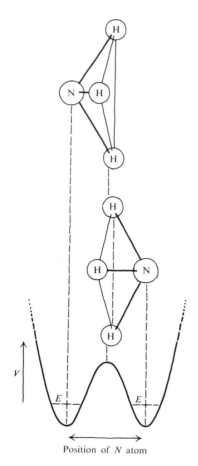

FIG. 8–12   The NH_3 molecule before and after inversion. The energy of the molecule in inversion is E, and the potential energy V of the molecule is shown as a function of position of the N atom. The H atoms remain fixed.

Position of N atom

"Because in most nuclei the floor of the crater is below the outer level, and only in the heaviest known nuclei is the floor sufficiently elevated to make such an escape possible."

—GEORGE GAMOW, *Mr. Tompkins Explores the Atom.**

PROBLEMS

8–1 Consider the one-dimensional functions $\psi_1 = ax$ and $\psi_2 = ibx^2$.

a) What must be the value of a in order that ψ_1 is normalized between 0 and L?

b) What must be the value of b in order that ψ_2 is normalized between 0 and L?

c) Are the *normalized* ψ_1 and ψ_2 functions orthogonal in the interval 0 to L? Prove.

* From *Mr. Tompkins Explores the Atom*, Cambridge University Press, London, 1945. Reprinted through permission of the publisher.

8–2 If ψ_m and ψ_n are orthonormal eigenfunctions of the Hamiltonian operator for the single particle in the one-dimensional box,

a) show that the linear-combination functions

$$\psi_{\mathrm{I}} = a(\psi_m + \psi_n) \quad \text{and} \quad \psi_{\mathrm{II}} = c(\psi_m - \psi_n)$$

are orthogonal in the interval $x = 0$ to $x = L$.

b) Determine the required values of a and c in order that ψ_{I} and ψ_{II} may also be individually normalized.

8–3 a) Show that if $\psi_m, \psi_n, \psi_p,$ and ψ_q are members of a set of wave functions which are orthonormal in a given space interval, the two linear-combination functions $(a\psi_m - b\psi_n)$ and $(f\psi_p + g\psi_q)$ are orthogonal in the same interval ($a, b, f,$ and g are constants).

b) What must be the relationship between a and b and between f and g in order that the two linear-combination functions are *normalized* in the same space interval?

8–4 Consider the two linear-combination functions which contain *one* function ψ_n in common:

$$\psi_{\mathrm{I}} = a\psi_m + b\psi_n, \qquad \psi_{\mathrm{II}} = f\psi_p + g\psi_n,$$

where $\psi_m, \psi_n,$ and ψ_p are members of an orthonormal set of eigenfunctions. Prove that ψ_{I} and ψ_{II} cannot be orthogonalized in the same interval unless b or g is zero.

8–5 Prove that the operator x is Hermitian.

8–6 Show that the linear-momentum operator $\hat{p}_x$ is a Hermitian operator for a single particle in three-dimensional space. That is, show that

$$\int\int\int \psi_m^* \hat{p}_x \psi_n \, dx \, dy \, dz = \int\int\int \psi_n \hat{p}_x^* \psi_m^* \, dx \, dy \, dz.$$

Make use of the relationship for integration by parts: $uv = \int v \, du + \int u \, dv$.

8–7 Show, for the particle in the three-dimensional box, that $\psi(x, y, z)$ must be normalized within the limits $(0 \leqslant x \leqslant a, 0 \leqslant y \leqslant b, 0 \leqslant z \leqslant c)$ of the box if each of the individual φ functions is normalized within its respective one-dimensional limits.

8–8 Sketch a boundary-surface probability density plot for ψ_{123}^2 versus x, y and z for a particle in a cubic box in the (123) quantum state.

8–9 Consider a single particle of mass m in a cubic box of edge L. Suppose that the system is perturbed slightly by pushing in one wall so that the new dimensions are $L \times L \times 0.99L$.

a) Calculate the first seven energy levels, in units of $h^2/8mL^2$.

b) Sketch an energy plot (similar to Fig. 8–1) for the perturbed system, indicating the degeneracy of each of the new levels.

c) Given the results of (a) and (b), how would you expect the perturbation of any symmetrical system to affect degeneracy?

8–10 Show in detail how Eq. (8–73) is obtained from Eq. (8–72).

8–11 Refer to the finite barrier described in Fig. 8–10 and start with a single particle initially in region I with energy E which is *greater* than V_0.

    a) Write the Schrödinger equations for each of the three regions.

    b) Write ψ-functions for each of the three regions.

    c) Write the continuity equations which apply at $x = 0$ and $x = a$.

    d) Since E is greater than V_0, *classical* theory implies that the particle should not be stopped by the barrier. Express the probability, in terms of the constants used in writing the ψ equations, that the particle is reflected back at $x = 0$.

8–12 The potential barriers encountered by an electron moving through the atoms of a crystal are about 10 eV high and 2×10^{-8} cm thick. Calculate the probability that an electron moving with a thermal energy kT will tunnel through such barriers at 300° K. (k = Boltzmann constant.)

8–13 Consider the barrier of Fig. 8–10 with $V_0 = E$, so that $\beta = 0$.

    a) Show that $C + Dx$, rather than Eq. (8–57), is then the general solution of Eq. (8–54).

    b) Using Eqns. (8–56) and (8–58), which are still valid, and the continuity conditions (8–59) through (8–62), find the transmission coefficient of the barrier.

    c) Show that the same answer for the transmission coefficient results from treating Eq. (8–73) as a limiting equation with β approaching zero.

In Chapter 4 we treated the classical linear harmonic oscillator in some detail. For our model we chose (Fig. 4–5) a single mass m constrained to move in a straight line subject to a restoring force proportional to the displacement y from the equilibrium position. The frequency ν of vibration was shown to be $\nu = (1/2\pi)(\kappa/m)^{1/2}$ where κ is the force constant, and the potential energy was given as $V = \kappa y^2/2$. In the classical model the total energy is constant and may have any positive value. In this chapter, we'll explore the quantum mechanical solution of the same oscillator. We immediately note that the oscillator is different from previous systems we have examined in that the potential energy V is not constant with position. In Fig. 9–1 the potential energy curve for the linear harmonic oscillator (parabolic well) is compared to that for the particle in the one dimensional box (square well). Because of the general similarity in shapes of the two curves, we'll expect a similarity in shapes of the ψ functions for the two systems.

9–1 SCHRÖDINGER EQUATION FOR
THE LINEAR HARMONIC OSCILLATOR

Since the total energy E of the oscillator is constant, the system is conservative, and we may write the Schrödinger equation as

$$\frac{-\hbar^2}{2m} \frac{d^2\psi(y)}{dy^2} + \frac{\kappa y^2}{2} \psi(y) = E\psi(y).$$

Substitution of κ from Eq. (4–16) gives

$$\frac{-\hbar^2}{2m} \frac{d\psi(y)}{dy^2} + 2\pi^2 m\nu^2 y^2\psi(y) = E\psi(y),$$

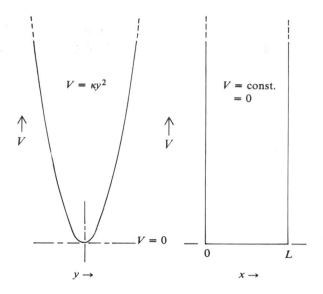

FIG. 9–1.   A comparison of the parabolic potential energy well of the linear harmonic oscillator
with the square well of the particle in the one-dimensional box.

which can be expressed more compactly as

$$\frac{d^2\psi(y)}{dy^2} + (\lambda - \alpha^2 y^2)\psi(y) = 0, \qquad (9\text{–}1)$$

where $\lambda = 2mE/\hbar^2$ and $\alpha = 2\pi m\nu/\hbar$. It is now mathematically convenient to
express the wave equation in terms of a new *dimensionless* variable ξ, such that

$$\xi \equiv \alpha^{1/2} y,$$

and to replace $\psi(y)$ with $\psi(\xi)$, to which it is equal. Then

$$\frac{d\psi(y)}{dy} = \frac{d\psi(\xi)}{d\xi} \cdot \frac{d\xi}{dy} = \alpha^{1/2} \frac{d\psi(\xi)}{d\xi},$$

and

$$\frac{d^2\psi(y)}{dy^2} = \frac{d}{d\xi}\left[\frac{d\psi(y)}{dy}\right]\frac{d\xi}{dy} = \frac{d}{d\xi}\left[\alpha^{1/2}\frac{d\psi(\xi)}{d\xi}\right]\frac{d\xi}{dy} = \alpha\frac{d^2\psi(\xi)}{d\xi^2}$$

so that Eq. (9–1) can be written

$$\alpha \frac{d^2\psi(\xi)}{d\xi^2} + (\lambda - \alpha\xi^2)\psi(\xi) = 0,$$

or

$$\frac{d^2\psi(\xi)}{d\xi^2} + \left(\frac{\lambda}{\alpha} - \xi^2\right)\psi(\xi) = 0. \qquad (9\text{–}2)$$

9–2 AN ASYMPTOTIC SOLUTION

We must now look for well-behaved $\psi(\xi)$ functions which satisfy Eq. (9–2), that is, which are continuous, single-valued and *finite* for all values of ξ from $-\infty$ to ∞. As it turns out, no simple mathematical solution is satisfactory, so that we'll find it convenient to resort to a power-series method. But first, in order to ensure that $\psi(\xi)$ remains finite as $|\xi| \to \infty$ and to get a rough idea of form of $\psi(\xi)$, let's specifically examine the asymptotic solution for very large values of $|\xi|$, and then determine what modifications are necessary to make the solution suitable for *all* values of $|\xi|$. For any arbitrary constant value of E, λ/α eventually becomes negligible compared to ξ^2 as $|\xi|$ increases, so that Eq. (9–2) becomes, in asymptotic form,

$$\frac{d^2\psi(\xi)}{d\xi^2} = \xi^2\psi(\xi), \qquad (|\xi| \to \infty).$$

We'll now show through substitution that the above asymptotic wave equation is satisfied by either the positive or negative exponential form

$$\psi(\xi) = ae^{\pm\xi^2/2},$$

where a is a constant.
To do this, we differentiate to obtain

$$\frac{d\psi(\xi)}{d\xi} = \pm\xi ae^{\pm\xi^2/2}$$

and

$$\frac{d^2\psi(\xi)}{d\xi^2} = \xi^2 ae^{\pm\xi^2/2} \pm ae^{\pm\xi^2/2}.$$

But as $|\xi| \to \infty$, the second term in the above equation becomes negligible compared to the first so that

$$\frac{d^2\psi(\xi)}{d\xi^2} = \xi^2 ae^{\pm\xi^2/2} = \xi^2\psi(\xi), \qquad (|\xi| \to \infty)$$

and the asymptotic wave equation is thus shown to be satisfied. However, of the two solutions, $ae^{+\xi^2/2}$ and $ae^{-\xi^2/2}$, only the negative exponential form is well-behaved, since the positive solution tends to infinity as $|\xi|$ increases. Thus, a satisfactory form of wave function for large values of $|\xi|$ is

$$\psi(\xi) = ae^{-\xi^2/2}, \qquad (|\xi| \to \infty).$$

9–3 SEARCH FOR THE GENERAL SOLUTION

But we must now find $\psi(\xi)$ functions which are also satisfactory for *finite* values of ξ in the interval $-\infty < \xi < \infty$, not just at the limits. In an attempt to construct a generally acceptable wave function, we can try modifying the asymptotic function by substituting a still-to-be-determined adjusting function $H(\xi)$ for the constant a:

$$\psi(\xi) = e^{-\xi^2/2}H(\xi). \tag{9–3}$$

What sort of function must $H(\xi)$ be? As $|\xi| \to \infty$, it will be necessary that $H(\xi)$ be slowly varying compared to $e^{-\xi^2/2}$ in order to ensure that $e^{-\xi^2/2}H(\xi)$ vanishes.

In order to determine an appropriate form for $H(\xi)$, we first construct a more easily solved wave equation by substituting $\psi(\xi)$ from Eq. (9–3) into the earlier wave equation (9–2). Differentiation of $\psi(\xi)$ gives:

$$\frac{d\psi(\xi)}{d\xi} = -\xi e^{-\xi^2/2}H + e^{-\xi^2/2}\frac{dH}{d\xi}$$

and

$$\frac{d^2\psi(\xi)}{d\xi^2} = -e^{-\xi^2/2}H + \xi^2 e^{-\xi^2/2}H - \xi e^{-\xi^2/2}\frac{dH}{d\xi} \cdot$$

$$- \xi e^{-\xi^2/2}\frac{dH}{d\xi} + e^{-\xi^2/2}\frac{d^2H}{d\xi^2}$$

$$\frac{d^2\psi(\xi)}{d\xi^2} = e^{-\xi^2/2}\left[-H + \xi^2 H - 2\xi\frac{dH}{d\xi} + \frac{d^2H}{d\xi^2}\right].$$

Substitution of $\psi(\xi)$ and $d^2\psi(\xi)/d\xi^2$ into Eq. (9–2) yields

$$e^{-\xi^2/2}\left[-H + \xi^2 H - 2\xi\frac{dH}{d\xi} + \frac{d^2H}{d\xi^2}\right] + \frac{\lambda e^{-\xi^2/2}}{\alpha}H - \xi^2 e^{-\xi^2/2}H = 0.$$

Division by $e^{-\xi^2/2}$ followed by cancellation of terms gives

$$\frac{d^2H}{d\xi^2} - 2\xi\frac{dH}{d\xi} + \left(\frac{\lambda}{\alpha} - 1\right)H = 0, \tag{9-4}$$

which is a well-known eigenvalue equation called the *Hermite differential equation*.

9-4 SERIES SOLUTION OF THE HERMITE DIFFERENTIAL EQUATION

We must now try to solve Eq. (9-4) for eigenfunctions $H(\xi)$ in order to determine acceptable forms for $\psi(\xi) = e^{-\xi^2/2}H(\xi)$. The solution will also generate the eigenvalues $-[(\lambda/\alpha) - 1]$ from which the corresponding eigenvalue energies can be calculated. As it turns out, Eq. (9-4) is typical of many differential equations arising in applied mathematics which cannot be solved in closed form, that is, in a simple way in terms of elementary functions. In such cases, it is natural to seek a solution in the form of an infinite power series. Let's then assume that it is possible to express the solution of Eq. (9-4) in the form:

$$H(\xi) = \sum_{n=0}^{\infty} a_n\xi^n = a_0 + a_1\xi + a_2\xi^2 + a_3\xi^3 + \cdots \tag{9-5}$$

Our next task is to determine the necessary coefficients $a_0, a_1, a_2, \ldots$, in the proposed series. To do this, we'll substitute the $H(\xi)$ series into Eq. (9-4) and demand that the coefficients be such that the resulting equation is satisfied for all values of ξ. Differentiation of $H(\xi)$ from Eq. (9-5) gives

$$\frac{dH}{d\xi} = \sum_{n=0}^{\infty} na_n\xi^{n-1} = 1a_1 + 2a_2\xi + 3a_3\xi^2 + \cdots,$$

and

$$\frac{d^2H}{d\xi^2} = \sum_{n=0}^{\infty}(n-1)na_n\xi^{n-2} = 1\cdot 2a_2 + 2\cdot 3a_3\xi + 3\cdot 4a_4\xi^2 + 4\cdot 5a_5\xi^3 + \cdots,$$

so that substitution of the series forms for H, $dH/d\xi$, and $d^2H/d\xi^2$ into Eq. (9-4) gives,

$$
\begin{aligned}
1\cdot 2a_2 + 2\cdot 3a_3\xi &\quad + 3\cdot 4a_4\xi^2 &\quad + 4\cdot 5a_5\xi^3 + \ldots \\
- 2\cdot 1a_1\xi &\quad - 2\cdot 2a_2\xi^2 &\quad - 2\cdot 3a_3\xi^3 - \ldots \\
+\left(\frac{\lambda}{\alpha} - 1\right)a_0 + \left(\frac{\lambda}{\alpha} - 1\right)a_1\xi + \left(\frac{\lambda}{\alpha} - 1\right)a_2\xi^2 &+ \left(\frac{\lambda}{\alpha} - 1\right)a_3\xi^3 + \ldots = 0.
\end{aligned}
$$

In order that the above expression be equal to zero for all values of ξ, including $\xi = 0$, it is necessary that the coefficients of individual powers of ξ vanish separately. To see this, consider the above series written as

$$b_0 + b_1\xi + b_2\xi^2 + b_3\xi^3 + \ldots = 0,$$

where the b terms are the sums of the coefficients of individual powers of ξ. If $\xi = 0$ in the above series, b_0 must be zero for the series to be zero. By taking the first derivative of the series and setting $\xi = 0$ we also show that b_1 must be zero. Similarly, by taking the nth derivative and setting $\xi = 0$, we show that b_n must be zero. Let's now sum the coefficients for the various powers of ξ:

$$\xi^0: \quad 1 \cdot 2a_2 + \left(\frac{\lambda}{\alpha} - 1\right)a_0 = 0,$$

$$\xi^1: \quad 2 \cdot 3a_3 + \left(\frac{\lambda}{\alpha} - 1 - 2 \cdot 1\right)a_1 = 0,$$

$$\xi^2: \quad 3 \cdot 4a_4 + \left(\frac{\lambda}{\alpha} - 1 - 2 \cdot 2\right)a_2 = 0,$$

$$\xi^3: \quad 4 \cdot 5a_5 + \left(\frac{\lambda}{\alpha} - 1 - 2 \cdot 3\right)a_3 = 0,$$

so that for the coefficients of ξ^n it is required that

$$\xi^n: \quad (n + 1)(n + 2)a_{n+2} + \left(\frac{\lambda}{\alpha} - 1 - 2n\right)a_n = 0,$$

from which

$$a_{n+2} = \frac{-\left(\frac{\lambda}{\alpha} - 1 - 2n\right)}{(n + 1)(n + 2)} a_n. \tag{9–6}$$

Eq. (9–6) is called a *recursion formula*. It generates and dictates the relative values of the coefficients in the $H(\xi)$ series in terms of the two arbitrary constants a_0 and a_1. That is, we can calculate $a_2, a_4, a_6, \ldots$, in terms of a_0; and we can calculate $a_3, a_5, a_7, \ldots$, in terms of a_1. We may thus treat the $H(\xi)$ series as the sum of two independent series, one even and one odd:

$$H(\xi) = a_0\left(1 + \frac{a_2}{a_0}\xi^2 + \frac{a_2}{a_0}\cdot\frac{a_4}{a_2}\xi^4 + \frac{a_2}{a_0}\cdot\frac{a_4}{a_2}\cdot\frac{a_6}{a_4}\xi^6 + \ldots\right)$$

$$+ a_1\left(\xi + \frac{a_3}{a_1}\xi^3 + \frac{a_3}{a_1}\cdot\frac{a_5}{a_3}\xi^5 + \frac{a_3}{a_1}\cdot\frac{a_5}{a_3}\cdot\frac{a_7}{a_5}\xi^7 + \ldots\right), \tag{9–7}$$

where each of the coefficients is determined by the recursion formula from a_0 or a_1. If a_0 is arbitrarily set at zero only the odd series remains; if a_1 is set at zero only the even series remains. The appearance of two arbitrary constants is entirely expected in the solution of any second-order differential equation.

As we'll show soon, boundary conditions will fix one of the constants at zero and final normalization will determine the other.

For arbitrary values of λ/α (arbitrary energies) each of the above series contains an infinite number of terms. We'll first show that $H(\xi)$ with an infinite number of terms is *not* a well-behaved function because it causes the total wave function $\psi(\xi)$ to diverge rather than vanish at $|\xi| \to \infty$. To show that $\psi(\xi)$ diverges, we first examine for comparison the Maclaurin series expansion of the related function e^{ξ^2}:

$$ e^{\xi^2} = 1 + \xi^2 + \frac{\xi^4}{2!} + \frac{\xi^6}{3!} + \ldots + \frac{\xi^n}{\left(\frac{n}{2}\right)!} + \frac{\xi^{n+2}}{\left(\frac{n}{2}+1\right)!} + \ldots \quad (9\text{-}8) $$

Let's look particularly at the later terms (large values of n) in the infinite e^{ξ^2} series, which are the dominant terms for large values of $|\xi|$, and examine the ratio of coefficients of successive powers of ξ. Considering the last two terms, for *large* values of n,

$$ \frac{a_{n+2}}{a_n} = \frac{\left(\frac{n}{2}\right)!}{\left(\frac{n}{2}+1\right)!} = \frac{\left(\frac{n}{2}\right)!}{\left(\frac{n}{2}\right)!\left(\frac{n}{2}+1\right)} \cong \frac{2}{n}. $$

In comparison, the ratio of coefficients of successive powers of ξ in either the even or odd series of $H(\xi)$ [Eq. (9-7)] is given through the recursion formula, for large values of n, as

$$ \frac{a_{n+2}}{a_n} = \frac{-\frac{\lambda}{\alpha} + 1 + 2n}{(n+1)(n+2)} \cong \frac{2n}{n^2} \cong \frac{2}{n}, $$

which is *identical* to the ratio for large values of n in the e^{ξ^2} series. This means that the higher terms in the even series of $H(\xi)$ must differ from those in the e^{ξ^2} series by a constant c. Furthermore, the higher terms in the odd series of $H(\xi)$ must differ by ξ times another constant c'. Thus, as $|\xi| \to \infty$, in which case the later high-n terms become the dominant terms, Eq. (9-7) can be written in terms of Eq. (9-8) as

$$ H(\xi) = ce^{\xi^2} + c'\xi e^{\xi^2}, \qquad |\xi| \to \infty, $$

so that at high values of $|\xi|$, the complete wave function $\psi(\xi)$ becomes

$$ \psi(\xi) = e^{-\xi^2/2}H(\xi) = (c + c'\xi)e^{\xi^2/2}, \qquad |\xi| \to \infty, $$

which as it stands is unacceptable, since $\psi(\xi)$ obviously diverges rather than vanishes as $|\xi| \to \infty$.

There is only one way to prevent $\psi(\xi)$ from diverging and to thus produce acceptable wave functions. *We must limit $H(\xi)$ to a finite number of terms*, in which case $\psi(\xi)$ at large values of $|\xi|$ will be dominated by the term $e^{-\xi^2/2}$ and will thus approach zero as required. We can easily cut off the series if we first define either a_0 or a_1 to be zero and then truncate the remaining odd or even series at any selected value of n by equating the numerator in the recursion formula for a_{n+2}/a_n to zero. For example, in order for the series to be terminated after the ξ^v term ($n = v$), that is, for a_{v+2} and all following coefficients to be zero, it is necessary from Eq. (9–6) that

$$\frac{\lambda}{\alpha} - 1 - 2v = 0,$$

or

$$\frac{\lambda}{\alpha} = 1 + 2v, \tag{9-9}$$

where

$$v = 0, 2, 4, \ldots, \text{ if } a_1 \equiv 0$$

$$v = 1, 3, 5, \ldots, \text{ if } a_0 \equiv 0.$$

9-5 QUANTIZED VIBRATIONAL ENERGIES AND FUNCTIONS

Substitution of our original definitions for λ and α into Eq. (9–9) followed by rearrangement gives

$$E = (v + \tfrac{1}{2})h\nu \tag{9-10}$$

where

$$v = 0, 1, 2, 3, \ldots$$

The non-negative integers v, called *vibrational quantum numbers*, obviously restrict the allowed energies of the simple harmonic oscillator to a set of discrete privileged values. Once more we encounter the now familiar result in which the imposition of boundary conditions forces a system's observable energies into a discrete spectrum of sharp eigenvalues. At the same time, acceptable eigenfunctions are limited to a discrete set of polynomials $H_v(\xi)$ of degree v, each polynomial having only even or only odd powers of ξ, as v is even or odd. Since there is only one polynomial $H_v(\xi)$ corresponding to each energy level E_v, the vibrational energy levels are nondegenerate.

The $H_v(\xi)$ polynomials which satisfy the Hermite differential equation (9–4) are called *Hermite polynomials*. The Hermite polynomial for any selected value of v is evaluated from Eq. (9–7) by first discarding the odd series if v is even or the even series if v is odd. We next evaluate the coefficients in the remaining series. Rather than choose an arbitrary initial coefficient a_0 or a_1, it is simpler in practice to arbitrarily define the coefficient of the highest power term as $a_v = 2^v$, and then evaluate the coefficients of the lower-power terms from the recursion formula, where $\lambda/\alpha = 1 + 2v$, as is required to terminate the series. For example, for $v = 4$,

$$H_4(\xi) = a_0 + a_2\xi^2 + a_4\xi^4.$$

Let $a_4 = 2^4 = 16$. Then, from the recursion formula,

$$a_n = a_{n+2}\frac{(n + 1)(n + 2)}{\left(-\dfrac{\lambda}{\alpha} + 1 + 2n\right)} = a_{n+2}\frac{(n + 1)(n + 2)}{(2n - 2v)}.$$

Thus,

$$a_2 = a_4\frac{(3)(4)}{(-4)} = -48, \quad \text{and} \quad a_0 = a_2\frac{(1)(2)}{(-8)} = +12,$$

so that,

$$H_4(\xi) = 12 - 48\xi^2 + 16\xi^4.$$

The first five Hermite polynomials so calculated are

$$H_0(\xi) = 1$$

$$H_1(\xi) = 2\xi$$

$$H_2(\xi) = -2 + 4\xi^2$$

$$H_3(\xi) = -12\xi + 8\xi^3$$

$$H_4(\xi) = 12 - 48\xi^2 + 16\xi^4$$

$$H_5(\xi) = 120\xi - 160\xi^3 + 32\xi^5$$

The arbitrariness in the selection of the value of a_v in the $H_v(\xi)$ polynomial disappears when the final $\psi_v(\xi)$ eigenfunction is normalized, that is, when we incorporate a normalization constant into $\psi_v(\xi)$ in order to ensure that

$$\int_{-\infty}^{\infty}\left[\psi_v(\xi)\right]^2 d\xi = 1.$$

Using the above set of Hermite polynomials in Eq. (9–3), the final normalized vibrational eigenfunctions can be shown* to be

$$\psi_v(\xi) = \left(\frac{1}{2^v v!\,\pi^{1/2}}\right)^{1/2} e^{-\xi^2/2} H_v(\xi),$$

---

* Pilar, F. L., *Elementary Quantum Chemistry*, McGraw-Hill Book Co., New York (1968) p. 105.

or, in terms of the original variable y,

$$\psi_v(y) = \left(\frac{\alpha^{1/2}}{2^v v! \, \pi^{1/2}} \right)^{1/2} e^{-\alpha y^2/2} H_v(y)$$

The first five $\psi_v(\xi)$ eigenfunctions are shown graphically in Fig. 9–2 super-posed on their corresponding eigenvalue energy levels within the classical potential energy well. The expression $V = (h\nu/2)\xi^2$ is obtained by substitution of $y = \xi/\alpha^{1/2}$, $\alpha = 4\pi^2 m\nu/h$, and κ from Eq. (4–16) into Eq. (4–20).

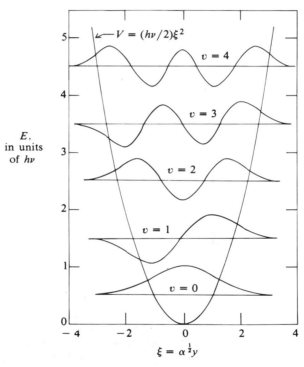

FIG. 9–2. The first five $\psi_v(\xi)$ eigenfunctions for the simple harmonic oscillator, superposed at their corresponding eigenvalue energy levels on the classical potential energy curve. Values for $\psi_v(\xi)$ are in scale relative to one another but not to the energy ordinate.

Note that the quantum mechanical solution of the simple harmonic oscilla-tor is in full accord with Planck's earlier spacing requirement $h\nu$ between adjacent energy levels. On the other hand, there was nothing in Planck's model for black-body radiation which would prevent individual oscillators from having energies or corresponding momenta equal to zero. But the actual fact is, as we have previously discovered in our treatment of the particle in the one-dimensional box, that the uncertainty principle does not allow a state of exactly defined position (the equilibrium position) and simultaneously exactly defined momentum (zero in the case of Planck's oscillators). Thus the *zero-point energy* $h\nu/2$, which is the minimum energy allowed to the oscillator in the quantum mechanical solution, corresponds to the uncertainty in momentum required by the localization of oscillation in the confined region of the potential energy well.

As we shall see in Chapter 15, the linear harmonic oscillator is a very useful approximation to vibrational motion in molecules. Even at $0°K$, when all translational molecular motion has ceased, molecules still retain zero-point energy contributions of $h\nu/2$ per vibrational mode.

Examination of the wave shapes in Fig. 9–2 reveals that the functions for the linear harmonic oscillator are generally similar to those for the particle in the box, with the very noticeable difference that the harmonic oscillator functions extend beyond the classical boundaries. Apparently, the quantum mechanical solution allows the particle to exist at classically forbidden displacements for which the total energy is less than the potential energy, that is, for which $E < V(\xi)$. This classically unexpected behavior is a further example of barrier penetration or tunneling which we have previously discussed in Chapter 8.

It is also interesting to compare the probability density $\psi_v^2(\xi)$ distributions with classical expectations, as shown in Fig. 9–3. The classical probability for a given displacement is inversely proportional to the velocity of the mass, and thus reaches a minimum at $\xi = 0$ where the velocity is greatest and approaches infinity at the classical boundaries of motion where the mass reverses direction. On the other hand, the quantum mechanical probability distribution is very different, particularly at lower values of v. For all energy levels $\psi_v^2(\xi)$ shows v nodes within the classically allowed displacement and vanishes rapidly outside this region. Note, however, that the quantum-mechanical and classical distributions begin to look more nearly alike as v increases. The equivalence of classical and quantum-mechanical results when quantum numbers are large is stated by the *Bohr correspondence principle*:

In the limit of large quantum numbers, quantum mechanics and classical mechanics must yield identical results.

But such an idea was fantastic; it was like saying that a swing may swing with a sweep of one yard, or two yards, or three, or four, and so on, but not with a sweep of one and a quarter yards or any other in-between value. Even a child would realize how fantastic that sort of thing would be. Yet it did lead to the proper answer . . . .

—BANESH HOFFMAN, *The Strange Story of the Quantum*, Dover Publications, Inc., New York, 1959. Reprinted through permission of the publisher.

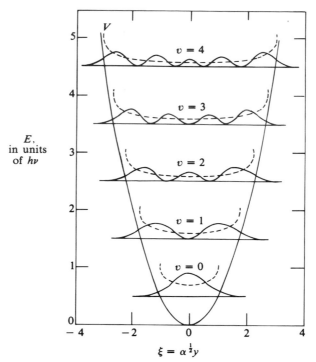

FIG. 9–3. Probability density functions $\psi_v^2(\xi)$ for the simple harmonic oscillator superposed in relative scale on corresponding eigenvalue energy levels. Dashed lines correspond to the classical probability functions at the same energies.

PROBLEMS

In order to evaluate integrals in several of the following problems, it is helpful to review the concept of *even* and *odd* functions. A function $f(x)$ is said to be *even* when

$$f(-x) = f(x).$$

For example, x^2, e^{x^2}, and $\cos x$ are even functions. That is, $(-x)^2 = x^2$, $e^{(-x)^2} = e^{x^2}$, and $\cos(-x) = \cos x$. The graph of an even function is symmetric about the ordinate, as shown in Fig. 9–4(a), so that

$$\int_{-a}^{a} f(x)\, dx = 2\int_{0}^{a} f(x)\, dx \quad \text{(even function)}.$$

A function $g(x)$ is said to be *odd* if

$$g(-x) = -g(x).$$

Thus, x, x^3, and $\sin x$ are odd. That is, $(-x) = -x$, $(-x)^3 = -x^3$, and $\sin(-x) = -\sin x$. The graph of an odd function must pass through the origin, as shown in Fig. 9–4(b), and the curve on the negative side is an inverted reflection of the curve on the

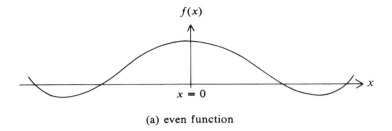

(a) even function

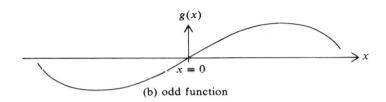

(b) odd function

Fig. 9–4. Graphs of even and odd functions about the origin.

positive side. Therefore

$$\int_{-a}^{0} g(x)\,dx = -\int_{0}^{a} g(x)\,dx, \quad \text{so that} \quad \int_{-a}^{a} g(x)\,dx = 0 \quad \text{(odd function)}.$$

9–1 Write the Maclaurin-series expansion (See Prob. 3–8) for e^x, and, letting $x = \xi^2$, write the expression for e^{ξ^2}.

9–2 Determine the form of the Hermite polynomial $H_6(\xi)$ and the normalized form of $\psi_6(\xi)$ for the linear harmonic oscillator.

9–3 Verify the normalization constant $(1/2^v v!\,\pi^{1/2})^{1/2}$ in $\psi_v(\xi)$ for the first excited state of the linear harmonic oscillator for which $v = 1$.

9–4 Calculate the expectation (average) values for ξ, ξ^2, and $|\xi|$ for a linear harmonic oscillator in the ground state. What is the value of ξ at the classical amplitude A?

9–5 Calculate the expectation values $\overline{T}$ and $\overline{V}$ for the kinetic and potential energies of the linear harmonic oscillator in the ground state and show that the sum of the two energies is equal to the zero-point energy.

9–6 The potential energy function for a three-dimensional harmonic oscillator is

$$V(x, y, z) = \tfrac{1}{2}\kappa_x x^2 + \tfrac{1}{2}\kappa_y y^2 + \tfrac{1}{2}\kappa_z z^2,$$

where κ_x, κ_y, and κ_z are the force constants in each of the coordinate directions.

a) By analogy to the solution of the Schrödinger equation for the particle in the three-dimensional box, write an expression for the energy eigenvalues in terms of the quantum numbers v_x, v_y, and v_z and the three fundamental vibrational frequencies v_x, v_y, and v_z.

b) Write the form of the complete wave function $\psi_{v_x, v_y, v_z}(x, y, z)$.

c) Write the equation for the energy eigenvalues for an *isotropic* three-dimensional harmonic oscillator in which the force constants and the fundamental frequencies are the same in all three directions.

d) Determine the degeneracies of the first four energy levels of the isotropic three-dimensional harmonic oscillator.

9–7 a) Estimate, from the probability distribution given in Fig. 9–3, the uncertainty in position Δy for the linear harmonic oscillator at zero-point energy.

b) Note that at $\xi = 0$ $(y = 0)$ the energy of the oscillator is entirely kinetic energy, and its momentum is either a maximum or a minimum, depending on its direction of motion. Calculate the uncertainty in momentum Δp for the oscillator at zero-point energy.

c) From the results of parts a and b, show that $\Delta p \, \Delta y \cong h$.

THE HYDROGEN ATOM I.
SOLUTION OF THE
SCHRÖDINGER EQUATION

In our earlier treatment of the hydrogen atom in Chapter 3, we noted that the Bohr model, even with several subsequent refinements, was not entirely satisfactory. In particular, the model failed to provide an acceptable interpretation of relative line intensities in the hydrogen spectrum and also failed to provide a satisfactory basis for interpretation of the energies and spectra of more complex atoms. In this chapter, we'll consider the hydrogen atom from a quantum-mechanical point of view.

10–1 THE SCHRÖDINGER EQUATION FOR THE HYDROGEN ATOM

In all of our previous treatments, we have been concerned with the motion, energy, and location of a *single* particle. In fact, we have stated the postulates of quantum mechanics, given in Chapter 6, in terms of a single particle rather than in terms of a *system* of particles. The conceptual extension of quantum mechanics from one particle to a system of many particles is simple. For example, if we postulate a state function $\Psi(x, y, z, t)$ or $\psi(x, y, z)$ for each particle in a system of j particles, then through appropriate combination we are able in principle to construct a single function

$$\Psi(x_1, y_1, z_1, x_2, y_2, z_2, \ldots, x_j, y_j, z_j, t),$$

and, in conservative systems,

$$\psi(x_1, y_1, z_1, x_2, y_2, z_2, \ldots, x_j, y_j, z_j),$$

which completely describes the *entire system*. We then apply those postulates which we have already developed for a single particle to the wave function of the entire system in order to obtain values of energies and other observables for the system, and to determine the probability of a *particular configuration of the system* as a function of the coordinate variables in that specific system.

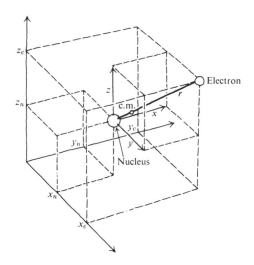

FIG. 10–1 The hydrogen atom or hydrogen-like ion in rectangular coordinate space. The point marked c.m. is the center of mass of the atom and is at the position x_M, y_M, z_M.

In order to show how the postulates of quantum mechanics are extended to multiparticle systems, let's specifically consider the free hydrogen atom or hydrogen-like ion, which is a *two-particle* system consisting of an electron of mass m_e and charge $-e$ and a nucleus of mass m_n and charge $+Ze$. For the hydrogen atom itself, $Z = 1$. A few other hydrogen-like ions which may be included in our model are

$$\text{He}^+ \text{ ion: }\; Z = 2,$$

$$\text{Li}^{2+} \text{ ion: }\; Z = 3,$$

$$\text{Be}^{3+} \text{ ion: }\; Z = 4.$$

Such ions are called *hydrogen-like* ions because in each case there is only one extranuclear electron, just as for the hydrogen atom. The potential energy of the *free* atom is entirely internal and depends on the distance r between the two particles. We have shown previously (Sec. 3–3) that

$$V = - Ze^2/r. \tag{10–1}$$

Recall that V is arbitrarily defined as zero when r is infinite. Since the potential energy is not an explicit function of time, that is, since t does not appear in Eq. (10–1), the coulombic force field is a *conservative* force field and we may therefore use the time-independent form of the Schrödinger equation,

$$\hat{H}\psi_T(x_e, y_e, z_e, x_n, y_n, z_n) = E_T \psi_T(x_e, y_e, z_e, x_n, y_n, z_n)$$

or

$$(\hat{T} + V)\psi_T = E_T \psi_T \tag{10–2}$$

in order to determine the wave function, ψ_T, for the *total* system. Note that ψ_T is a function of *six* coordinate variables, three for the electron (x_e, y_e, z_e) and three for the nucleus (x_n, y_n, z_n). In our treatment in this chapter, we'll ignore rest-mass energy so that E_T is the *total nonrelativistic energy*, and includes both the translational energy *and* the energy of relative motion of the two particles within the atom. The system is sketched in Fig. 10–1.

Because the total kinetic energy of the system is given as the sum of two terms, one for each of the particles, the kinetic energy operator $\hat{T}$ must also contain two terms, one for the electron and one for the nucleus. That is,

$$\hat{T} = -\frac{\hbar^2 \nabla_e^2}{2m_e} - \frac{\hbar^2 \nabla_n^2}{2m_n}$$

or

$$\hat{T} = -\frac{\hbar^2}{2m_e}\left(\frac{\partial^2}{\partial x_e^2} + \frac{\partial^2}{\partial y_e^2} + \frac{\partial^2}{\partial z_e^2}\right) - \frac{\hbar^2}{2m_n}\left(\frac{\partial^2}{\partial x_n^2} + \frac{\partial^2}{\partial y_n^2} + \frac{\partial^2}{\partial z_n^2}\right). \quad (10\text{--}3)$$

Substitution of Eqs. (10–1) and (10–3) into Eq. (10–2) yields the total non-relativistic Schrödinger equation for the hydrogen atom or hydrogen-like ion:

$$-\frac{\hbar^2}{2m_e}\left(\frac{\partial^2\psi_T}{\partial x_e^2} + \frac{\partial^2\psi_T}{\partial y_e^2} + \frac{\partial^2\psi_T}{\partial z_e^2}\right)$$

$$-\frac{\hbar^2}{2m_n}\left(\frac{\partial^2\psi_T}{\partial x_n^2} + \frac{\partial^2\psi_T}{\partial y_n^2} + \frac{\partial^2\psi_T}{\partial z_n^2}\right) - \frac{Ze^2}{r}\psi_T = E_T\psi_T. \quad (10\text{--}4)$$

We can now separate Eq. (10–4) into two simpler equations if we proceed as follows.

1. Define a new set of coordinates x_M, y_M, z_M as the coordinates of the center of mass of the system, so that

$$x_M = \frac{m_e x_e + m_n x_n}{m_e + m_n}, \qquad y_M = \frac{m_e y_e + m_n y_n}{m_e + m_n}, \qquad z_M = \frac{m_e z_e + m_n z_n}{m_e + m_n}.$$

$$(10\text{--}5)$$

2. Define a second new set of coordinates x, y, z as the coordinates of the electron, *assuming the nucleus at the origin*, so that

$$x = x_e - x_n, \qquad y = y_e - y_n, \qquad z = z_e - z_n. \quad (10\text{--}6)$$

3. Assume that the total wave function ψ_T may be expressed as the product of two contributing wave functions, one dependent only on the coordinates of the center of mass, and the other dependent only on the coordinates of the electron relative to the nucleus, that is,

$$\psi_T = \psi_t(x_M, y_M, z_M)\psi(x, y, z). \quad (10\text{--}7)$$

Incorporation of Eqs. (10–5), (10–6), and (10–7) into Eq. (10–4), with appropriate algebraic manipulation, yields

$$-\frac{\hbar^2}{2\psi_t(m_e + m_n)}\left(\frac{\partial^2\psi_t}{\partial x_M^2} + \frac{\partial^2\psi_t}{\partial y_M^2} + \frac{\partial^2\psi_t}{\partial z_M^2}\right)$$

$$-\frac{\hbar^2}{2\psi}\left(\frac{1}{m_e} + \frac{1}{m_n}\right)\left(\frac{\partial^2\psi}{\partial x^2} + \frac{\partial^2\psi}{\partial y^2} + \frac{\partial^2\psi}{\partial z^2}\right) - \frac{Ze^2}{r} = E_T, \quad (10\text{–}8)$$

where $r = \sqrt{x^2 + y^2 + z^2}$. Since the second and third terms in Eq. (9–8) are independent of x_M, y_M, and z_M, and since the sum of all three terms on the left is equal to a constant, it follows that the first term on the left is itself a constant, which we'll call E_t, so that

$$\frac{-\hbar^2}{2\psi_t(m_e + m_n)}\left(\frac{\partial^2\psi_t}{\partial x_M^2} + \frac{\partial^2\psi_t}{\partial y_M^2} + \frac{\partial^2\psi_t}{\partial z_M^2}\right) = E_t. \quad (10\text{–}9)$$

Since $m_e + m_n = M$, the mass of the *atom*, we may rearrange Eq. (10–9) to

$$\frac{-\hbar^2}{2M}\left(\frac{\partial^2\psi_t}{\partial x_M^2} + \frac{\partial^2\psi_t}{\partial y_M^2} + \frac{\partial^2\psi_t}{\partial z_M^2}\right) = E_t\psi_t. \quad (10\text{–}10)$$

By similar argument, the sum of the second and third terms in Eq. (10–8) must be a constant, which we'll call E, so that

$$\frac{-\hbar^2}{2\psi}\left(\frac{1}{m_e} + \frac{1}{m_n}\right)\left(\frac{\partial^2\psi}{\partial x^2} + \frac{\partial^2\psi}{\partial y^2} + \frac{\partial^2\psi}{\partial z^2}\right) - \frac{Ze^2}{r} = E, \quad (10\text{–}11)$$

where $E = E_T - E_t$. But

$$1/m_e + 1/m_n = 1/\mu,$$

where μ is the reduced mass of the atom, so that we may rearrange Eq. (10–11) to

$$\frac{-\hbar^2}{2\mu}\left(\frac{\partial^2\psi}{\partial x^2} + \frac{\partial^2\psi}{\partial y^2} + \frac{\partial^2\psi}{\partial z^2}\right) - \frac{Ze^2\psi}{r} = E\psi. \quad (10\text{–}12)$$

But, note that Eq. (10–10) is simply the Schrödinger equation for a free particle of mass M, where E_t is the *translational kinetic energy* associated with the free movement of the center of mass of the atom through space (compare with Eq. 8–27). Since we have already solved this problem it will not interest us here.

On the other hand, Eq. (10–12) is exactly the form of Schrödinger equation we would write for the motion of a fictitious electron of assigned mass μ moving in the coulombic potential energy field of a positively charged, *stationary* nucleus, where E is the energy of relative motion of the electron with respect to the nucleus. *In the remainder of our treatment of the hydrogen atom and hydrogen-like ions, we'll ignore the translational energy E_t and will be concerned only with E, the energy of relative motion.* That is, we'll view the atom from a frame of reference in which the center of mass is at rest.

Just as it was apparent in our previous treatment of the corrected Bohr model for the hydrogen atom (Section 3–4), it is once more apparent from Eq. (10–12) that we may mathematically account for the contribution of nuclear motion to the energy of relative motion of the atom if we assign to the electron the fictitious reduced mass $\mu = m_e m_n / (m_e + m_n)$, rather than the mass m_e of the electron alone. Thus, in our present model, which is formally identical in terms of energy to a model in which both particles revolve about a common center of mass, we'll imagine the single electron, of *assigned* reduced mass μ and charge $-e$, to move at a distance r about a stationary nucleus of charge $+ Ze$. This simple model is shown in Fig. 10–2.

The problem which now remains is that of solving Eq. (10 12), which is an eigenvalue equation for a conservative system, and which leads to the determination of a series of functions $\psi(x, y, z)$, each related to a particular stationary-state energy of relative motion for the hydrogen atom. Once each of the wave functions $\psi(x, y, z)$ is known, we'll immediately be able to write $\psi^* \psi$, which is the probability density distribution for the electron in the state characterized by the function ψ. That is, the solution of Eq. (10–12) will tell us where the electron *most probably exists* relative to the nucleus and what the allowed eigenvalue energies of relative motion are. Furthermore, through application of Postulates III and V we may calculate either exact or average values of each observable property of the atom. Once we have written the Schrödinger equation, then, we have *in principle* stated the complete problem. Unfortunately, the most frustrating part of quantum mechanics lies in the

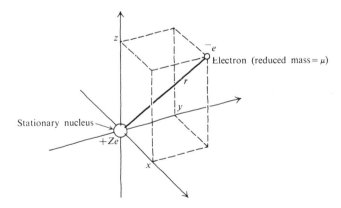

FIG. 10–2   A model for the relative motion of the electron and the nucleus in the hydrogen atom and hydrogen-like ions.

mathematical efforts required to solve equations such as Eq. (10–12). In fact, direct solutions have not been obtained for atoms more complex than hydrogen. We are fortunate, however, in being able to solve Eq. (10–12) directly, since we are able to separate the variables *in this specific case*.

10–2 SEPARATION OF VARIABLES IN THE EQUATION OF RELATIVE MOTION

Because of the spherical symmetry of the potential-energy field (V depends only on r), it is convenient to transform Eq. (10–12) into an equation involving spherical polar coordinates (r, θ, ϕ) rather than rectangular coordinates. The relationship between the two coordinate systems is shown in Fig. 10–3. Appropriate substitution of the transformation equations (given in Fig. 10–3) into Eq. (10–12) is a lengthy process* which finally leads to the spherical polar form of the Schrödinger equation:

$$\frac{-\hbar^2}{2\mu r^2}\left[\frac{\partial}{\partial r}\left(r^2 \frac{\partial \psi}{\partial r}\right) + \frac{1}{\sin\theta}\frac{\partial}{\partial\theta}\left(\sin\theta\frac{\partial\psi}{\partial\theta}\right) + \frac{1}{\sin^2\alpha}\frac{\partial^2\psi}{\partial\phi^2}\right] - \frac{Ze^2}{r}\psi = E\psi,$$

(10–13)

where ψ is now a function of r, θ, and ϕ.

Recall that in the case of the single particle in the three-dimensional box we were able to separate successfully the three-dimensional Schrödinger amplitude equation into three independent and soluble one-dimensional eigenvalue equations by writing the wave function $\psi(x, y, z)$ as the product of three wave functions, each dependent on only one of the coordinates (Eq. 8–28). We'll now solve Eq. (10–13) using the same technique. That is, assume that $\psi(r, \theta, \phi)$ may be expressed as the product of three independent functions, each dependent on only one of the coordinate positions,

$$\psi(r, \theta, \phi) = R(r)\Theta(\theta)\Phi(\phi).$$

(10–14)

The function $R(r)$ depends only on r, the function $\Theta(\theta)$ depends *only* on θ, and the function $\Phi(\phi)$ depends *only* on ϕ. Let's substitute Eq. (10–14) into Eq. (10–13) and determine whether or not a suitable separation results. Substitution of Eq. (10–14) with simplified notation into Eq. (10–13) yields

$$\frac{-\hbar^2}{2\mu r^2}\left[\Theta\Phi\frac{d}{dr}\left(r^2\frac{dR}{dr}\right) + R\Phi\frac{1}{\sin\theta}\frac{d}{d\theta}\left(\sin\theta\frac{d\Theta}{d\theta}\right)\right.$$

$$\left. + R\Theta\frac{1}{\sin^2\theta}\frac{d^2\Phi}{d\phi^2}\right] - \frac{Ze^2}{r}R\Theta\Phi = ER\Theta\Phi. \quad (10\text{–}15)$$

* See for example C. W. Sherwin, *Introduction to Quantum Mechanics*, Holt, Rinehart and Winston, New York, 1959, Appendix V.

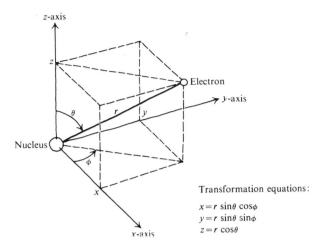

Transformation equations:

$$x = r \sin\theta \cos\phi$$
$$y = r \sin\theta \sin\phi$$
$$z = r \cos\theta$$

Fig. 10–3.   The relationships among the spherical polar coordinates (r, θ, ϕ) and the rectangular coordinates (x, y, z) of the electron relative to the nucleus.

Note that each of the derivatives in Eq. (10–15) is now a *total* derivative rather than a partial derivative, since in each term the operand contains only that variable with respect to which the derivative is taken. If we multiply the entire equation by $\sin^2\theta / R\Theta\Phi$ and rearrange, we may convert Eq. (10–15) to

$$\frac{\sin^2\theta}{R}\frac{d}{dr}\left(r^2\frac{dR}{dr}\right) + \frac{\sin\theta}{\Theta}\frac{d}{d\theta}\left(\sin\theta\,\frac{d\Theta}{d\theta}\right)$$

$$+ \frac{1}{\Phi}\frac{d^2\Phi}{d\phi^2} + \frac{2\mu r^2 \sin^2\theta}{\hbar^2}\left(E + \frac{Ze^2}{r}\right) = 0. \quad (10\text{--}16)$$

But note that the third term of Eq. (10–16) is the only term involving ϕ. Since this term is independent of variations in r and θ, and the other terms are independent of variations in ϕ,

$$\frac{1}{\Phi}\cdot\frac{d^2\Phi}{d\phi^2}$$

must be a constant, and the sum of the other terms in the equation must be the negative of this same constant. Calling the constant A, we may write

$$\frac{d^2\Phi}{d\phi^2} = A\Phi. \quad (10\text{--}17)$$

Equation (10–17) is an eigenvalue equation whose satisfactory solution will yield Φ as eigenfunctions and A as eigenvalues. Substitution of the constant A

back into Eq. (10–16) followed by division by $\sin^2 \theta$ yields

$$\frac{1}{R} \cdot \frac{d}{dr}\left(r^2 \frac{dR}{dr}\right) + \frac{1}{\Theta \sin \theta} \cdot \frac{d}{d\theta}\left(\sin \theta \frac{d\Theta}{d\theta}\right) + \frac{A}{\sin^2 \theta} + \frac{2\mu r^2}{\hbar^2}\left(E + \frac{Ze^2}{r}\right) = 0.$$

(10–18)

But the first and fourth terms of Eq. (10–18) depend only on r and are independent of θ, and the second and third terms depend only on θ and are independent of r. It then follows that, since the sum of all four terms is a constant (zero), the sum of the set of terms involving r and the sum of the set of terms involving θ must each independently be equal to constants. Furthermore, the value of the constant for the r terms must be the negative of the value of the constant for the θ terms, since the sum of the two constants must be equal to zero. Thus, arbitrarily letting the constant for the θ terms equal $(-B)$, we may write

$$\frac{1}{\Theta \sin \theta}\frac{d}{d\theta}\left(\sin \theta \frac{d\Theta}{d\theta}\right) + \frac{A}{\sin^2 \theta} = -B$$

(10–19)

or

$$\frac{1}{\sin \theta}\frac{d}{d\theta}\left(\sin \theta \frac{d\Theta}{d\theta}\right) + \frac{A\Theta}{\sin^2 \theta} = -B\Theta.$$

(10–20)

Equation (10–20) is an eigenvalue equation whose solution will yield Θ as eigenfunctions and $-B$ as eigenvalues.

In addition, since the sum of the terms involving r in Eq. (10–18) must be equal to B, we may write

$$\frac{1}{R}\frac{d}{dr}\left(r^2 \frac{dR}{dr}\right) + \frac{2\mu r^2}{\hbar^2}\left(E + \frac{Ze^2}{r}\right) = B.$$

(10–21)

Multiplication by $R\hbar^2/2\mu r^2$ yields

$$\frac{\hbar^2}{2\mu r^2}\frac{d}{dr}\left(r^2 \frac{dR}{dr}\right) + ER + \left(\frac{Ze^2}{r}\right)R = \left(\frac{B\hbar^2}{2\mu r^2}\right)R,$$

(10–22)

and rearrangement of Eq. (10–22) gives

$$\frac{\hbar^2}{2\mu r^2}\frac{d}{dr}\left(r^2 \frac{dR}{dr}\right) + \left(\frac{Ze^2}{r}\right)R - \left(\frac{B\hbar^2}{2\mu r^2}\right)R = -ER,$$

(10–23)

which is an eigenvalue equation whose satisfactory solution yields R as eigenfunctions and $-E$ as eigenvalues.

Thus our attempt to separate the Schrödinger equation involving r, θ, and ϕ (Eq. 10–13) into three separate eigenvalue equations, each involving only one of the coordinates, has been successful. In fact, *such a separation is always successful when V is a function of r only.* Solution of Eq. (10–17) will yield the allowed forms for the function $\Phi(\phi)$, solution of Eq. (10–20) will in turn yield the allowed forms for the function $\Theta(\theta)$, and final solution of Eq. (10–23) will yield the allowed forms for the function $R(r)$.

In mathematically completing the solutions, we'll first solve the Φ eigenvalue equation, Eq. (10–17), which is now rather straightforward, in order to obtain the allowed values for A, which, in turn, must be used in the solution of the Θ eigenvalue equation, Eq. (10–20). The values of B obtained from the solution of Eq. (10–20) must then be used in the solution of the R eigenvalue equation, Eq. (10–23).

10–3 SOLUTION OF THE Φ EQUATION

Let's first solve Eq. (10–17) in order to determine Φ and A. The form of the Φ equation,

$$\frac{d^2\Phi}{d\phi^2} = A\Phi, \qquad (10\text{–}17')$$

by now should be familiar. An acceptable eigenfunction, which may be verified easily by substitution, is

$$\Phi(\phi) = \alpha e^{A^{1/2}\phi}, \qquad (10\text{–}24)$$

where α is a constant. In order that Φ be continuous and single-valued, it is necessary that Φ have the same value at ϕ as at $(\phi + 2\pi)$ since these are identical angular positions. Otherwise, two different values of Φ at the same position would lead to two different values of $\psi = \Phi\Theta R$ and hence to two different values of $\psi^*\psi$, the probability density, at the same position, which is not physically acceptable. Let's then set Φ at $\phi = 0$ equal to Φ at $\phi = 2\pi$:

$$\Phi(0) = \Phi(2\pi),$$

or

$$\alpha e^{A^{1/2}(0)} = \alpha e^{A^{1/2}(2\pi)}, \qquad (10\text{–}25)$$

which apparently cannot be so if $A^{1/2}$ is a real number, since $2\pi \neq 0$. Equation (10–25) may be true only if Φ is a periodic function involving an *imaginary* exponent, that is, if $A^{1/2}$ is imaginary. Therefore, let's arbitrarily write

$$A^{1/2} = im, \qquad \text{or} \qquad A = -m^2,$$

where m is a constant. Substitution for $A^{1/2}$ in Eq. (10–25) yields

$$e^{im(0)} = e^{im(2\pi)},$$

from which, through Euler's formula, we obtain

$$1 = \cos(m2\pi) + i\sin(m2\pi),$$

which requires that

$$m = 0, \pm 1, \pm 2, \pm 3, \ldots, \pm \infty.$$

There are, therefore, an infinite number of discrete solutions to the Φ-equation, each of the form

$$\Phi(\phi) = \alpha e^{im\phi}, \tag{10-26}$$

where m is restricted to integers as given above.

Now let's determine the constant α through normalization. We may ensure that the total wave function $\psi(\phi, \theta, r)$ is normalized if we normalize $\Phi(\phi)$, $\Theta(\theta)$, and $R(r)$ individually. That is, since in spherical polar coordinates $d\tau = r^2 \sin \theta \, dr \, d\theta \, d\phi$, the normalization condition for $\psi = R\Theta\Phi$ is

$$\int_0^\infty \int_0^\pi \int_0^{2\pi} R^* R \Theta^* \Theta \Phi^* \Phi r^2 \sin \theta \, dr \, d\theta \, d\phi = 1$$

or

$$\int_0^\infty R^* R r^2 \, dr \cdot \int_0^\pi \Theta^* \Theta \sin \theta \, d\theta \cdot \int_0^{2\pi} \Phi^* \Phi \, d\phi = 1 \tag{10-27}$$

which is guaranteed if each of the integrals is unity. Then, for normalized Φ functions, the constant α may be determined easily from the condition

$$\int_0^{2\pi} \Phi^*(\phi)\Phi(\phi) \, d\phi = 1.$$

Substitution of Eq. (10–26) into the above equation yields

$$\alpha^2 \int_0^{2\pi} e^{-im\phi} e^{im\phi} \, d\phi = \alpha^2 \int_0^{2\pi} d\phi = \alpha^2 2\pi = 1,$$

from which

$$\alpha = 1/\sqrt{2\pi},$$

so that Eq. (10–27) may be written, for *normalized* Φ functions, as

$$\Phi_m(\phi) = \frac{1}{\sqrt{2\pi}} e^{im\phi}. \tag{10-28}$$

The added subscript m reminds us that the form of the Φ function depends on the value of the integer m. In effect, m serves to quantize the eigenvalues A in Eq. (10–17) such that $A = -m^2$. The integer m is called the *magnetic quantum number*, for reasons to be discussed later. Several specific normalized Φ_m functions for selected values of m are given in Table 10–1.

TABLE 10–1

Selected Normalized $\Phi_m(\phi)$
Functions

m	$\Phi_m(\phi)$
0	$1/\sqrt{2\pi}$
+1	$(1/\sqrt{2\pi})e^{i\phi}$
−1	$(1/\sqrt{2\pi})e^{-i\phi}$
+2	$(1/\sqrt{2\pi})e^{i2\phi}$
−2	$(1/\sqrt{2\pi})e^{-i2\phi}$

10–4 SOLUTION OF THE Θ EQUATION

The detailed solution of the Θ equation, Eq. (10–20), is rather lengthy and mathematically complicated so that we'll at times only indicate the general steps in the procedure and refer to more complete treatments. We begin by substituting $-m^2$ for A in Eq. (10–20) and rearranging, which yields

$$\frac{1}{\sin\theta}\frac{d}{d\theta}\left(\sin\theta\frac{d\Theta}{d\theta}\right) - \frac{m^2\Theta}{\sin^2\theta} + B\Theta = 0 \qquad (10\text{–}29)$$

In order to write the equation in what will turn out to be a more suitable form, it is convenient to define the new independent variable z (*which is not the cartesian coordinate*) as

$$z \equiv \cos\theta$$

so that

$$1 - z^2 = 1 - \cos^2\theta = \sin^2\theta. \qquad (10\text{–}30)$$

Let's also replace $\Theta(\theta)$ with an equivalent function of z, $P(z)$, defined so that

$$P(z) \equiv \Theta(\theta). \qquad (10\text{–}31)$$

Then, using the chain rule,

$$\frac{d\Theta}{d\theta} = \frac{dP}{d\theta} = \frac{dP}{dz}\cdot\frac{dz}{d\theta} = \frac{dP}{dz}\frac{d(\cos\theta)}{d\theta} = -\sin\theta\frac{dP}{dz}, \qquad (10\text{–}32)$$

from which, since $\Theta = P$, we may also write the operator

$$\frac{d}{d\theta} = -\sin\theta\frac{d}{dz}. \qquad (10\text{–}33)$$

Substitution of Eqs. (10–31), (10–32), and (10–33) into Eq. (10–29) yields

$$\frac{1}{\sin\theta}\left(-\sin\theta\frac{d}{dz}\right)\left[\sin\theta\left(-\sin\theta\frac{dP}{dz}\right)\right] - \frac{m^2P}{\sin^2\theta} + BP = 0,$$

or

$$\frac{d}{dz}\left(\sin^2\theta\,\frac{dP}{dz}\right) + \left(B - \frac{m^2}{\sin^2\theta}\right)P = 0. \tag{10–34}$$

Further substitution of Eq. (10–30) into Eq. (10–34) yields

$$\frac{d}{dz}\left[(1-z^2)\frac{dP}{dz}\right] + \left(B - \frac{m^2}{1-z^2}\right)P = 0,$$

and differentiation of the term in brackets gives

$$(1-z^2)\frac{d^2P}{dz^2} - 2z\frac{dP}{dz} + \left(B - \frac{m^2}{1-z^2}\right)P = 0. \tag{10–35}$$

If we now arbitrarily define the constant B in terms of another constant l as

$$B = l(l+1)$$

we can rewrite Eq. (10–35) as

$$(1-z^2)\frac{d^2P}{dz^2} - 2z\frac{dP}{dz} + \left[l(l+1) - \frac{m^2}{1-z^2}\right]P = 0, \tag{10–36}$$

Eq. (10–36) is a well known differential equation called the *associated Legendre equation**, whose solutions $P(z)$ are in the forms of polynomial series each of which must be restricted to a finite number of terms if $P(z)$ is to remain everywhere finite and accordingly well-behaved. This restriction is similar to that previously imposed on the Hermite polynomial series in our treatment of the simple harmonic oscillator in Chapter 9, and results in the requirement that l be restricted to non-negative integer values, that is, $l = 0, 1, 2, 3, \ldots, \infty$. The solutions, P, are given as the *associated Legendre polynomials* $P_l^{|m|}$ of *degree* l and *order* $|m|$, as follows:

$$P = P_l^{|m|}(z) = (1-z^2)^{|m|/2}\frac{d^{|m|}}{dz^{|m|}}P_l(z). \tag{10–37}$$

The subscript l and the superscript $|m|$ in $P_l^{|m|}(z)$ are meant to signify that the forms of the associated Legendre polynomials, which are functions of z, depend on the value of the positive integer l, which is called the *azimuthal quantum number*, and on the absolute value $|m|$ of the integer m, which we have already defined in the previous section. Note that $P_l^{|m|}(z)$ is defined in terms of a simpler function, $P_l(z)$, which is called the *Legendre polynomial*. The Legendre polynomials are also functions of z, whose forms depend on the value of the positive integer l and are given by

$$P_l(z) = \frac{1}{2^l(l!)}\frac{d^l}{dz^l}(z^2-1)^l. \tag{10–38}$$

---

* For a detailed discussion of Legendre equations and polynomials, see H. Margenau and G. M. Murphy, *The Mathematics of Physics and Chemistry*, 2nd Ed., D. Van Nostrand Co., New York, 1956, Chap. 3.

Note also that even though the value of the integer m was not restricted to any maximum value in the previous solution of the Φ equation, *the present solution of the Θ equation requires that $|m| \leqslant l$ if the functions are not to vanish* (see Problem 10–3). That is,

$$l = 0, 1, 2, 3, \ldots, \infty$$

$$|m| = 0, 1, 2, \ldots, \pm l, \quad \text{or,} \quad m = 0, \pm 1, \pm 2, \ldots, \pm l.$$

Some of the Legendre polynomials $P_l(z)$ for selected values of l are

$$l = 0, \qquad P_0(z) = 1,$$

$$l = 1, \qquad P_1(z) = z,$$

$$l = 2, \qquad P_2(z) = \tfrac{3}{2}z^2 - \tfrac{1}{2},$$

$$l = 3, \qquad P_3(z) = \tfrac{5}{2}z^3 - \tfrac{3}{2}z.$$

The $\Theta(\theta)$ eigenfunctions are identical to the solutions of the associated Legendre equation, that is, identical to the associated Legendre polynomials as given by Eq. (10–37). However, in order to normalize the functions, we must further introduce the constant β such that

$$\Theta(\theta)_{\text{normalized}} = \beta P_l^{|m|}(z). \tag{10–39}$$

The value of β must be determined from the normalization condition on ψ given by Eq. (10–27), that is

$$\int_0^\pi \Theta^*\Theta \sin\theta \; d\theta = 1, \text{ or,}$$

$$\beta^2 \int_{+1}^{-1} P_l^{|m|} P_l^{|m|} \, (-d\cos\theta) = \beta^2 \int_{-1}^{+1} P_l^{|m|} P_l^{|m|} \, (d\cos\theta) = 1.$$

But $\cos\theta = z$, so that

$$\beta^2 \int_{-1}^{+1} P_l^{|m|} P_l^{|m|} \, dz = 1. \tag{10–40}$$

By appropriate substitution and integration it may be shown* that

$$\beta = \left[\frac{(2l+1)}{2} \frac{(l-|m|)!}{(l+|m|)!} \right]^{1/2}, \tag{10–41}$$

so that the normalized Θ functions may be written, by substitution of Eq. (10–40) into Eq. (10–39), as

$$\Theta_{lm}(\theta) = \left[\frac{(2l+1)}{2} \frac{(l-|m|)!}{(l+|m|)!} \right]^{1/2} P_l^{|m|}(z). \tag{10–42}$$

* See for example, L. Pauling and E. B. Wilson, *Introduction to Quantum Mechanics*, McGraw-Hill, New York, 1935, App. VI.

TABLE 10–2

Selected Normalized $\Theta_{lm}(\theta)$ Functions

l	m	$\Theta_{lm}(\theta)$
0	0	$\sqrt{2}/2$
1	0	$(\sqrt{6}/2)\cos\theta$
1	±1	$(\sqrt{3}/2)\sin\theta$
2	0	$(\sqrt{10}/4)(3\cos^2\theta - 1)$
2	±1	$(\sqrt{15}/2)\sin\theta\cos\theta$
2	±2	$(\sqrt{15}/4)\sin^2\theta$

We have now added the subscripts l and m to Θ to indicate that the Θ function depends on the values of the quantum numbers l and m.

Although Equations (10–37), (10–38), and (10–42) seem to be rather formidable, the Θ functions reduce to relatively simple forms on substitution of specific values of the quantum numbers l and m. Let's, for example, determine $\Theta_{21}(\theta)$, for which $l = 2$ and $m = +1$. From Eq. (10–38),

$$P_2(z) = \frac{3z^2}{2} - \frac{1}{2},$$

which is substituted in Eq. (10–37) to give

$$P_2^1(z) = (1 - z^2)^{1/2}3z = 3(1 - \cos^2\theta)^{1/2}\cos\theta$$

$$= 3\sin\theta\cos\theta,$$

which, in turn, is substituted into Eq. (10–42) to yield the final normalized function:

$$\Theta_{21}(\theta) = \frac{\sqrt{15}}{2}\sin\theta\cos\theta.$$

Several normalized functions for selected values of l and m are given in Table 10–2.

10–5 SOLUTION OF THE *R* EQUATION

In solving Eq. (10–23) for the $R(r)$ functions, or *radial functions*, we'll use a method which is similar to that we have used for the Θ equation. Through a lengthy procedure involving a transformation of the independent variable and examination of the forms of solution as the variable approaches infinity and then zero, we'll eventually manipulate Eq. (10–23) into the form of an equation whose solutions are already known.

Let's begin by substituting $l(l + 1)$ for B in Eq. (10–23) and then rearranging to

$$\frac{1}{r^2}\frac{d}{dr}\left(r^2\frac{dR}{dr}\right) - \frac{l(l + 1)R}{r^2} + \frac{2\mu ER}{\hbar^2} + \frac{2\mu Ze^2R}{\hbar^2 r} = 0. \quad (10\text{–}43)$$

Recalling that E is constant and *negative* for the *bound* electron, we now define two new constants α and λ, related to the energy E, such that

$$\alpha^2 = -2\mu E/\hbar^2, \qquad (10\text{–}44)$$

and

$$\lambda = \mu Z e^2/\hbar^2\alpha. \qquad (10\text{–}45)$$

Substitution of Eqs. (10–44) and (10–45) into Eq. (10–43) yields

$$\frac{1}{r^2}\frac{d}{dr}\left(r^2\frac{dR}{dr}\right) - \frac{l(l+1)R}{r^2} - \alpha^2 R + \frac{2\alpha\lambda R}{r} = 0. \qquad (10\text{–}46)$$

In order eventually to change the equation to a more familiar form, we find it convenient now to transform to the new independent variable ρ where

$$\rho \equiv 2\alpha r, \qquad (10\text{–}47)$$

and the new function $S(\rho)$, where

$$S(\rho) \equiv R(r), \qquad (10\text{–}48)$$

which are to be substituted into Eq. (10–46). In order to transform the differential terms, we again use the chain rule,

$$\frac{dR(r)}{dr} = \frac{dS(\rho)}{dr} = \frac{dS(\rho)}{d\rho}\frac{d\rho}{dr} = 2\alpha\frac{dS(\rho)}{d\rho}, \qquad (10\text{–}49)$$

and from Eq. (10–49), since $R(r) = S(\rho)$,

$$\frac{d}{dr} = 2\alpha\frac{d}{d\rho}. \qquad (10\text{–}50)$$

Substitution of Eqs. (10–47), (10–48), (10–49), and (10–50) into Eq. (10–46) yields

$$\left(\frac{4\alpha^2}{\rho^2}\right)2\alpha\frac{d}{d\rho}\left[\left(\frac{\rho^2}{4\alpha^2}\right)2\alpha\frac{dS(\rho)}{d\rho}\right] - \frac{4\alpha^2}{\rho^2}l(l+1)S(\rho) - \alpha^2 S(\rho) + \left(\frac{2\alpha}{\rho}\right)2\alpha\lambda S(\rho) = 0,$$

and division by $4\alpha^2$ gives

$$\frac{1}{\rho^2}\frac{d}{d\rho}\left[\rho^2\frac{dS(\rho)}{d\rho}\right] + \left[\frac{-l(l+1)}{\rho^2} - \frac{1}{4} + \frac{\lambda}{\rho}\right]S(\rho) = 0, \qquad (10\text{–}51)$$

where ρ, which is directly proportional to r, may assume positive values between zero and infinity. Completion of the differentiation indicated in Eq. (10–51) yields

$$\frac{d^2S(\rho)}{d\rho^2} + \frac{2}{\rho}\frac{dS(\rho)}{d\rho} + \left[\frac{-l(l+1)}{\rho^2} - \frac{1}{4} + \frac{\lambda}{\rho}\right]S(\rho) = 0. \qquad (10\text{–}52)$$

In our attempt to find a solution for Eq. (10–52) let's first look at the form of the equation as ρ approaches infinity. As $\rho \to \infty$, Eq. (10–52) becomes

$$\frac{d^2S(\rho)_\infty}{d\rho^2} = \frac{S(\rho)_\infty}{4}, \qquad (\rho \to \infty) \quad (10\text{–}53)$$

which is an *asymptotic equation* which may be used to obtain the *asymptotic solution* $S(\rho)_\infty$. The solution of Eq. (10–53) may be verified by substitution to be

$$S(\rho)_\infty = c e^{\pm\rho/2}, \qquad (\rho \to \infty) \quad (10\text{–}54)$$

where c is an arbitrary constant in the asymptotic solution, but may otherwise *generally* be a function of ρ, $F(\rho)$, such that $F(\rho) \to c$ as $\rho \to \infty$. That is, in general

$$S(\rho) = e^{\pm\rho/2}F(\rho) \qquad (0 \leqslant \rho \leqslant \infty)$$

It can be shown* that it makes no difference whether we now choose either the positive exponent or the negative exponent in the above equation for $S(\rho)$. Each leads to the same final solution. Let's then select

$$S(\rho) = e^{-\rho/2}F(\rho), \qquad (10\text{–}55)$$

where $F(\rho)$ is presumably a simpler function than $S(\rho)$. Now let's transform Eq. (10–52) into an equation involving $F(\rho)$ rather than $S(\rho)$. Differentiating Eq. (10–55), we obtain

$$\frac{dS(\rho)}{d\rho} = e^{-\rho/2}\frac{dF(\rho)}{d\rho} - \tfrac{1}{2}e^{-\rho/2}F(\rho), \qquad (10\text{–}56)$$

and further differentiation and collection of terms yields

$$\frac{d^2S(\rho)}{d\rho^2} = e^{-\rho/2}\left[\frac{d^2F(\rho)}{d\rho^2} - \frac{dF(\rho)}{d\rho} + \frac{F(\rho)}{4}\right]. \qquad (10\text{–}57)$$

Substitution of Eqs. (10–55), (10–56), and (10–57) into Eq. (10–52) yields

$$e^{-\rho/2}\left[\frac{d^2F(\rho)}{d\rho^2} - \frac{dF(\rho)}{d\rho} + \frac{F(\rho)}{4}\right] + \frac{2}{\rho}e^{-\rho/2}\left[\frac{dF(\rho)}{d\rho} - \frac{F(\rho)}{2}\right]$$

$$+ \left[\frac{-l(l+1)}{\rho^2} - \frac{1}{4} + \frac{\lambda}{\rho}\right]e^{-\rho/2}F(\rho) = 0,$$

and division by $e^{-\rho/2}$ followed by rearrangement leads to

$$\frac{d^2F(\rho)}{d\rho^2} + \left(\frac{2}{\rho} - 1\right)\frac{dF(\rho)}{d\rho} + \left[\frac{\lambda}{\rho} - \frac{l(l+1)}{\rho^2} - \frac{1}{\rho}\right]F(\rho) = 0, \quad (10\text{–}58)$$

where ρ may vary from zero to infinity.

---

* B. F. Gray, G. Hunter, and H. O. Pritchard, *J. Chem. Phys.*, **38**, 2790 (1963).

In order to examine the solution to Eq. (10–58) as $\rho \to 0$, we now define still further another new function $L(\rho)$, such that

$$F(\rho) = \rho^s L(\rho), \tag{10–59}$$

where the *constant* s is to be determined from the nature of the equation at $\rho = 0$. Differentiation of Eq. (10–59) yields

$$\frac{dF(\rho)}{d\rho} = \rho^s \frac{dL(\rho)}{d\rho} + s\rho^{s-1}L(\rho), \tag{10–60}$$

and further differentiation gives

$$\frac{d^2F(\rho)}{d\rho^2} = \rho^s \frac{d^2L(\rho)}{d\rho^2} + 2s\rho^{s-1}\frac{dL(\rho)}{d\rho} + s(s-1)\rho^{s-2}L(\rho). \tag{10–61}$$

Substitution of Eqs. (10–59), (10–60), and (10–61) into Eq. (10–58), followed by collection of terms, yields

$$\rho^s \frac{d^2L(\rho)}{d\rho^2} + (2s\rho^{s-1} + 2\rho^{s-1} - \rho^s)\frac{dL(\rho)}{d\rho}$$

$$+ [s(s-1)\rho^{s-2} + 2s\rho^{s-2} - s\rho^{s-1} + \lambda\rho^{s-1} - l(l+1)\rho^{s-2} - \rho^{s-1}]L(\rho) = 0.$$

Division by ρ^{s-2} with further collection of terms yields

$$\rho^2 \frac{d^2L(\rho)}{d\rho^2} + [(2s + 2 - \rho)\rho]\frac{dL(\rho)}{d\rho}$$

$$+ [s(s+1) - l(l+1) + (\lambda - s - 1)\rho]L(\rho) = 0. \tag{10–62}$$

Note that Eq. (10–62) must be satisfied by *all* values of ρ, including $\rho = 0$. In order to determine the value of the constant s let's then substitute $\rho = 0$ into Eq. (10–62), which yields

$$s(s+1) - l(l+1) = 0,$$

which, in turn, is satisfied by the two roots

$$s = l \quad \text{or} \quad s = -(l+1).$$

But the negative solution, $s = -(l+1)$, must be rejected since, according to Eq. (10–59), a negative value of s would require $F(\rho)$ to approach infinity (to diverge) as ρ approaches zero and would consequently cause divergence of $R(r)$ and $\psi(\phi, \theta, r)$ as $\rho \to 0$. Then, accepting the solution $s = l$, we may rewrite Eq. (10–59) as

$$F(\rho) = \rho^l L(\rho). \tag{10–63}$$

Let's now substitute l for s in Eq, (10–62):

$$\frac{\rho^2\, d^2L(\rho)}{d\rho^2} + [(2l + 2 - \rho)\rho]\frac{dL(\rho)}{d\rho}$$

$$+ [l(l + 1) - l(l + 1) + (\lambda - l - 1)\rho]L(\rho) = 0.$$

Division by ρ, followed by collection of terms, yields the differential equation

$$\frac{\rho\, d^2L(\rho)}{d\rho^2} + [2(l + 1) - \rho]\frac{dL(\rho)}{d\rho} + (\lambda - l - 1)L(\rho) = 0. \quad (10\text{–}64)$$

Before we proceed with the solution of Eq. (10–64), let's summarize what we have done so far:

1. In order to examine the asymptotic solution as $r \to \infty$, we have defined $R(r)$ in terms of a new function $S(\rho)$, such that

$$R(r) = S(\rho), \quad \text{where } \rho = 2\alpha r. \qquad (10\text{–}48')$$

2. The asymptotic solution as $\rho \to \infty$ led us to define $S(\rho)$ in terms of the simpler function $F(\rho)$, such that

$$S(\rho) = e^{-\rho/2}F(\rho). \qquad (10\text{–}55')$$

3. In order to learn more of the structure of the function, we further examined the form of solution as $\rho \to 0$, we then defined $F(\rho)$ in terms of a still simpler function, $L(\rho)$ as

$$F(\rho) = \rho^s L(\rho). \qquad (10\text{–}59')$$

4. The solution of the $L(\rho)$ equation at $\rho = 0$ required that $s = l$, so that

$$F(\rho) = \rho^l L(\rho). \qquad (10\text{–}63')$$

We may thus now write the original function $R(r)$ as

$$R(r) = S(\rho) = e^{-\rho/2}F(\rho) = \rho^l e^{-\rho/2}L(\rho). \qquad (10\text{–}65)$$

We can put Eq. (10–64) into a more familiar form if we now define the constants l and λ in terms of the two new constants j and k, where $j = 2l + 1$, and $k = \lambda + l$. Note that the constant j is required to be an odd positive integer since the solution of the $\Theta(\theta)$ equation requires that $l = 0, 1, 2, 3, \ldots, \infty$. Substitution of j and k into Eq. (10–64) yields

$$\frac{\rho^2 d^2L(\rho)}{d\rho^2} + (j + 1 - \rho)\frac{dL(\rho)}{d\rho} + (k - j)L(\rho) = 0 \qquad (10\text{–}66)$$

Eq. (10–66) is a well-known differential equation called the *associated*

Laguerre equation* whose solutions $L(\rho)$ are in the forms of polynomial series, each of which must be restricted to a finite number of terms if the function $L(\rho)$ is to be well-behaved and accordingly vanish as $\rho \to \infty$, rather than diverge as it would were it an infinite series. This boundary restriction mathematically requires that k be a positive integer, that is, $k = 1, 2, 3, \ldots, \infty$. The solutions of Eq. (10–66) are the *associated Laguerre poly-nomials*, of degree $k - j$ and order j:

$$L(\rho) = L_k^j(\rho) = \frac{d^j}{d\rho^j} L_k(\rho), \tag{10-67}$$

where $L_k(\rho)$, the *Laguerre polynomials* of degree k, are given by

$$L_k(\rho) = e^\rho \frac{d^k}{d\rho^k} \rho^k e^{-\rho}. \tag{10-68}$$

Some of the Laguerre polynomials obtained through substitution of selected k values in Eq. (10–68) are

$$k = 1, \qquad L_1(\rho) = 1 - \rho, \tag{10-69}$$

$$k = 2, \qquad L_2(\rho) = 2 - 4\rho + \rho^2, \tag{10-70}$$

$$k = 3, \qquad L_3(\rho) = 6 - 18\rho + 9\rho^2 - \rho^3. \tag{10-71}$$

Since the solution of the associated Laguerre equation demands that k be a positive integer, and since we have previously shown in the solution of the Θ equation that l must be zero or a positive integer, it follows that λ *must also be a positive integer*, since $\lambda = k - l$. We'll now redefine the integer λ as n, the *principle quantum number*, so that

$$k = n + l, \tag{10-72}$$

and the associated Laguerre polynomials which satisfy Eq. (10–66) may be written as

$$L(\rho) = L_k^j(\rho) = L_{n+l}^{2l+1}(\rho). \tag{10-73}$$

Substitution of Eq. (10–73) into Eq. (10–65) yields, for *normalized $R(r)$* functions,

$$R(r) = \gamma \rho^l e^{-\rho/2} L_{n+l}^{2l+1}(\rho), \tag{10-74}$$

where γ is introduced as a normalization constant (to be given shortly) and

---

* For a detailed discussion of Laguerre equations and polynomials, see H. Margenau and G. M. Murphy, *The Mathematics of Physics and Chemistry*, 2nd Ed., D. Van Nostrand, New York, 1956, pp. 77, 126–132.

where $\rho = 2\alpha r$. Note that from Eq. (10–45), where $\lambda = n$, we may write

$$\alpha = \mu Z e^2 / n \hbar^2 = Z / n a_0,$$

where

$$a_0 = \hbar^2 / \mu e^2. \tag{10–75}$$

The term a_0 has been defined previously as the radius of the first orbit in the Bohr hydrogen atom (Eq. 3–30). Then, we may write the variable ρ as

$$\rho = 2\alpha r = (2Z / n a_0) r. \tag{10–76}$$

Now let's examine the allowed values for the integer n which we have called the principal quantum number. If the associated Laguerre polynomial is not to vanish, it is necessary, according to Eq. (10–67), that

$$j \leqslant k$$

(see Problem 10–7), which on substitution of $j = 2l + 1$ and $k = n + l$ becomes

$$2l + 1 \leq n + l,$$

so that

$$l + 1 \leq n \quad \text{or} \quad l \leq n - 1.$$

Since l is allowed the values $0, 1, 2, 3, \ldots$, the principal quantum number n obviously may not have a value of zero but is restricted to the *positive* integers given by

$$n = 1, 2, 3, \ldots, \infty,$$

and since the value of l may not exceed $n - 1$, l is restricted to the values

$$l = 0, 1, 2, \ldots, (n - 1).$$

As the final step in determining a completely satisfactory R equation, we must normalize the expression given by Eq. (10–74). The value of γ must be determined from the normalization condition on ψ as given by Eq. (10–27). That is,

$$\int_0^\infty R^*(r) R(r) r^2 dr = 1.$$

Substitution of Eq. (10–74), where $r = \rho/2\alpha$ and $dr = d\rho/2\alpha$, gives

$$\frac{\gamma^2}{(2\alpha)^3} \int_0^\infty \rho^{2l} e^{-\rho} \left[L_{n+1}^{2l+1}(\rho) \right]^2 \rho^2 \, d\rho = 1,$$

and integration* yields

$$\frac{\gamma^3}{(2\alpha)^3} \frac{2n \left[(n + l)! \right]^3}{(n - l - 1)!} = 1,$$

---

* L. Pauling and E. B. Wilson, *Introduction to Quantum Mechanics*, McGraw-Hill, New York, 1935, App. VII.

from which, since $\alpha = Z/na_0$,

$$\gamma = \pm \left\{ \left(\frac{2Z}{na_0} \right)^3 \frac{(n-l-1)!}{2n[(n+l)!]} \right\}^{1/2}.$$

If we select the negative root for γ in order to make the radial function positive,

$$R_{nl}(r) = -\left\{ \left(\frac{2Z}{na_0} \right)^3 \frac{(n-l-1)!}{2n[(n+l)!]^3} \right\}^{1/2} \rho^l e^{-\rho/2} L_{n+l}^{2l+1}(\rho), \quad (10\text{–}77)$$

where

$$\rho = (2Z/na_0)r.$$

In order to evaluate a specific normalized radial function for selected values of n and l, use the following procedure:

1. Determine the form of the Laguerre polynomial, $L_{n+l}(\rho)$, by substituting the value of $n + l = k$ into Eq. (10–68) or by using directly the polynomials given in Equations (10–69), (10–70), or (10–71).

2. Determine the form of the associated Laguerre polynomial, $L_k^j(\rho) = L_{n+l}^{2l+1}(\rho)$, by substituting the resultant $L_{n+l}(\rho)$ from procedure 1 into Eq. (10–67).

3. Substitute $L_{n+l}^{2l+1}(\rho)$, which is a simple polynomial, into Eq. (10–77).

Several normalized radial functions $R_{nl}(r)$ are given in Table 10–3 for selected values of the quantum numbers n and l.

10–6 THE PRINCIPAL QUANTUM NUMBER n
AND ALLOWED ENERGY LEVELS

Before we combine the Φ_m function, the Θ_{lm} function, and the R_{nl} function into the general wave function ψ_{nlm}, let's examine the dependence of the energy E of the atom on the principal quantum number n. If we substitute n for λ in Eq. (10–45), we obtain

$$\alpha = \mu Z e^2 / n\hbar^2. \qquad (10\text{–}78)$$

Substitution of α from Eq. (10–78) into Eq. (10–44), followed by rearrangement, yields

$$E = -2\pi^2 \mu Z^2 e^4 / n^2 h^2, \qquad (10\text{–}79)$$

which is in *exact agreement* with the corrected Bohr equation, Eq. (3–28). According to Eq. (10–79), the energy of relative motion of the hydrogen atom or hydrogen-like ion is quantized and depends on the value of the principal quantum number n, but is independent of the values of the other two quantum numbers l and m. Since the principal quantum number n appears

TABLE 10–3

Selected Normalized $R_{nl}(r)$ Functions for
Hydrogen-Like Atoms

n	l	$R_{nl}(r)$
1	0	$2\left(\dfrac{Z}{a_0}\right)^{3/2} e^{-Zr/a_0}$
2	0	$\left(\dfrac{Z}{2a_0}\right)^{3/2}\left(2 - \dfrac{Zr}{a_0}\right) e^{-Zr/2a_0}$
2	1	$\dfrac{1}{\sqrt{3}}\left(\dfrac{Z}{2a_0}\right)^{3/2}\left(\dfrac{Zr}{a_0}\right) e^{-Zr/2a_0}$
3	0	$\dfrac{2}{3}\left(\dfrac{Z}{3a_0}\right)^{3/2}\left(3 - \dfrac{2Zr}{a_0} + \dfrac{2Z^2r^2}{9a_0^2}\right) e^{-Zr/3a_0}$
3	1	$\dfrac{2\sqrt{2}}{9}\left(\dfrac{Z}{3a_0}\right)^{3/2}\left(\dfrac{2Zr}{a_0} - \dfrac{Z^2r^2}{3a_0^2}\right) e^{-Zr/3a_0}$
3	2	$\dfrac{4}{27\sqrt{10}}\left(\dfrac{Z}{3a_0}\right)^{3/2}\left(\dfrac{Z^2r^2}{a_0^2}\right) e^{-Zr/3a_0}$

only in the radial wave function $R(r)$, and does not appear in $\Theta(\theta)$ nor in $\Phi(\phi)$, we conclude that the energy of relative motion of the atom is related generally to the distance of the electron from the nucleus and does not depend on any particular angular orientation. This is certainly not surprising in consideration of the spherically symmetrical form of the potential energy equation (Eq. 10–1).

The atom may have its lowest energy in quantum states for which $n = 1$. But when $n = 1$, l may have *only* the value zero, and m may also have *only* a value of zero. We may designate this first allowed quantum state as the (100) state. In general, we'll use the form (*nlm*) to designate quantum states. The energy in the lowest quantum state is

$$E_{(nlm)} = E_{(100)} = \frac{-2\pi^2\mu e^4}{h^2}$$

$$= \frac{-2\pi^2(9.104 \times 10^{-28}\ \text{g})(4.803 \times 10^{-10}\ \text{esu})^4}{(6.626 \times 10^{-27}\ \text{erg-sec})^2}$$

$$= -2.178 \times 10^{-11}\ \text{erg}. \tag{10–80}$$

The atom may have its next highest energy when $n = 2$, in which case both l and m may have more than one value. The allowed values for the three quantum numbers, when $n = 2$, are listed in Table 10–4.

TABLE 10-4

n	l	m	Quantum state
2	0	0	(200)
2	1	0	(210)
2	1	+1	(211)
2	1	−1	(21-1)

Since the energy of a quantum state depends only on the value of n, the second quantum state is *fourfold degenerate*. Thus

$$E_{(200)} = E_{(210)} = E_{(211)} = E_{(21-1)} = \frac{E_{(100)}}{4} = -0.545 \times 10^{-11} \text{ erg.}$$

Obviously, higher quantum states will exhibit even greater degeneracies. Consider, for example, the degenerate states corresponding to the third energy level (Table 10–5). The third energy level is apparently *ninefold degenerate*.

TABLE 10-5

n	l	m	Quantum state
3	0	0	(300)
3	1	0	(310)
3	1	+1	(311)
3	1	−1	(31-1)
3	2	0	(320)
3	2	+1	(321)
3	2	−1	(32-1)
3	2	+2	(322)
3	2	−2	(32-2)

Except for very small corrections that result from relativistic considerations, which we'll accommodate later by incorporation of an additional *electron-spin* quantum number, the stationary-state energy levels generated from Eq. (10-79) account fully for the atomic spectrum of hydrogen. Furthermore, knowing the number of degenerate quantum states corresponding to any particular stationary-state energy helps one to interpret spectral line intensities in terms of statistics. That is, although the basic probability that an electron exists at a given energy level is primarily dependent on the magnitude of the energy level itself, the greater the number of possible degenerate quantum states corresponding to a given energy level, the greater is the probability that the electron may assume that level. For a given energy transition, then, it also follows that the number of spectral photons emitted (spectral intensity) will be proportional to the degeneracy of the initial higher energy state. Thus the concept of degeneracy of energy levels in the hydrogen atom, in conjunction with the statistics of thermal equilibrium, affords a partial explanation of differences in relative line intensities in the spectrum of atomic hydrogen. Recall that the Bohr model was unable to even partially account for relative line intensities.

The interpretation of line intensities is not quite so simple as it may appear from the above discussion. A more thorough quantum-mechanical study of the transitions between quantum states indicates that certain transitions are *forbidden*. Such restrictions are summarized in what are called the *selection rules*. In the case of atomic spectra, the selection rules state that n may change by any integer, l must change by ± 1, and m may change by ± 1 or not at all.

10–7 SOME COMMENTS CONCERNING THE TOTAL WAVE FUNCTION ψ

The total normalized wave function for relative motion in the hydrogen atom or hydrogenlike ion is expressed, according to Eq. (10–14), as

$$\psi_{nlm}(r, \theta, \phi) = R_{nl}(r) \cdot \Theta_{lm}(\theta) \cdot \Phi_m(\phi), \qquad (10\text{–}81)$$

where normalized functions for $R_{nl}(r)$ are given by Eq. (10–77) and in Table 10–3, those for $\Theta_{lm}(\theta)$ are given by Eq. (10–42) and in Table 10–2, and those for $\Phi_m(\phi)$ are given by Eq. (10–28) and in Table 10–1.

It is interesting to note that just as for the single particle in the three-dimensional box, where ψ was a function of n_x, n_y, and n_z, we have found that *three* quantum numbers n, l and m are needed in order to define the wave functions for the hydrogen atom. In fact, we may state generally that in a given system, *each degree of freedom requires the introduction of one quantum number*. Where the particle is able to move in any one of three coordinate directions, three quantum numbers arise.

Rather than present at this time the resultant forms for the ψ_{nlm} wave function, we'll find it profitable to first discuss the physical meanings and implications of the quantum numbers we have introduced. Recall that m was introduced to quantize the allowed values for A in the solution of Eq. (10–17), l was introduced to quantize the allowed values for $(-B)$ in Eq. (10–20), and n was introduced to quantize the allowed values for $(-E)$ in Eq. (10–23). The three quantum numbers which naturally arise in the nonrelativistic quantum-mechanical solution of the wave function for relative motion in the hydrogen atom or hydrogen-like ion are summarized in Table 10–6.

We have been able to associate the principal quantum number n with the energy of relative motion of the atom. In the next chapter, we'll develop corresponding relationships for the azimuthal quantum number l and the magnetic quantum number m.

TABLE 10–6

A Summary of the Quantum Numbers

Quantum number	Name	Allowed values	Eigenvalue quantized
n	Principal	$1, 2, 3, \ldots, \infty$	$E = \dfrac{-2\pi^2\mu Z^2 e^4}{n^2 h^2}$
l	Azimuthal	$0, 1, 2, \ldots, (n-1)$	$B = l(l+1)$
m	Magnetic	$0, \pm 1, \pm 2, \ldots, \pm l$	$A = -m^2$

To sum up the meaning of wave mechanics it can be stated that: A wave must be associated with each corpuscle and only the study of the wave's propagation will yield information to us on the successive positions of the corpuscle in space . . . and this wave is no myth; its wavelength can be measured and its interferences predicted. It has thus been possible to predict a whole group of phenomena without their actually having been discovered. And it is on this concept of the duality of waves and corpuscles in Nature, expressed in a more or less abstract form, that the whole recent development of theoretical physics has been founded and that all future developments of this science will apparently have to be founded.

—LOUIS de BROGLIE, Nobel Lecture, Stockholm, 1929.*

PROBLEMS

10–1 The Φ_m functions given by Eq. (10–28) are not the only forms of solution to the Φ equation. $d^2\Phi/d\phi^2 = -m^2\phi$.

a) Write a sine function $\Phi_{\sin}$ and also a cosine function $\Phi_{\cos}$ each of which is a solution to the Φ equation.

b) Normalize $\Phi_{\sin}$ and $\Phi_{\cos}$ over the range $0 \leqslant \phi \leqslant 2\pi$.

c) Using Euler's formulas, write the alternate exponential functions, Φ_{+1} and Φ_{-1} as given in Table 10–1, as linear combinations of $\Phi_{\sin}$ and $\Phi_{\cos}$ functions in which $m = \pm 1$.

10–2 By substitution of $l = 4$ in Eq. (10–38), determine the form for the Legendre polynomial $P_4(z)$.

10–3 a) Show that if $l = 1$ and $|m| = 2$, the associated Legendre polynomial vanishes, that is, $P_l^{|m|} = 0$.

b) Note, from the list of selected Legendre polynomials given in Section 10–4, that the highest power of z in the equation for $P_l(z)$ is given by z^l. According to Eq. (10–37), will the associated Legendre polynomial $P_l^{|m|}(z)$ vanish whenever $|m| > l$? Why?

---

* For a complete translation, see *Nobel Lectures, Physics,* 1922–1941, Elsevier Publishing Company, Amsterdam, 1965. © Nobel Foundation. Used with permission.

10-4 Show that the associated Legendre polynomial $P_l^{|m|}(z)$ reduces to the Legendre polynomial $P_l(z)$ when $|m| = 0$.

10-5 Using the equation for the Legendre polynomial $P_3(z)$ given in Section 10-4, in conjunction with Eq. (10-42), determine the form, in terms of θ, of the normalized $\Theta_{lm}(\theta)$ function when $l = 3$ and $|m| = 0$.

10-6 Through substitution of $k = 4$ in Eq. (10-68), evaluate the Laguerre polynomial, $L_4(\rho)$.

10-7 a) Show that if $n = 1$ and $l = 1$, the associated Laguerre polynomial vanishes, that is,

$$L_i^j(\rho) = L_{n+l}^{2l+1}(\rho) = 0.$$

b) Note, from the selected list of Laguerre polynomials, $L_k(\rho)$, given in Equations (10-69), (10-70), and (10-71), that the highest power of ρ in the polynomial is given by ρ^k. Show that the associated Laguerre polynomial vanishes *whenever* $j > k$.

10-8 Consider the state for a hydrogen-like ion of nuclear charge Z in which $n = 2$ and $l = 0$.

a) Determine the form of the Laguerre polynomial

$$L_k(\rho) = L_{n+l}(\rho) = L_2(\rho)$$

through use of Eq. (10-68). Compare your answer with $L_2(\rho)$ given in Eq. (10-70).

b) Determine the form of the *associated* Laguerre polynomial

$$L_k^j(\rho) = L_{n+l}^{2l+1}(\rho) = L_2^1(\rho)$$

by substitution of $L_2(\rho)$ from (a) into Eq. (10-67) followed by differentiation.

c) Substitute $L_2^1(\rho)$ from (b), along with $\rho = 2Zr/na_0$, into Eq. (10-77) in order to obtain the normalized radial wave function $R_{20}(r)$. Compare your answer with the corresponding function given in Table 10-3.

10-9 What is the degree of degeneracy of the fourth energy level of the hydrogen atom? List all the degenerate quantum states.

10-10 In an equilibrium distribution of hydrogen atoms, virtually all of the electrons are in the ground state, for which $n = 1$. Using Eq. (2-5), which is called the *Boltzmann equation*,

a) Calculate the ratio of the probability that an electron is in the first excited state as compared to the probability that it is in the ground state at $300°K$.

b) Calculate the ratio of the probability that an electron is in the $n = 10$ state as compared to the probability that it is in the $n = 9$ state at $300°K$.

c) In general, what must be the approximate magnitude of the energy difference between a given excited state and the ground state if the excited state is to have a reasonable probability of existence?

10-11 Show that the ψ_{nlm} function is normalized if the R_{nl}, Θ_{lm}, and Φ_m functions are each independently normalized.

THE HYDROGEN ATOM II.
ANGULAR MOMENTUM
AND ORBITAL SHAPES

We have shown in the previous chapter that the quantized energy of relative motion of the hydrogen atom is entirely dependent on the value of the principal quantum number n. Through use of the operator postulate, we'll show in this chapter that the azimuthal quantum number l is associated in a similar fashion with the total angular momentum of the atom and that the magnetic quantum number m is associated with the orientation of the angular momentum vector in an external magnetic field.

11–1 ANGULAR MOMENTUM OPERATORS

Whenever a mass m moves in a curved path with a velocity v, it has a magnitude of angular momentum L, which is given classically as

$$L = rmv = rp, \qquad (11–1)$$

where p is the linear momentum and r is the radius of curvature. In vector algebra the angular momentum is represented by a vector arrow of length rp, which is perpendicular to both the radius vector and the linear velocity vector (Fig. 11–1). If the motion is circular, r is a constant and the angular momentum vector is represented by an arrow projecting perpendicularly from the center of the circle. To determine the direction of the L vector, we may use a "right-hand rule," such that when the curled fingers of the right hand point in the direction of the v vector, the thumb points in the direction of the L vector. It may be shown that the total angular momentum L of a system may also be expressed in terms of its vector components L_x, L_y, and L_z in the rectangular coordinate directions x, y, and z, such that

$$L_x = yp_z - zp_y, \qquad (11–2)$$

$$L_y = zp_x - xp_z, \qquad (11–3)$$

$$L_z = xp_y - yp_x, \qquad (11–4)$$

199

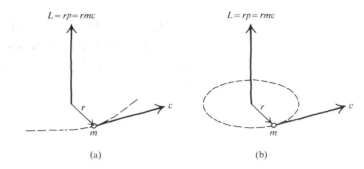

FIG. 11–1   Vector representation of the angular momentum L for a mass m moving about an axis with a linear velocity v, at a radius of curvature r. (a) Any curved motion. (b) Circular motion.

where p_x, p_y, and p_z are the component linear momenta in the x, y, and z component directions. According to the operator rules postulated in Chapter 6, we may now formulate the quantum-mechanical operators for L_x, L_y, and L_z by simply replacing the p terms in Eqs. (11–2), (11–3), and (11–4) with corresponding linear-momentum operators, so that

$$\hat{L}_x = \frac{\hbar}{i} \left(y \frac{\partial}{\partial z} - z \frac{\partial}{\partial y} \right),$$ (11–5)

$$\hat{L}_y = \frac{\hbar}{i} \left(z \frac{\partial}{\partial x} - x \frac{\partial}{\partial z} \right),$$ (11–6)

$$\hat{L}_z = \frac{\hbar}{i} \left(x \frac{\partial}{\partial y} - y \frac{\partial}{\partial x} \right).$$ (11–7)

But in our present treatment of the hydrogen atom, we have chosen to work in spherical polar coordinates. By applying the transformation equations given in Fig. 10–3, we can (after lengthy manipulation) convert the angular momentum operators given by Eqs. (11–5), (11–6), and (11–7) into forms involving r, θ, and ϕ, so that

$$\hat{L}_x = \frac{\hbar}{i} \left(-\sin \phi \frac{\partial}{\partial \theta} - \cot \theta \cos \phi \frac{\partial}{\partial \phi} \right),$$ (11–8)

$$\hat{L}_y = \frac{\hbar}{i} \left(\cos \phi \frac{\partial}{\partial \theta} - \cot \theta \sin \phi \frac{\partial}{\partial \phi} \right),$$ (11–9)

$$\hat{L}_z = \frac{\hbar}{i} \frac{\partial}{\partial \phi},$$ (11–10)

and since, by vector algebra,

$$\widehat{L^2} = \widehat{L_x^2} + \widehat{L_y^2} + \widehat{L_z^2},$$ (11–11)

it can be shown that

$$\widehat{L^2} = -\hbar^2 \left[\frac{1}{\sin \theta} \frac{\partial}{\partial \theta} \left(\sin \theta \frac{\partial}{\partial \theta} \right) + \frac{1}{\sin^2 \theta} \frac{\partial^2}{\partial \phi^2} \right]. \qquad (11\text{–}12)$$

11–2 ORBITAL ANGULAR MOMENTUM AND THE AZIMUTHAL QUANTUM NUMBER l

Let's now multiply Eq. (10–20) by $-\hbar^2 \Phi$:

$$-\hbar^2 \left[\frac{1}{\sin \theta} \cdot \frac{\partial}{\partial \theta} \left(\sin \theta \frac{\partial \Theta \Phi}{\partial \theta} \right) + \frac{A \Phi \Theta}{\sin^2 \theta} \right] = (B\hbar^2)\Theta \Phi. \qquad (11\text{–}13)$$

In incorporating Φ into the first term on the left of Eq. (11–13) we note that Φ is independent of θ, so that we may choose to include it as part of the term with respect to which the θ derivatives are taken. Further substitution for $A\Phi$ from Eq. (10–17) and of $l(l+1)$ for B yields

$$-\hbar^2 \left[\frac{1}{\sin \theta} \frac{\partial}{\partial \theta} \left(\sin \theta \frac{\partial \Theta \Phi}{\partial \theta} \right) + \frac{\Theta d^2 \Phi}{\sin^2 \theta \, d\phi^2} \right] = \left[l(l+1)\hbar^2 \right]\Theta \Phi. \qquad (11\text{–}14)$$

Since Θ is not a function of ϕ, the second term in brackets on the left may be rewritten to include Θ as part of the function with respect to which the ϕ derivative is taken. Thus

$$-\hbar^2 \left[\frac{1}{\sin \theta} \frac{\partial}{\partial \theta} \left(\sin \theta \frac{\partial \Theta \Phi}{\partial \theta} \right) + \frac{1}{\sin^2 \theta} \frac{\partial^2 \Theta \Phi}{\partial \phi^2} \right] = \left[l(l+1)\hbar^2 \right]\Theta \Phi,$$
$$(11\text{–}15)$$

or, in operator symbolism,

$$-\hbar^2 \left[\frac{1}{\sin \theta} \cdot \frac{\partial}{\partial \theta} \left(\sin \theta \frac{\partial}{\partial \theta} \right) + \frac{1}{\sin^2 \theta} \cdot \frac{\partial^2}{\partial \phi^2} \right]\Theta \Phi = \left[l(l+1)\hbar^2 \right]\Theta \Phi.$$
$$(11\text{–}16)$$

But, according to Eq. (11–12), the operator in Eq. (11–16), which is an eigenvalue equation, is $\widehat{L^2}$. Substitution of Eq. (11–11) into Eq. (11–16) yields

$$\widehat{L^2}\Theta \Phi = \left[l(l+1)\hbar^2 \right]\Theta \Phi. \qquad (11\text{–}17)$$

Furthermore, since the operator $\widehat{L^2}$ does not differentially operate on r, and since $R(r)$ does not depend on θ or on ϕ, we can introduce R into the operand on both sides of Eq. (11–17) in such a way that

$$\widehat{L^2}\Theta \Phi R = \left[l(l+1)\hbar^2 \right]\Theta \Phi R, \qquad (11\text{–}18)$$

and since the total wave function ψ is equal to $\Theta\Phi R$, we may write

$$\widehat{L^2}\psi = \left[l(l + 1)\hbar^2\right]\psi. \qquad (11\text{-}19)$$

Equation (11-19) is an eigenvalue equation in which the operator is $\widehat{L^2}$, the operand is ψ, the wave function for the hydrogen atom, and the eigenvalues, according to Postulate IIIb, are the allowed stationary-state values of L^2, the square of the magnitude of orbital angular momentum of the hydrogen atom. Since l is restricted to integral values, L^2 and L are then quantized such that

$$L^2 = l(l + 1)\hbar^2, \qquad (11\text{-}20)$$

and

$$L = \sqrt{l(l + 1)}\ \hbar, \qquad (11\text{-}21)$$

where $l = 0, 1, 2, \ldots, (n - 1)$. The values of L calculated from Eq. (11-21) are the *only exact values of L which may be observed experimentally*. The negative square root was not included in writing Eq. (11-21) because negative magnitudes for vectors are meaningless.

Thus, just as the principal quantum number n serves to quantize the allowed energy of relative motion of the hydrogen atom, *the azimuthal quantum number l serves to quantize the allowed magnitude of orbital angular momentum of the hydrogen atom*, which may be thought of conveniently as the angular momentum of a fictitious electron of *reduced mass* μ revolving with a linear velocity v at a radius r about a stationary nucleus (see Section 3-4).

In comparison, recall that in the Bohr model the angular momentum of the atom was quantized *by postulation* as

$$L = \mu v r = n\hbar \quad \text{(Bohr)}. \qquad (11\text{-}22)$$

It is worth noting that the Sommerfeld extension of the Bohr theory required quantization of angular momentum according to Eq. (11-21), rather than Eq. (11-22). Note also that in the Schrödinger model, the principal quantum number is related to the *energy* of relative motion and not directly to the angular momentum.

If we are willing to sacrifice some conceptual rigor, we can gain helpful insight into the relationship between angular momentum and magnetic properties by imagining an idealized model in which the electron moves in a classical planar orbit, as illustrated in Fig. 11-2. For the direction of electron revolution indicated, the angular momentum vector L of length $\sqrt{l(l + 1)}\ \hbar$ is pointed upward. Furthermore, whenever a charged particle revolves about an axis, a magnetic field is generated along that axis. For example, if we wind copper wire around an iron core and cause electrons to flow in the wire, a magnetic field is generated in the core. Such a device is called an electromagnet. In determining the direction of the magnetic field, *when using electrons*, we use a "left-hand rule," such that if the curled fingers of the left hand point in the

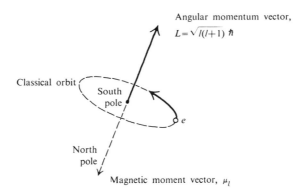

FIG. 11–2   A simplified pictorialization of the relationship between the orbital angular momentum vector and the magnetic moment vector for an electron in a classical orbit.

direction of the electron flow around the axis, the left thumb points toward the *north pole* of the electromagnet. Thus, in Fig. 11–2, the magnetic moment vector μ_l is pointed in a direction exactly opposite to that of the orbital angular momentum vector L. As expected, it can be shown that the magnitude of the magnetic moment vector is also quantized by the azimuthal quantum number l (specifically, it can be shown that $\mu_l = eL/2mc$).

It is customary to designate l quantum states by letters, according to the following scheme:

Value of l	0	1	2	3	4	5	...
Orbital notation	s	p	d	f	g	h	...

Thus, when the electron is in a quantum state in which $n = 1$ and $l = 0$, it is said to be in the 1s *orbital*; when it is in a state in which $n = 3$ and $l = 1$, it is said to be in the 3p orbital. *We define an orbital as a one-electron wave function.* The letters s, p, d, and f are derived from the spectroscopic terms: *sharp, principal, diffuse,* and *fundamental.*

Finally, it is interesting to note that for all states in which $l = 0$, that is, for all s orbitals, the orbital angular momentum has a value of zero according to Eq. (11–21). The orbital motion of an electron having a finite mass and finite velocity without angular momentum is difficult to visualize in classical terms.

11–3 THE *z*-COMPONENT OF ANGULAR MOMENTUM AND THE MAGNETIC QUANTUM NUMBER *m*

Although we were able to define the allowed stationary-state values for the magnitude of orbital angular momentum of the electron according to Eq. (11–21), we are not able to state the *orientation* of the angular-momentum vector with respect to an external reference. If there is no external magnetic field, any orientation of the angular momentum vector L, which is directed oppositely to the magnetic moment vector μ_l, should have the same energy. In the presence of an external magnetic field, however, different orientations of the magnetic moment vector and the associated antiparallel angular momentum vector with respect to the external field direction represent different energies. Consider the two particular orientations of the angular momentum vector and its associated magnetic moment vector shown in Fig. 11–3. We have again chosen to simplify the representation by pictorializing planar orbits for the electrons. In addition, we'll impose an *external magnetic field H* on the system and will arbitrarily define the z-axis as the field direction. By convention, the field arrow H points in the direction *toward which an imaginary isolated north pole would migrate.* Thus, the more nearly the arrow of the magnetic moment vector μ_l aligns itself with the field direction, the lower will be the energy of the orientation. In terms of L, the more nearly the angular momentum vector L aligns itself with the external magnetic field direction, the *higher* will be the energy of the system. For comparative

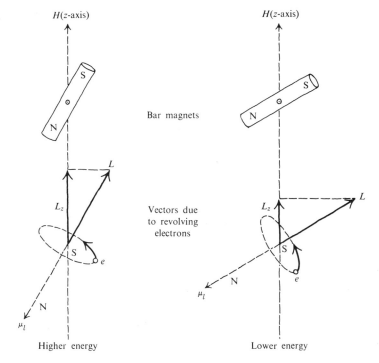

FIG. 11–3   Relation between orientation and energy for angular momentum vectors L and magnetic moment vectors μ_l, in the presence of an external magnetic field H.

purpose, two pivoted bar magnets in the same relative orientations as the two vectors are also shown in Fig. 11–3. A convenient measure of the orientation of the vector with respect to the z-axis (and hence of the relative energy of the orientation) is the length of the z-component L_z of the angular momentum vector. The component L_z is the projection of the L vector on the z-axis. In Fig. 11–3, although L is identical in the high-energy and low-energy orientations, L_z is larger in the higher-energy orientation.

We'll now show that the energy related to the *orientation* of the angular momentum vector in an external field is quantized. More specifically, we'll show that L_z is quantized. If we take the derivative of Eq. (10–28) with respect to ϕ and then multiply by $\hbar$, we obtain

$$\hbar \frac{d\Phi}{d\phi} = \hbar \frac{im}{\sqrt{2}} e^{im\varphi}, \tag{11–23}$$

which on rearrangement and substitution of Eq. (10–28) gives

$$\left(\frac{\hbar}{i} \frac{d}{d\phi} \right) \Phi = m\hbar\Phi. \tag{11–24}$$

Note that the operator in the above equation is $\hat{L}_z$ as given by Eq. (11–10), so that we may write

$$\hat{L}_z \Phi = m\hbar\Phi. \tag{11–25}$$

Since the operator $\widehat{L_z^2}$ does not differentially operate on θ or r, and since neither $\Theta(\theta)$ nor $R(r)$ contains ϕ, we can introduce ΘR into the operand on both sides of Eq. (11–25) in such a way that

$$\hat{L}_z \Phi \Theta R = m\hbar \Phi \Theta R$$

and since $\psi = \Phi \Theta R$, we may write

$$\hat{L}_z \psi = m\hbar\psi \tag{11–26}$$

Equation (11–26) is an eigenvalue equation in which the operator is $\hat{L}_z$, the operand is ψ, the wave function for the hydrogen atom, and the eigenvalues, according to the operator postulate, are the allowed stationary-state values for L_z, the z-component of angular momentum. Since m is restricted to integral values, L_z is quantized such that

$$L_z = m\hbar, \tag{11–27}$$

where $m = -l, -(l-1), \ldots, 0, \ldots, (l-1), l$.

The *only exact experimental values* for L_z which may be observed are those calculated from Eq. (11–27). Thus, it is now apparent that the magnetic *quantum number m serves to quantize the allowed levels for the z-component of angular momentum.* That is, the quantum number m serves to quantize the orientation of the orbital angular momentum vector in the presence of an external magnetic field.

The statement of n, l, and m for the quantum state of an electron in the hydrogen atom allows the immediate calculation of its total energy, $E = -2\pi^2 me^4/n^2 h^2$, the magnitude of its orbital angular momentum, $L = \sqrt{l(l+1)}\,\hbar$, and the z-component of its angular momentum, $L_z = m\hbar$. According to quantum mechanics, the values calculated from the above equations are the *only possible observable exact values* for E, L, and L_z.

In the presence of an external magnetic field, it may be shown that the classical angular momentum vector *precesses* about the field direction axis (z-axis) in the same way that a spinning top precesses in a gravitational field. The two motions are compared in Fig. 11–4. In each case, precession occurs in the same direction as the rotation of the top or the revolution of the electron.

The allowed orientations of the angular momentum vector for a d electron ($l = 2$) are shown in Fig. 11–5. The length of the vector in all cases is $L = \sqrt{2(2+1)}\,\hbar = \sqrt{6}\,\hbar$. Since $l = 2$, m may assume the values -2, -1, 0, 1, and 2, and L_z may thus have the five values, according to Eq. (11–27),

$$-2\hbar, \ -\hbar, \ 0, \ +\hbar, \ +2\hbar.$$

The position of the angular momentum vector L in the precessional cone is undefined.

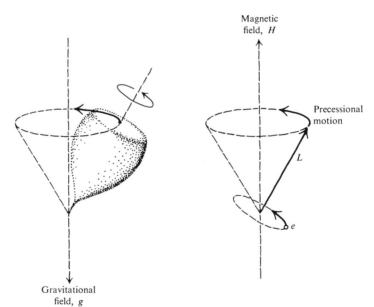

FIG. 11–4   A comparison of the precessional motion of a top in a gravitational field with the precessional motion of the orbital angular momentum vector of the electron in an external magnetic field.

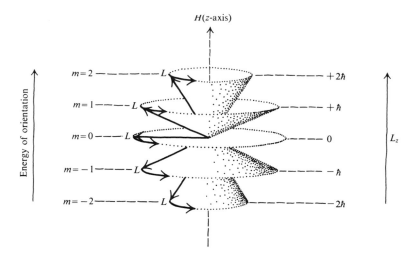

FIG 11–5   Allowed orientations of the angular momentum vector L for a d electron ($l = 2$) in an external magnetic field. The L vectors, each of length $\sqrt{2(2+1)}\ \hbar$, precess about the z-axis.

11–4 COMMUTATION OF OPERATORS AND SIMULTANEOUS MEASUREMENTS

We have shown that it is possible in any state of the hydrogen atom to calculate simultaneous exact values for the total energy E, the magnitude L of orbital angular momentum, and the z-component L_z of orbital angular momentum. On the other hand, the indefinite position of the L vector in its precessional cone about the z-axis reflects a total lack of knowledge concerning the values of L_x and L_y. In this section we'll show that the above results can be predicted from a knowledge of commutative properties of operators.

Whenever two different operators $\hat{A}$ and $\hat{B}$ operate on the same function the operations conventionally follow a sequence from right to left. Thus, the operation $\hat{A}\hat{B}\Psi$ requires that $\hat{B}$ first operate on Ψ and that $\hat{A}$ then operate on the resultant $\hat{B}\Psi$. It is important that we observe the proper sequence since often

$$\hat{A}\hat{B}\Psi \neq \hat{B}\hat{A}\Psi$$

For example, if $\hat{A}$ is $9 \cdot$ ("multiply by nine") and $\hat{B}$ is $\sqrt{\ }$ ("take the square root"), then

$$\hat{A}\hat{B}\Psi = 9 \cdot (\sqrt{\ })\Psi = 9\sqrt{\Psi}\ ,$$

whereas

$$\hat{B}\hat{A}\Psi = (\sqrt{\ })9 \cdot \Psi = 3\sqrt{\Psi}\ ,$$

so that the two results are not equal.

If for two given operators it should happen that the result is the same regardless of the order in which the operations are performed,

$$\hat{A}\hat{B} = \hat{B}\hat{A},$$

and the operators are said to *commute*. The *commutator* $[\hat{A}, \hat{B}]$ of two operators is defined as

$$[\hat{A}, \hat{B}] = \hat{A}\hat{B} - \hat{B}\hat{A}.$$

For example, the commutator of the two operators $9 \cdot$ and $\sqrt{}$ is

$$9(\sqrt{}) - (\sqrt{})9 \cdot = 9\sqrt{} - 3\sqrt{} = 6\sqrt{}.$$

Obviously, if two operators commute, their commutator is zero ("multiply by zero").

We'll now show that

if a system in a given state is to have simultaneous exact values for two different dynamical variables, it is necessary that the operators for these variables commute.

To prove this, consider the operators $\hat{A}$ and $\hat{B}$ for two different dynamical variables. According to Postulate III, the only possible exact values for the variables which may be observed in a state Ψ are given as eigenvalues A and B in the equations

$$\hat{A}\Psi = A\Psi$$

and

$$\hat{B}\Psi = B\Psi.$$

Thus, if A and B are each obtained as definite exact values in the same state, Ψ must be an eigenfunction of both $\hat{A}$ and $\hat{B}$. Let's operate on the first of the above equations with $\hat{B}$ and on the second with $\hat{A}$. Since A and B are constants, they commute with each other and with all linear quantum-mechanical operators, so that

$$\hat{B}\hat{A}\Psi = \hat{B}A\Psi = A\hat{B}\Psi = AB\Psi$$

and

$$\hat{A}\hat{B}\Psi = \hat{A}B\Psi = B\hat{A}\Psi = BA\Psi = AB\Psi.$$

Thus

$$\hat{B}\hat{A}\Psi = \hat{A}\hat{B}\Psi$$

and the operators must therefore commute. It can be shown* further that if any two operators commute, it is possible to find a set of functions which are

---

* H. Eyring, J. Walter, G. E. Kimball, *Quantum Chemistry*, John Wiley & Sons, New York, 1944, p. 35.

simultaneous eigenfunctions of both operators. Conversely, if two given operators do not commute, the system cannot have exact values for each of the corresponding dynamical variables, but only for one.

Let's consider a few examples. The linear momentum p_x and location x of a particle cannot each be exactly known in the same state of a system since $\hat{p}_x$ and $\hat{x}$ do not commute. That is:

$$\hat{p}_x \hat{x} - \hat{x}\hat{p}_x = \frac{\hbar}{i} \frac{\partial}{\partial x} x \cdot - x \frac{\hbar}{i} \frac{\partial}{\partial x}$$

$$= \frac{\hbar}{i} \left(\frac{\partial}{\partial x} x \cdot - x \frac{\partial}{\partial x} \right)$$

$$= \frac{\hbar}{i} \left(x \frac{\partial}{\partial x} + 1 \cdot - x \frac{\partial}{\partial x} \right) = \frac{\hbar}{i} \neq 0.$$

In fact, a general statement of the Heisenberg uncertainty principle, which we have previously discussed in Chapter 7, can be derived* in terms of the commutator of the two operators involved. Note that it is possible in a three dimensional system to obtain simultaneous exact values for p_x and y, the linear momentum in one direction and location in *another*, since $\hat{p}_x$ and $\hat{y}$ do commute; that is

$$\hat{p}_x \hat{y} - \hat{y}\hat{p}_x = \frac{\hbar}{i} \left(\frac{\partial}{\partial x} y \cdot - y \frac{\partial}{\partial x} \right) = 0.$$

Now let's explore the significance of similar relationships, which are called *commutation rules*, among the angular momentum operators. Using Equations (11–5), (11–6), and (11–7) it is easy to show that

$$\hat{L}_x \hat{L}_y - \hat{L}_y \hat{L}_x = i\hbar \hat{L}_z$$

$$\hat{L}_y \hat{L}_z - \hat{L}_z \hat{L}_y = i\hbar \hat{L}_x$$

$$\hat{L}_z \hat{L}_x - \hat{L}_x \hat{L}_z = i\hbar \hat{L}_y$$

For example, let's prove the first of the above three relations:

$$\hat{L}_x \hat{L}_y = \frac{\hbar}{i} \left(y \frac{\partial}{\partial z} - z \frac{\partial}{\partial y} \right) \frac{\hbar}{i} \left(z \frac{\partial}{\partial x} - x \frac{\partial}{\partial z} \right)$$

$$= \left(\frac{\hbar}{i} \right)^2 \left(y \frac{\partial}{\partial x} + yz \frac{\partial^2}{\partial z \partial x} - yx \frac{\partial^2}{\partial z^2} - z^2 \frac{\partial^2}{\partial y \partial x} + zx \frac{\partial^2}{\partial y \partial z} \right)$$

---

* H. Margenau and G. M. Murphy, *The Mathematics of Physics and Chemistry*, 2nd Ed., D. Van Nostrand Co., Princeton, N.J., 1956, p. 349.

and similarly

$$\hat{L}_y \hat{L}_x = \frac{\hbar}{i} \left(z \frac{\partial}{\partial x} - x \frac{\partial}{\partial z} \right) \frac{\hbar}{i} \left(y \frac{\partial}{\partial z} - z \frac{\partial}{\partial y} \right)$$

$$= \left(\frac{\hbar}{i} \right)^2 \left(zy \frac{\partial^2}{\partial x \partial z} - z^2 \frac{\partial^2}{\partial x \partial y} - xy \frac{\partial^2}{\partial z^2} + x \frac{\partial}{\partial y} + xz \frac{\partial^2}{\partial z \partial y} \right).$$

Since the order of differentiation is immaterial, subtraction of the above two expressions gives

$$\hat{L}_x \hat{L}_y - \hat{L}_y \hat{L}_x = \frac{\hbar}{i} \left[\frac{\hbar}{i} \left(y \frac{\partial}{\partial x} - x \frac{\partial}{\partial y} \right) \right] = i\hbar \hat{L}_z,$$

which is the required proof.

We immediately conclude that, since none of the component angular-momentum operators commutes with any other, it is impossible to simultaneously measure exact values for more than one component of orbital angular momentum in any given state. On the other hand, it is easily seen from Equations (11–10) and (11–12) that

$$\widehat{L^2 \hat{L}_z} - \widehat{\hat{L}_z L^2} = 0$$

and it follows from symmetry that

$$\widehat{L^2 \hat{L}_x} - \widehat{\hat{L}_x L^2} = 0,$$

and

$$\widehat{L^2 \hat{L}_y} - \widehat{\hat{L}_y L^2} = 0,$$

which means that, although in a given state we cannot simultaneously know exactly any two of the components of orbital angular momentum, we can obtain exact values for both L^2, the square of the magnitude of orbital angular momentum (and hence L itself) and *any one* of the three components, L_x, L_y, or L_z. Although the choice of coordinate directions is arbitrary, in constructing eigenfunctions we usually select L_z as the component to be defined exactly because the expression for $\hat{L}_z$ is relatively simple in polar coordinates. Refer once more to Figure 11–5 and note that in any given state both L and L_z are defined exactly. Note also that the position of the L vector in the precessional cone is not defined, that is, L_x and L_y are not known.

Finally, using Eq. (11–1) we may express the total energy E in terms of L as

$$E = \frac{p^2}{2m} + V = \frac{L^2}{2mr^2} + V$$

so that $\hat{H}$, the operator for total energy, can be written in a conservative system with a spherically-symmetric potential-energy field as

$$\hat{H} = \frac{\widehat{L^2}}{2mr^2} + V(r).$$

It is now easy to show that $\widehat{L^2}$ and $\hat{L}_z$ each commute with $\hat{H}$, since $\widehat{L^2}$ commutes with itself and with $\hat{L}_z$ and since both $\widehat{L^2}$ and $\hat{L}_z$ involve taking derivatives with respect to ϕ or θ and thus have no effect on r or $V(r)$. That is,

$$\widehat{L^2}\hat{H} = \widehat{L^2}\left[\frac{\widehat{L^2}}{2mr^2} + V(r) \right] = \left[\frac{\widehat{L^2}}{2mr^2} + V(r) \right]\widehat{L^2}$$

and

$$\hat{L}_z\hat{H} = \hat{L}_z\left[\frac{\widehat{L^2}}{2mr^2} + V(r) \right] = \left[\frac{\widehat{L^2}}{2mr^2} + V(r) \right]\hat{L}_z$$

We then conclude, as we have indeed already shown, that it is possible to obtain simultaneous exact values for E, L, and L_z for any state of the hydrogen atom.

11-5 THE HYDROGEN-LIKE WAVE FUNCTIONS

Some of the wave functions which we have developed in the previous chapter for the hydrogen atom are again presented in Table 11–1. We have arbitrarily restricted the tabulated solutions to the first two energy levels and shall investigate these in some detail. For convenience, the total wave function

$$\psi_{nlm}(r, \theta, \phi) = R_{nl}(r) \cdot \Theta_{lm}(\theta) \cdot \Phi_m(\phi) \tag{11–28}$$

is presented as the product of two parts: a *radial function*, $R_{nl}(r)$, and an *angular function*, $\Theta_{lm}(\theta)\Phi_m(\phi)$. The radial function $R_{nl}(r)$ depends only on r and has the same form for all states which have the same values for their n quantum numbers and for their l quantum numbers. For example, note from Table 11–1 that all of the 2p orbitals have the same form for $R(r)$. The angular part of the wave function, $\Theta_{lm}(\theta) \cdot \Phi_m(\phi)$, is also called the *spherical harmonic* and shows the combined dependence of ψ on θ and ϕ. We'll consider the significance of the $2p_x$ and $2p_y$ hybrid functions which are also included in Table 11–1 when we later discuss quantum states having angular dependence.

11-6 SPHERICALLY SYMMETRICAL STATES: s ORBITALS

It is apparent from Table 11–1 that the angular functions $\Theta(\theta)\Phi(\phi)$ for the 1s and 2s orbitals are not dependent on the values of θ or ϕ. It also follows from Eq. (11–28) that the *total* wave function ψ is independent of both θ and ϕ for

TABLE 11–1

Normalized Hydrogen-Like Wave Functions for the First Two Energy Levels

$$\psi_{nlm}(r, \theta, \phi) = R_{nl}(r) \cdot \Theta_{lm}(\theta) \cdot \Phi_m(\phi)$$

n	l	m	$R_{nl}(r)$ (radial function)	$\Theta_{lm}(\theta) \cdot \Phi_m(\phi)$ (angular function)	Symbol for wave function or orbital
1	0	0	$2\left(\dfrac{Z}{a_0}\right)^{3/2} e^{-Zr/a_0}$	$\left(\dfrac{1}{4\pi}\right)^{1/2}$	1s
2	0	0	$\left(\dfrac{Z}{2a_0}\right)^{3/2}\left(2 - \dfrac{Zr}{a_0}\right)e^{-Zr/2a_0}$	$\left(\dfrac{1}{4\pi}\right)^{1/2}$	2s
2	1	0	$\dfrac{1}{\sqrt{3}}\left(\dfrac{Z}{2a_0}\right)^{3/2}\left(\dfrac{Zr}{a_0}\right)e^{-Zr/2a_0}$	$\left(\dfrac{3}{4\pi}\right)^{1/2}\cos\theta$	$2p_0$ or $2p_z$
2	1	+1	$\dfrac{1}{\sqrt{3}}\left(\dfrac{Z}{2a_0}\right)^{3/2}\left(\dfrac{Zr}{a_0}\right)e^{-Zr/2a_0}$	$\left(\dfrac{3}{8\pi}\right)^{1/2}\sin\theta e^{i\phi}$	$2p_{+1}$
2	1	−1	$\dfrac{1}{\sqrt{3}}\left(\dfrac{Z}{2a_0}\right)^{3/2}\left(\dfrac{Zr}{a_0}\right)e^{-Zr/2a_0}$	$\left(\dfrac{3}{8\pi}\right)^{1/2}\sin\theta\, e^{-i\phi}$	$2p_{-1}$

Equivalent Hybrid Functions for $2p_{+1}$ and $2p_{-1}$

n	l	m	$R_{nl}(r)$	angular	symbol
2	1	±1 hybrid	$\dfrac{1}{\sqrt{3}}\left(\dfrac{Z}{2a_0}\right)^{3/2}\left(\dfrac{Zr}{a_0}\right)e^{-Zr/2a_0}$	$\left(\dfrac{3}{4\pi}\right)^{1/2}\sin\theta\cos\phi$	$2p_x$
2	1	±1 hybrid	$\dfrac{1}{\sqrt{3}}\left(\dfrac{Z}{2a_0}\right)^{3/2}\left(\dfrac{Zr}{a_0}\right)e^{-Zr/2a_0}$	$\left(\dfrac{3}{4\pi}\right)^{1/2}\sin\theta\sin\phi$	$2p_y$

the 1s and 2s states. Similarly, independence of θ and ϕ may be shown for the wave functions of *all* s orbitals, that is, the 3s orbital, the 4s orbital, and so forth. Thus, the wave functions $\psi_{1s}, \psi_{2s}, \psi_{3s}, \ldots$ or, in more simple notation, 1s, 2s, 3s, . . . , depend only on r and are spherically symmetrical. Plots of ψ versus r for the 1s orbital and the 2s orbital are shown in Fig. 11–6. Since, for s orbitals, $R(r)$ is directly proportional to ψ, a corresponding plot of $R(r)$ versus r would have the same form as that for the ψ versus r curves shown in Fig. 11–6.

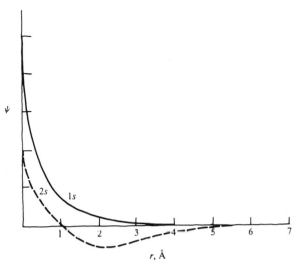

FIG. 11-6   The ψ wave functions for the electron in the 1s orbital and in the 2s orbital of the hydrogen atom. ψ is given in relative units.

In themselves, plots of ψ versus r or $R(r)$ versus r are not fully instructive. We are usually more interested in the probability distribution of the electron about the nucleus, which is given by plots of $\psi^*\psi$ versus r. Probability density distributions are given for the 1s and 2s orbitals in Fig. 11-7. Recall that $\psi^*\psi$ is the probability *per unit volume* of finding the electron at a given point in space. For both the 1s orbital and the 2s orbital, $\psi^*\psi$ is maximum in the

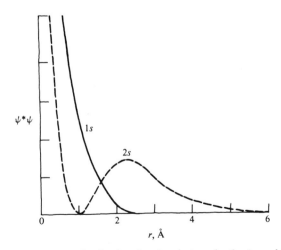

FIG. 11-7   Probability density distribution for the electron in the 1s and 2s orbitals of the hydrogen atom.

center of the atom. Three-dimensional plots of $\psi^*\psi$ versus r are shown as cloud density patterns and boundary surface plots in Fig. 11–8 for 1s and 2s orbitals.

In locating electrons in atoms, we are often concerned with the probability of finding the electron at some particular value of r *irrespective of direction*, rather than at a particular value of r, θ, and ϕ. Since in spherical polar coordinates the volume element $d\tau$ is given as $r^2 \sin \theta \, dr \, d\theta \, d\phi$, the probability of finding an electron in the region of space where its r coordinate is between r and $r + dr$, its θ coordinate is between θ and $\theta + d\theta$, and its ϕ coordinate is between ϕ and $\phi + d\phi$ is

$$|\psi|^2 \, d\tau = R^2(r)|\Theta(\theta)|^2|\Phi(\phi)|^2 r^2 \sin \theta \, dr \, d\theta \, d\phi. \qquad (11\text{–}29)$$

Now suppose we wish to express the probability that an electron exists in the differential annular shell between a sphere of radius r and a concentric sphere of radius $r + dr$, as shown in Fig. 11–9. In effect, we must integrate Eq. (11–29) over all values of θ and ϕ while holding r constant. The resultant probability $P(r)$ is called the *radial probability* and is given by

$$P(r) = r^2 R^2(r) \, dr \cdot \int_0^\pi |\Theta(\theta)|^2 \sin \theta \, d\theta \cdot \int_0^{2\pi} |\Phi(\phi)|^2 \, d\phi \qquad (11\text{–}30)$$

But, since the $\Theta(\theta)$ and $\Phi(\phi)$ functions are separately normalized for all orbitals, each of the above integrals is unity and

$$P(r) = r^2 R^2(r) \, dr. \qquad (11\text{–}31)$$

Equation (11–31) gives the probability of finding the electron *in any orbital* in a shell of thickness dr at a distance r from the nucleus. Since dr is a differential quantity, so is $P(r)$ a differential quantity. A *finite* probability may be expressed if we define the *radial distribution function* $D(r)$ as the probability *per unit radius* of finding the electron in a spherical shell at a distance r from the nucleus. Thus

$$D(r) = P(r)/dr = r^2 R^2(r). \qquad (11\text{–}32)$$

Plots of the radial distribution function $r^2 R^2(r)$ for the 1s and 2s orbitals are shown in Fig. 11–10. For the 1s orbital, which represents the ground state, the most probable radius is given as $a_0 = 0.529$ Å, *which is exactly the same as that calculated from the earlier Bohr theory.*

Thus, the quantum-mechanical model for the hydrogen atom allows the 1s electron to have some probability for existence, however small, at *all* values of r. However, there is a maximum probability per unit radius for finding the electron at $r = a_0$. In effect, quantum mechanics serves to smear out the neat orbits of the older Bohr theory.

For the 2s orbital there are two maxima or peaks in the radial distribution curve: one just below 3 Å and a much smaller peak at about 0.4 Å.   The

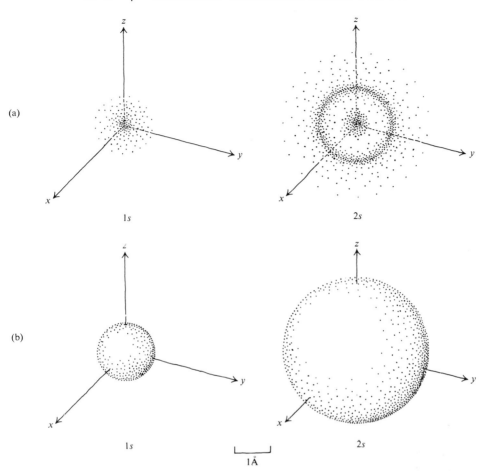

FIG. 11–8 Probability density ($\psi^*\psi$) as a function of position for the electron in the 1s and 2s orbitals of hydrogen. (a) A cross section in which cloud density is proportional to $\psi^*\psi$. (b) A boundary surface which excludes all space elements in which $\psi^*\psi$ is less than one-tenth of the maximum value.

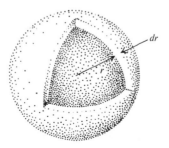

FIG. 11–9 The probability that an electron exists between the radii r and $r + dr$ is given as $P(r) = r^2 R^2(r)\, dr$.

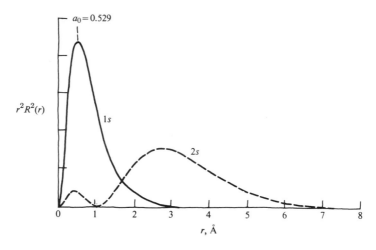

FIG. 11–10   Radial distribution functions $r^2 R^2(r)$ for the electron in the 1s and 2s orbitals of hydrogen.

unexpected appearance of the smaller peak much closer to the nucleus is referred to as *penetration*. In multi-electronic atoms, penetration effects are important in states with low values of l and contribute significantly to the bonding energy between the electron and the nucleus. States with high penetration have low energy and high stability.

11–7   STATES WITH ANGULAR DEPENDENCE

An examination of the wave functions given in Table 11–1 indicates that the ψ functions for those states for which l is greater than zero are not symmetrical about the nucleus but depend on the values of θ and ϕ. (Recall that states for which l is greater than zero are states which have angular momentum.) For example, for the (210) quantum state, or the $2p_0$ orbital, ψ depends on r and on θ, but not on ϕ, and is thus symmetrical about the z-axis.

Boundary-surface plots of $\psi^*\psi$ versus r, θ, and ϕ for several orbitals having angular momentum are given in Fig. 11–11. The contours shown are considerably more complicated than those for the 1s and 2s orbitals which we have considered previously, and the lack of spherical symmetry is evident. The probability term plotted in Fig. 11–11 is

$$\psi^*\psi = |\psi|^2 = |R|^2 \, |\Theta\Phi|^2,$$

or, since none of the radial functions are complex,

$$\psi^*\psi = R^2 \, |\Theta\Phi|^2. \tag{11–33}$$

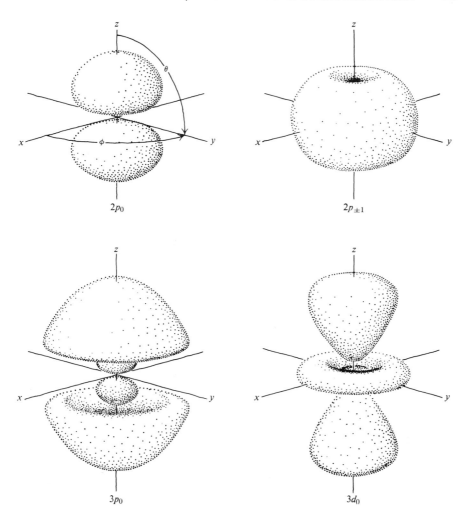

FIG. 11–11   Boundary surface plots of the probability distributions ($\psi^*\psi$ versus r, θ, and ϕ) for several orbitals with angular momentum. The boundary surfaces exclude all space in which the probability density is less than one-tenth of the maximum.

Note, however, from Tables 10–1 and 11–1, that $\Phi_m(\phi)$ is a complex function for all values of m other than zero, which means that we must express all angular functions for which $m \neq 0$ in terms of their absolute values $|\Theta\Phi|$ if we are to use them in Eq. (11–33). Since the R^2 term in Eq. (11–33) is spherically symmetrical, it is often more convenient to consider *angular* dependence of the probability distribution in terms of the square of the *absolute* angular function $|\Theta\Phi|^2$. In addition, the absolute angular function

$$|\Theta_{10}\Phi_0| = \left(\frac{3}{4\pi}\right)^{1/2} \cos\theta$$

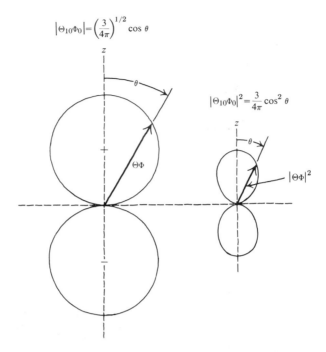

$$|\Theta_{10}\Phi_0|^2 = \frac{3}{4\pi}\cos^2\theta$$

FIG. 11–12 Two-dimensional polar graphs of $|\Theta\Phi|$ versus θ and $|\Theta\Phi^2|$ versus θ for p_0 or p_z orbitals. Since $|\Theta\Phi|$ is independent of ϕ for p_0 orbitals the two-dimensional plots appear the same for any value of ϕ. The values of $|\Theta\Phi|$ and $|\Theta\Phi^2|$ are given by the lengths of the radial coordinate.

$|\Theta\Phi|$ is often used without squaring for plots of angular dependence of wave functions. The advantage of using plots involving only the angular function and not R is that the forms for $|\Theta\Phi|$ are not dependent on the value of n, the principal quantum number. Thus, graphs of $|\Theta\Phi|$ versus θ and ϕ and of $|\Theta\Phi|^2$ versus θ and ϕ are the same for all p_0 orbitals (regardless of n), for all p_{+1} orbitals, for all p_{-1} orbitals, for all d_0 orbitals, and so on.

Two-dimensional polar graphs of $|\Theta\Phi|$ versus θ and of $|\Theta\Phi|^2$ versus θ are shown in Fig. 11–12 for the p_0 orbitals, which are not complex. Since $|\Theta\Phi|$ and $|\Theta\Phi|^2$ for p_0 orbitals are independent of ϕ, the corresponding three-dimensional graphs are symmetric about the z-axis, as shown in Fig. 11–13 (a). Note that in a *polar* graph the value of $|\Theta\Phi|$ is given as the length of the radial coordinate. Note especially that the surfaces shown in Fig. 11–13 are *not* boundary surfaces such as those presented in Fig. 11–11. Spherical polar graphs of $|\Theta\Phi|$ versus θ and ϕ are shown in Fig. 11–13(b) for the angular functions corresponding to the p_{+1} and p_{-1} orbitals. It may be shown easily (see Problem 11–15) that the *absolute* values of the angular functions for the p_{+1} and p_{-1} orbitals are identical and are given as

$$|\Theta_{1\,\pm1}\Phi_{\pm1}| = (3/8\pi)^{1/2}\sin\theta.$$

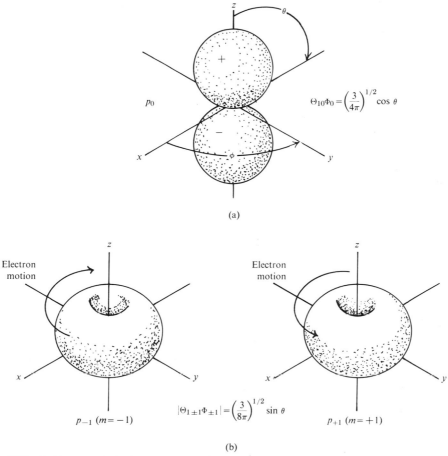

$$\Theta_{10}\Phi_0 = \left(\frac{3}{4\pi}\right)^{1/2} \cos\theta$$

(a)

Electron motion

Electron motion

$$|\Theta_{1\pm1}\Phi_{\pm1}| = \left(\frac{3}{8\pi}\right)^{1/2} \sin\theta$$

p_{-1} $(m = -1)$ p_{+1} $(m = +1)$

(b)

FIG. 11–13   (a) Spherical-polar graph of $|\Theta\Phi|$ versus θ and ϕ for the p_0 orbital. The value of $|\Theta\Phi|$ is given by the length of the radial coordinate. (b) Spherical-polar graphs of the absolute values $|\Theta\Phi|$ of the spherical harmonics for p_{+1} and p_{-1} orbitals.

Thus, for the p_{+1} and p_{-1} orbitals, the three-dimensional spherical polar graphs of $|\Theta\Phi|$ versus θ and ϕ are identical, as shown in Fig. 11–13 (b). Each of the donut-like plots is symmetric about the z-axis, since $|\Theta\Phi|$ is independent of ϕ for p_{+1} and p_{-1} orbitals. The only distinction we may make between the p_{+1} orbital and the p_{-1} orbital is with respect to the general direction of classical rotation of the electron about the z-axis. In the presence of an external magnetic field directed upward along the z-axis the p_{+1} orbital ($m = +1$ state) would represent the higher energy state if the electron revolved in a *generally* counterclockwise direction, that is, if its angular momentum vector pointed generally *upward* along the z-axis (right hand rule). It then follows that the electron would revolve in a *generally* clockwise direction in the p_{-1} orbital.

11–8 HYBRID ORBITALS

Because of their complex nature and the impossibility of distinguishing between the shapes of their spherical harmonic graphs, the $2p_{+1}$ and $2p_{-1}$ orbitals are inconvenient to use, particularly in the interpretation of the geometry of molecules. To avoid such difficulties, we can construct a new but equivalent set of *real* orthonormal wave functions through linear combination. Since the $2p_{+1}$ and $2p_{-1}$ orbitals are degenerate, any linear combination of these two wave functions will also be an eigenfunction which satisfies the Schrödinger equation. We find it convenient to define the following two new linear combinations of $2p_{+1}$ and $2p_{-1}$:

$$2p_x = \frac{1}{\sqrt{2}}(2p_{+1} + 2p_{-1}) \tag{11–34}$$

and

$$2p_y = \frac{-i}{\sqrt{2}}(2p_{+1} - 2p_{-1}). \tag{11–35}$$

The $2p_x$ function and the $2p_y$ function are called *hybrid* functions. The particular linear combinations given by Eqs. (11–34) and (11–35) are chosen to provide *real orthonormal* hybrid functions (prove, see Problem 11–9) with convenient directional properties. The wisdom of the choice will become clear after we perform the indicated substitutions and examine the resultant three-dimensional distributions. The original $2p_{+1}$ and $2p_{-1}$ wave functions are given from Table 11–1 as

$$2p_{+1} = R_{21}(r)(3/8\pi)^{1/2} \sin \theta e^{i\phi}, \tag{11–36}$$

and

$$2p_{-1} = R_{21}(r)(3/8\pi)^{1/2} \sin \theta e^{-i\phi}. \tag{11–37}$$

Substitution of Eqs. (11–36) and (11–37) into Eq. (11–34) yields

$$2p_x = \tfrac{1}{2}R_{21}(r)(3/4\pi)^{1/2} \sin \theta (e^{i\phi} + e^{-i\phi}). \tag{11–38}$$

But, according to Euler's formula (Eq. 5–9),

$$e^{i\phi} = \cos \phi + i \sin \phi, \tag{11–39}$$

and

$$e^{-i\phi} = \cos \phi - i \sin \phi, \tag{11–40}$$

so that

$$e^{i\phi} + e^{-i\phi} = 2 \cos \phi. \tag{11–41}$$

Substitution of Eq. (11–41) into Eq. (11–38) yields

$$2p_x = R_{21}(r)(3/4\pi)^{1/2} \sin\theta \cos\phi.$$

Thus, for any p_x orbital, for *any* value of n, the angular function is

$$(\Theta\Phi)_{\dot{p}_x} = (3/4\pi)^{1/2} \sin\theta \cos\phi. \qquad (11–42)*$$

The second hybrid function $2p_y$ is found by substitution of Eqs. (11–36) and (11–37) into Eq. (11–35):

$$2p_y = -i\tfrac{1}{2}R_{21}(r)(3/4\pi)^{1/2} \sin\theta(e^{i\phi} - e^{-i\phi}). \qquad (11–43)$$

Combination of Eqs. (11–39) and (11–40) yields

$$e^{i\phi} - e^{-i\phi} = 2i \sin\phi, \qquad (11–44)$$

and substitution of Eq. (11–44) into Eq. (11–43) yields

$$2p_y = R_{21}(r)(3/4\pi)^{1/2} \sin\theta \sin\phi. \qquad (11–45)$$

The angular function for the p_y hybrid orbital, for *any* value of n may therefore be written as

$$(\Theta\Phi)_{p_y} = (3/4\pi)^{1/2} \sin\theta \sin\phi. \qquad (11–46)*$$

Equations (11–42) and (11–46) are listed in Table 11–1 as alternative angular functions for the $(2\ 1\ \pm\ 1)$ quantum states.

Spherical-polar graphs for angular functions $\Theta\Phi$, and for *angular distribution functions* $(\Theta\Phi)^2$ are shown for p_x, p_y, and p_z orbitals in Fig. 11–14. Note that the lobes in the $\Theta\Phi$ plots have algebraic signs associated with them, whereas all lobes for $(\Theta\Phi)^2$, by virtue of squaring, are positive.

Normalized angular functions for d orbitals, as taken from the Θ functions given in Table 10–2 and the Φ functions given in Table 10–1, are presented in Table 11–2. Recall that for d orbitals, $l = 2$, so that m is allowed the values $-2, -1, 0, +1, +2$. A spherical polar graph of the angular function $\Theta\Phi$ for the d_0 orbital is given in Fig. 11–15 (a), and spherical polar graphs of the

---

* Note that identical forms for $(\Theta\Phi)_{p_x}$ and $(\Theta\Phi)_{p_y}$ would have been obtained had we originally accepted $(1/\sqrt{\pi})\sin m\phi$ and $(1/\sqrt{\pi})\cos m\phi$ as alternate *real* normalized solutions to Eq. 10–24. However, we would not have been able then to associate an unambiguous sign of m with each Φ function.

ΘΦ $(\Theta\Phi)^2$

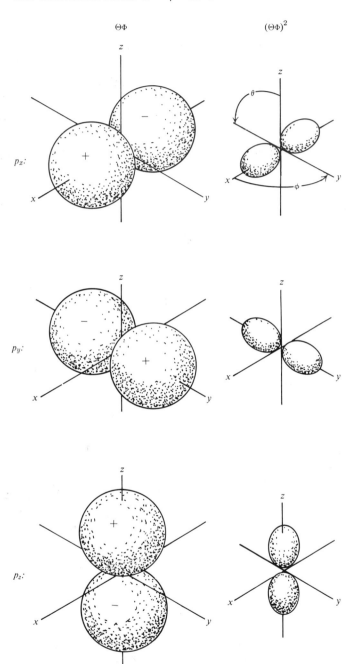

FIG. 11–14 Spherical-polar plots of the angular functions ΘΦ and the angular distribution functions $(\Theta\Phi)^2$ for the p_z orbital and for the hybrid p_x and p_y orbitals.

TABLE 11–2

Normalized Angular Functions for d Orbitals ($l = 2$)

l	m	$\Theta_{lm}(\theta)\Phi_m(\phi)$	Symbol
2	0	$\left(\dfrac{5}{16\pi}\right)^{1/2}(3\cos^2\theta - 1)$	d_0 or d_{z^2}
2	+1	$\left(\dfrac{15}{8\pi}\right)^{1/2}\sin\theta\cos\theta\, e^{i\phi}$	d_{+1}
2	−1	$\left(\dfrac{15}{8\pi}\right)^{1/2}\sin\theta\cos\theta\, e^{-i\phi}$	d_{-1}
Equivalent hybrid functions for d_{+1} and d_{-1}			
2	±1 hybrid	$\left(\dfrac{15}{4\pi}\right)^{1/2}\sin\theta\cos\theta\cos\phi$	d_{xz}
2	±1 hybrid	$\left(\dfrac{15}{4\pi}\right)^{1/2}\sin\theta\cos\theta\sin\phi$	d_{yz}
2	+2	$\left(\dfrac{15}{32\pi}\right)^{1/2}\sin^2\theta\, e^{i2\phi}$	d_{+2}
2	−2	$\left(\dfrac{15}{32\pi}\right)^{1/2}\sin^2\theta\, e^{-i2\phi}$	d_{-2}
Equivalent hybrid functions for d_{+2} and d_{-2}			
2	±2 hybrid	$\left(\dfrac{15}{16\pi}\right)^{1/2}\sin^2\theta\sin 2\phi$	d_{xy}
2	±2 hybrid	$\left(\dfrac{15}{16\pi}\right)^{1/2}\sin^2\theta\cos 2\phi$	$d_{x^2-y^2}$

absolute angular functions $|\Theta\Phi|$ are given for the d_{+1}, d_{-1}, d_{+2}, and d_{-2} orbitals in Fig. 11–15 (b) and (c).

As in the case of the p orbitals, the d_{+1} and d_{-1} plots are indistinguishable and are shown as a single $d_{\pm1}$ plot. In addition, the d_{+2} and d_{-2} graphs are also indistinguishable and are shown as a single $d_{\pm2}$ plot. In order to provide a better basis for later geometric interpretation, we may develop new linear combinations, as we have for the $p_{\pm1}$ orbitals, in order to define *two* new *real* and *orthonormal* hybrid functions *each* for the $d_{\pm1}$ and $d_{\pm2}$ orbitals. Although we'll not consider the mathematical details, the general hybridization procedure may

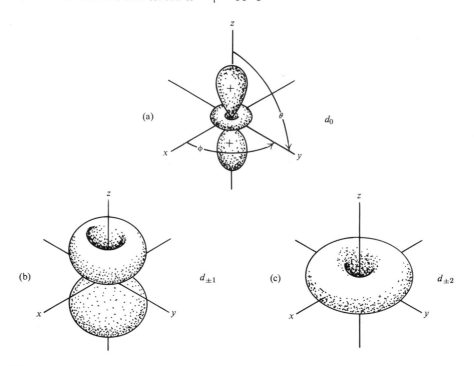

FIG. 11–15 Angular functions for d orbitals. (a) $\Theta\Phi$ versus θ and ϕ for the d_0 orbital. (b) $|\Theta\Phi|$ versus θ and ϕ for the d_{+1} and d_{-1} orbitals. (c) $|\Theta\Phi|$ versus θ and ϕ for the d_{+2} and d_{-2} orbitals.

be outlined as follows:

	Original functions			Hybrid functions
	d_0	$\xrightarrow{\text{No change}}$	$d_{z^2}\ (=d_0)$	
(Complex)	$\left.\begin{array}{c} d_{+1} \\ d_{-1} \end{array}\right\}$	$\xrightarrow{\text{Linear superposition}}$	$\left\{\begin{array}{c} d_{xz} \\ d_{yz} \end{array}\right.$	(real)
(Complex)	$\left.\begin{array}{c} d_{+2} \\ d_{-2} \end{array}\right\}$	$\xrightarrow{\text{Linear superposition}}$	$\left\{\begin{array}{c} d_{x^2-y^2} \\ d_{xy} \end{array}\right.$	(real)

The real angular functions for the four resulting hybrid orbitals, d_{xz}, d_{yz}, d_{xy}, and $d_{x^2-y^2}$, are given in Table 11–2 and spherical polar plots are shown in Fig. 11–16. Note that the d_0 orbital (which does not appear in Fig. 11–16) is not changed; that is, it is not hybridized. It is, however, given the new

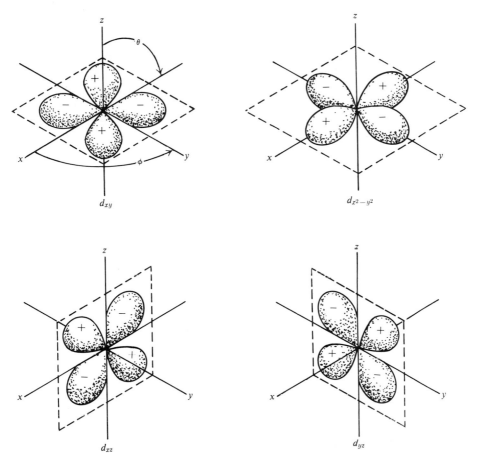

FIG. 11–16 The angular function $\Theta\Phi$ for the hybrid d orbitals. In each case, the lobes are bisected by the dotted planes.

designation d_{z^2}. It is not too difficult to visualize the spatial equivalence of the hybrid d orbitals (shown in Fig. 11–16) to the corresponding original d orbitals (shown in Fig. 11–15). We might, for example, as an approximation imagine that the $d_{\pm 1}$ orbital shown in Fig. 11–15 is produced as an envelope of rotation about the z-axis of either the d_{xz} or d_{yz} orbital shown in Fig. 11–16. Also, note that in forming the new hybrid d orbitals through linear superposition, we have given the hybrid orbitals *four* lobes rather than two as we had previously done for the p_x and p_y hybrid orbitals. The four lobes result from an insistence on orthogonality. The spatial representations provided by the hybrid d orbitals will be most helpful in the treatment of the geometry and stability of complex ions and molecules, particularly those involving transition-metal atoms.

11-9 RADIAL DEPENDENCE OF p ORBITALS AND d ORBITALS

In our previous discussions involving p and d orbitals we have been primarily concerned only with the angular part of the total wave function. We must remember, however, that the total probability density ψ^2 is the product of $|\Theta\Phi|^2$ multiplied by the radial function R^2. Curves of R^2 as a function of r for several p and d orbitals are given in Fig. 11-17. Note that the d orbitals show much less penetration than the p orbitals, which in turn show less penetration than s orbitals. We'll find that *in systems containing more than one electron*, d electrons are held less tightly by the nucleus than are p electrons, which in turn are held less tightly than s electrons.

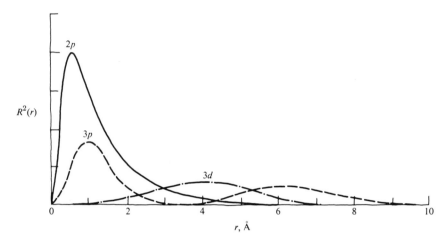

FIG. 11-17   The square $R^2(r)$ of the radial function for several states with angular dependence.

11-10 ELECTRON SPIN AND THE SPIN QUANTUM NUMBER

In spite of the remarkable success enjoyed by quantum mechanics in accounting for atomic spectra, there still remained in the mid-twenties certain troublesome features. For example, spectral lines for the alkali metals appear as closely spaced doublets rather than as the single lines predicted by nonrelativistic quantum-mechanical theory. As early as 1925, Wolfgang Pauli* attributed such spectral doublets to the existence of two closely spaced energy levels. Although Goudsmit and Uhlenbeck† and others advanced explanations for these levels in terms of intrinsic angular momentum of the electron, it was not until 1928 that

---

* W. Pauli, Jr., *Z. Physik*, **31**, 765 (1926).
† G. E. Uhlenbeck and S. Goudsmit, *Naturwiss.*, **13**, 953 (1925); *Nature*, **117**, 264 (1926).

Dirac* quantitatively accounted for the unexpected multiplicity of energy levels in terms of relativity. The mathematical treatment is complicated, so that we'll only outline a few of the most important features.

According to relativity theory, the total energy E of a particle not only must include terms for kinetic and potential energy, it must also include a term for the relativistic rest mass energy m_0c^2 (see Section 3–6). For example, the total energy of the particle in the one-dimensional box was previously given as $E = n^2h^2/8mL^2$ (Eq. 7–35). According to Eq. (3–37) the relativistic quantum-mechanical solution of the same problem requires that the total energy be very nearly expressed as

$$E = (n^2h^2/8mL^2) + m_0c^2, \qquad (11\text{–}47)$$

where m_0 is the rest mass of the particle. However, since usual experimental methods measure only *differences* in energy levels rather than energy levels themselves, the m_0c^2 term is not ordinarily observed experimentally.

In our previous treatment of nonrelativistic quantum-mechanical operators we have shown that the operator for the z component of orbital angular momentum is

$$\hat{L}_z = \frac{\hbar}{i} \cdot \frac{\partial}{\partial \phi} . \qquad (11\text{–}10')$$

In the *relativistic* treatment, Dirac showed that it was necessary to modify the operator for the z component of angular momentum to include a second term such that

$$\hat{J}_z = \frac{\hbar}{i} \cdot \frac{\partial}{\partial \phi} \pm \frac{\hbar}{2} , \qquad (11\text{–}48)$$

where $\hat{J}_z$ is the *relativistic* operator for the z component J_z of total angular momentum. We might rewrite Eq. (11-48) as

$$\hat{J}_z = \hat{L}_z \pm \frac{\hbar}{2} . \qquad (11\text{–}49)$$

Note that the additional term, $\pm\hbar/2$, in the operator provides for the existence of *two separate* allowed levels for the z component of angular momentum for any given value of m, instead of the single level which would be allowed if $\hat{J}_z$ were equal to $\hat{L}_z$ alone. One level is $\hbar/2$ *above* the z component of angular momentum allowed by nonrelativistic quantum mechanics and the other one is $\hbar/2$ *below* the level allowed by nonrelativistic quantum mechanics.

We may provide a gratuitous classical interpretation of the two new levels in terms of *intrinsic spin of the electron* on its own axis (although such an interpretation is not *required* by relativistic quantum mechanics). For the

---

* P. A. M. Dirac, *Proc. Roy. Soc. (London)*, **A117**, 610 (1928); **A118**, 351 (1928).

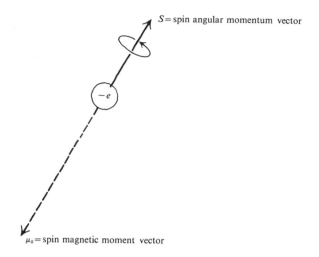

FIG. 11–18   Intrinsic spin of the electron on its own axis.

direction of spin shown in Fig. 11–18 the *spin angular momentum vector S* points upward (right-hand rule) and the spin magnetic moment vector μ_s points downward (left-hand rule).* Just as we have shown that the orbital angular momentum L is quantized according to the relationship

$$L = \sqrt{l(l + 1)}\ \hbar, \tag{11–21'}$$

where $l = 0, 1, 2, \ldots, n - 1$, we can derive an analogous equation to show that spin angular momentum S is given by

$$S = \sqrt{s(s + 1)}\ \hbar. \tag{11–50}$$

However, here the number s is restricted to the single value $\frac{1}{2}$. Tnat is, in spinning on its own axis, the electron may have only one allowed value for the magnitude of its total spin angular momentum vector S.

For the *z component* of spin angular momentum, however, the situation is quite different. We have previously shown that the z component of orbital angular momentum, L_z, is given by

$$L_z = m\hbar, \tag{11–27'}$$

where $m = -l, \ldots, 0, \ldots, +l$. By comparison, it can be shown that the z component of spin angular momentum is given as

$$S_z = m_s\hbar, \tag{11–51}$$

---

* It can be shown that $\mu_s \simeq (e/mc)S$. Thus, μ_s/S is twice as large as μ_l/L.

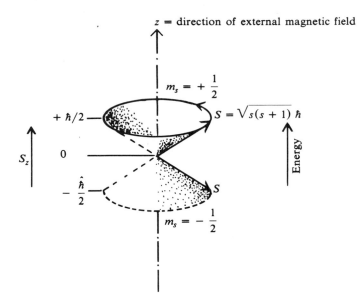

z = direction of external magnetic field

$m_s = + \dfrac{1}{2}$

$+ \hbar/2$

$S = \sqrt{s(s + 1)}\,\hbar$

S_z 0

$- \dfrac{\hbar}{2}$

S

$m_s = - \dfrac{1}{2}$

Energy

FIG. 11–19   The two allowed precessional cones for the spin angular momentum vector in an external magnetic field along the *z*-axis.

where m_s, the *spin quantum number*, is allowed the values $+\frac{1}{2}$, $-\frac{1}{2}$. Thus, for given values of *n*, *l*, and *m*, only two values are allowed for S_z: one in which $m_s = +\frac{1}{2}$, in which case the spin angular momentum vector *S* points *upward* in an external magnetic field directed upward, and one in which $m_s = -\frac{1}{2}$, in which case the spin angular momentum vector *S* points *downward* in an external magentic field directed upward. The relationships shown in Fig. 11–19 are analogous to those formerly considered for the orbital angular momentum vector. In fact, we assume that the operators for total and component spin angular momentum commute in the same way as those for orbital angular momentum. For example, the *S* vector may be considered to precess about the *z*-axis. However, only *two* precessional cones are allowed.

Thus, in the presence of an external magnetic field it is predicted that the spinning electron may assume either of two different closely spaced energy states. The Stern-Gerlach* experiment confirmed this prediction by showing that if a beam of atoms of an alkali metal is passed between the poles of a strong magnet, the beam is split into two parts which may be collected separately. In the absence of an external magnetic field, for quantum states in which *l* = 0 (states of zero orbital angular momentum), the states designated

* O. Stern, *Z. Physik*, **7**, 249 (1921); W. Gerlach and O. Stern, *Z. Physik*, **8**, 110 (1922); **9**, 349, 353 (1922).

by $m_s = +\frac{1}{2}$ and $m_s = -\frac{1}{2}$ have identical energies and are therefore degenerate. For those states which have orbital angular momentum (for which l is greater than zero) the classical electron rotates on its own axis in the magnetic field produced by its own orbital motion. In this case, the S vector is allowed one of two different orientations with respect to the L vector. From the electron's point of view, L and μ_l may be imagined to result from the relative revolution of the nucleus about the electron. Each orientation has a slightly different energy level. This slight difference in energy levels between the $m_s = +\frac{1}{2}$ and $m_s = -\frac{1}{2}$ states leads to the appearance of closely spaced doublets in the atomic spectrum. The interaction of the L vector and the S vector is shown in Fig. 11–20.

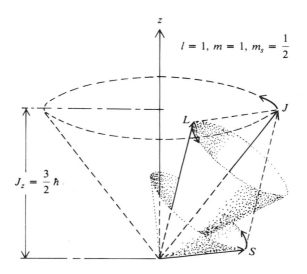

FIG. 11–20   The interaction of the magnetic fields associated with the L and S vectors exerts a torque on both magnetic dipoles, causing them to precess as shown about a common axis which is the direction of their *vectorial* sum J. In turn, the J vector precesses about the z axis with a z-component $J_z = L_z + S_z$. The higher-energy form of the two possible combinations is shown. In the lower-energy form, for which $m_s = -\frac{1}{2}$, the S cone is inverted so that J is smaller and J_z is reduced to $\hbar/2$.

Although relativity requires the introduction of a fourth quantum number (essentially due to the consideration of *time* as a fourth degree of freedom), we find it very convenient to adhere to the previous description for the state function ψ in terms of the quantum numbers n, l, and m and to account for electron spin by introducing a spin function α or β in the product of functions defining ψ, such that two total functions are possible. That is, for *spin up* ($m_s = +\frac{1}{2}$),

$$\psi_{nlmm_s} = R_{nl}\Theta_{lm}\Phi_m\alpha, \qquad (11\text{–}52)$$

whereas, for *spin down* ($m_s = -\frac{1}{2}$),

$$\psi_{nlmm_s} = R_{nl}\Theta_{lm}\Phi_m\beta. \tag{11–53}$$

11–11 SUMMARY OF THE QUANTUM STATES FOR THE ELECTRON IN THE HYDROGEN ATOM

We have shown that the quantum state of the electron in the hydrogen atom or in the hydrogen-like ion may be expressed entirely in terms of four quantum numbers: n, l, m, and m_s. The allowed values for the quantum numbers are

$$n = 1, 2, 3, \ldots, \infty,$$

$$l = 0, 1, 2, \ldots, (n-1),$$

$$m = -l, -(l-1), \ldots, 0, \ldots, (l-1), l,$$

$$m_s = +\tfrac{1}{2}, -\tfrac{1}{2}.$$

The allowed quantum states for the first three energy levels are summarized in Table 11–3. Note that there are two 1s states, two 2s states, six 2p states, two 3s states, six 3p states, and ten 3d states. Although in the case of the hydrogen

TABLE 11–3

Allowed Quantum States for the Electron in the First Three Energy Levels of the Hydrogen Atom or Hydrogen-Like Ion

Shell	n	l	m	m_s	Orbital
K	1	0	0	$+\frac{1}{2}, -\frac{1}{2}$	1s (two states)
L	2	0	0	$+\frac{1}{2}, -\frac{1}{2}$	2s (two states)
		1	-1	$+\frac{1}{2}, -\frac{1}{2}$	2p (six states)
			0	$+\frac{1}{2}, -\frac{1}{2}$	
			$+1$	$+\frac{1}{2}, -\frac{1}{2}$	
M	3	0	0	$+\frac{1}{2}, -\frac{1}{2}$	3s (two states)
		1	-1	$+\frac{1}{2}, -\frac{1}{2}$	3p (six states)
			0	$+\frac{1}{2}, -\frac{1}{2}$	
			$+1$	$+\frac{1}{2}, -\frac{1}{2}$	
		2	-2	$+\frac{1}{2}, -\frac{1}{2}$	3d (ten states)
			-1	$+\frac{1}{2}, -\frac{1}{2}$	
			0	$+\frac{1}{2}, -\frac{1}{2}$	
			$+1$	$+\frac{1}{2}, -\frac{1}{2}$	
			$+2$	$+\frac{1}{2}, -\frac{1}{2}$	

atom itself there is only one electron and it may occupy only one state at a time, we'll later find the descriptions of states for hydrogen to be most useful in the interpretation of allowed energy levels in more complex atoms. In multi-electron systems we will certainly expect extensive interactions among orbital angular momenta and spin angular momenta.

If you ask me: Now what are these particles, these atoms and molecules? I should have to admit that I know as little about it as where Sancho Panza's second donkey came from. However, to say something, even if not something momentous: They can at the most perhaps be thought of as more or less temporary creations within the wave field, whose structure and structural variety, in the widest sense of the word, are so clearly and sharply determined by means of wave laws as they recur always in the same manner; that much takes place as if they were a permanent material reality.

—ERWIN SCHRÖDINGER, in a lecture on September 4, 1952.*

PROBLEMS

11–1 Which of the following sets of operators commute?
   a) $3 \cdot$ and $\sqrt{}$
   b) $x \cdot$ and $y \cdot$
   c) $a \cdot$ and d^2/dx^2 (a = constant)
   d) y and $a(\partial/\partial y)$
   e) $x \cdot$ and $a(\partial/\partial y)$

11–2 Using $\hat{E}$ and $\hat{t}$ from Table 6–1, calculate the value of the commutator $[\hat{t}, \hat{E}]$. Compare to $[\hat{p}_x, \hat{x}]$ (Sec. 11–4) and state by analogy to Eq. (7–45) the uncertainty principle in terms of Δt and ΔE.

11–3 Is it possible in a one-dimensional system to locate a particle precisely and at the same time to know its energy precisely? That is, do $\hat{x}$ and $\hat{H}$ commute?

11–4 Prove that the commutator $[\hat{L}_y, \hat{L}_z]$ is $i\hbar\hat{L}_x$. Use Equations (11–6) and (11–7).

11–5 Prove that $\widehat{L^2}$ and $\hat{L}_z$, as given by equations (11–10) and (11–12) commute.

11–6 Show that the momentum of a particle in a given coordinate direction and its energy can be known precisely only if the potential energy does not vary in that direction.

11–7 a) Sketch the allowed orientations with respect to an external magnetic field for the orbital angular momentum vector of a p electron in the hydrogen atom. Show the field arrow and the vector arrow clearly.
   b) What is the length of the L vector?
   c) Which orientation represents the greatest energy?

---

* See W. Heisenberg, M. Born, E. Schrödinger, and P. Auger, *On Modern Physics*, Clarkson N. Potter, Inc., New York, 1960, p. 56.

11–8 Show that the 1s function and the 2s function for the hydrogen atom are orthogonal. (Hint: $d\tau = r^2 \sin\theta\, dr\, d\theta\, d\phi$)

11–9 Prove that the hybrid $2p_x$ and $2p_y$ orbitals for the hydrogen atom, as given by Eqs. (11–34) and (11–35) are orthonormal.

11–10 Show that the probability of finding a $2p_z$ electron anywhere in the xy-plane is zero.

11–11 Using the $2p_x$ function given in Table 11–1,

a) prepare a two-dimensional polar plot of $\Theta\Phi$ versus ϕ for the p_x orbital in the xy-plane,

b) prepare a two-dimensional polar plot of the angular distribution function $(\Theta\Phi)^2$ versus ϕ for the p_x orbital in the xy-plane.

11–12 Show that in a hydrogen-like ion, the radius at which there is maximum probability density of finding a 1s electron in *any direction* is a_0/Z, where Z is the nuclear charge and a_0 is the Bohr radius of the first hydrogen orbit. (Hint: differentiate to find maximum in radial distribution function.)

11–13 Using the mean-value postulate, show that for the 1s orbital of hydrogen, the *average* value of $1/r$ is $1/a_0$.

11–14 Calculate the finite value of r, in terms of a_0, at which a node occurs in the probability distribution for a 2s electron in hydrogen. Compare your answer with Fig. 11–7.

11–15 Prove that the absolute values of the $2p_{+1}$ and $2p_{-1}$ orbitals given in Table 11–1 are identical.

11–16 a) Show that the $2p_{+1}$ and $2p_{-1}$ orbitals given in Table 11–1 are each eigenfunctions of the operator $\hat{L}_z$. What are the eigenvalues for L_z in each case?

b) Are the $2p_x$ and $2p_y$ orbitals eigenfunctions of $\hat{L}_z$? Are they eigenfunctions of $\hat{L}_z^2$? What can be said concerning the allowed values of L_z associated with the $2p_x$ and $2p_y$ orbitals?

11–17 Using appropriate functions given in Table 11–2, prepare two-dimensional polar plots for:

a) $\Theta\Phi$ versus θ for the d_0 orbital in the yz-plane,

b) $\Theta\Phi$ versus ϕ for the d_{xy} orbital in the xy-plane.

11–18 Using the functions given in Table 11–2, show mathematically how the d_{xz} angular function and the d_{yz} angular function are derived from the angular functions for the d_{+1} and d_{-1} orbitals.

The quantum-mechanical solutions which we have considered thus far have been direct. That is, we have been able to solve directly the Schrödinger amplitude equations in order to evaluate wave functions and allowed stationary-state energies of systems. Even though the mathematics has at times been rather cumbersome, we have been able to determine directly the ψ functions for two-particle systems such as the hydrogen atom and hydrogen-like ions. However, in this chapter we'll find that the nonrelativistic Schrödinger equation for the next simplest atom, the helium atom, which contains *three* particles, cannot be solved directly. For all such systems we'll have to resort to approximation methods. Since most of the atomic and molecular systems with which we must deal are relatively complicated, it is then important that we develop a few basic approximation techniques.

12-1 THE HELIUM ATOM

The helium atom, as depicted in Fig. 12–1, is a three-particle system which consists of two electrons and a nucleus whose mass is 4.0026 amu (6.6461 × 10^{-24} g), and whose charge is $+2e$. Just as in the treatment of the hydrogen atom, the total Schrödinger equation may be separated into two equations, one involving the translational energy of the atom and the other involving the energy of relative motion of the electrons and the nucleus. We'll ignore the translational energy associated with the motion of the center of mass of the atom in space and will concentrate our attention on the relative motion of the particles within the atom and on the energy of relative motion. Furthermore, in our treatment of relative motion, we'll *assume that the nucleus is stationary*. Although this is not exactly true, the mass of the nucleus is so much larger (about four thousand times larger) than the combined mass of the electrons that the resultant error is not significant. The total potential energy of the

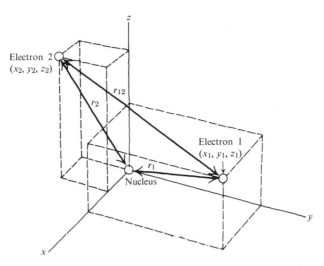

FIG. 12–1   The helium atom. The charge on the nucleus is $+2e$ and the charge on each of the electrons is $-e$.

atom is the sum of the potential-energy terms for each of the possible two-particle interactions. That is, it is the sum of:

1. The potential energy of attraction between the first electron and the nucleus, $-2e^2/r_1$, where r_1 is the distance between the first electron and the nucleus.
2. The potential energy of attraction between the second electron and the nucleus, $-2e^2/r_2$, where r_2 is the distance between the second electron and the nucleus.
3. The potential energy of repulsion between the two electrons, $+e^2/r_{12}$, where r_{12} is the distance between the two electrons.

In writing the above potential-energy terms we have again assumed that infinite separation between any pair of particles represents zero potential energy of interaction. The total potential energy V for the atom then may be expressed as

$$V = -\frac{2e^2}{r_1} - \frac{2e^2}{r_2} + \frac{e^2}{r_{12}}. \tag{12–1}$$

Since V is not explicitly dependent on time the force field is conservative, and we may use the nonrelativistic time-independent Schrödinger amplitude equation

$$\hat{H}\psi = E\psi, \tag{12–2}$$

to calculate the family of wave functions ψ and the corresponding energies of relative motion E. However, in writing the Hamiltonian operator for our model of the helium atom, we must take into account the sum of the kinetic energies of *two particles* (two electrons) and must, therefore, include a kinetic-energy operator term for *each* of the particles. Thus

$$\hat{H} = \frac{-\hbar^2}{2m_e} \nabla_1^2 - \frac{\hbar^2}{2m_e} \nabla_2^2 + V, \qquad (12\text{--}3)$$

where m_e is the mass of each electron. The first Laplacian operator, ∇_1^2, operates only on the coordinates (x_1, y_1, z_1) of the first electron, and the second Laplacian operator, ∇_2^2, operates only on the coordinates (x_2, y_2, z_2) of the second electron, that is,

$$\nabla_1^2 = \frac{\partial^2}{\partial x_1^2} + \frac{\partial^2}{\partial y_1^2} + \frac{\partial^2}{\partial z_1^2}, \qquad (12\text{--}4)$$

and

$$\nabla_2^2 = \frac{\partial^2}{\partial x_2^2} + \frac{\partial^2}{\partial y_2^2} + \frac{\partial^2}{\partial z_2^2}. \qquad (12\text{--}5)$$

Substitution of Eqs. (12–1) and (12–3) into Eq. (12–2) yields the complete Schrödinger amplitude equation for relative motion in the helium atom:

$$\frac{-\hbar^2}{2m_e} (\nabla_1^2 \psi + \nabla_2^2 \psi) - \frac{2e^2 \psi}{r_1} - \frac{2e^2 \psi}{r_2} + \frac{e^2 \psi}{r_{12}} = E\psi. \qquad (12\text{--}6)$$

Note that the Hamiltonian operator contains a Laplacian operator term for each electron and a potential energy term for each interacting pair. The general rule requiring one Laplacian operator term for each electron and one potential energy term for each interacting pair also holds for other more complex multi-electron systems. Following this rule, it is not at all difficult to write the correct Schrödinger amplitude equation for any atom we choose. The real difficulty arises in *solving* the eigenvalue equation which we so construct. Specifically, returning to the relatively simple Schrödinger equation (Eq. 12–6) for the helium atom, we note that if we were to neglect the electron repulsion term, $e^2\psi/r_{12}$, we would be able to separate variables and solve the eigenvalue equation directly (we'll do this shortly). As it stands, however, Eq. (12–6) has never been solved by any direct method. Rather, we are forced to rely on certain approximation methods, which we'll now outline as we attempt to calculate the ground-state energy of the helium atom.

12–2 GROUND-STATE ENERGY OF THE HELIUM ATOM. FIRST APPROXIMATION: IGNORE ELECTRON REPULSION

Since the presence of the electron repulsion term in Eq. (12–6) prevents us from obtaining a direct solution, we may, as a first approximation, naïvely assume that the electrons do not repel one another and proceed with the calculation. If we ignore the electron repulsion energy, the total potential energy of the free helium atom is given from Eq. (12–1) as

$$V = -\frac{2e^2}{r_1} - \frac{2e^2}{r_2},$$ (12–7)

and the Schrödinger amplitude equation is

$$\left[\frac{-\hbar^2}{2m_e}(\nabla_1^2 + \nabla_2^2) - \frac{2e^2}{r_1} - \frac{2e^2}{r_2}\right]\psi^0 = E^0\psi^0,$$ (12–8)

where the eigenvalues E^0 are the *stationary-state energies of relative motion of the helium atom assuming no electron repulsion* and the eigenfunctions ψ^0 are the *wave functions for the helium atom in which electron repulsion is ignored*. The associated nonrepulsion Hamiltonian operator $\hat{H}^0$ is then

$$\hat{H}^0 = \frac{-\hbar^2}{2m_e}(\nabla_1^2 + \nabla_2^2) - \frac{2e^2}{r_1} - \frac{2e^2}{r_2},$$ (12–9)

so that Eq. (12–8) may also be written as

$$\hat{H}^0\psi^0 = E^0\psi^0.$$ (12–10)

In order to separate the variables associated with the first electron from those associated with the second electron, we'll assume that the total wave function ψ^0 may be expressed as the product of two functions, such that the first function, φ_1, depends only on the coordinates of the first electron and the second function, φ_2, depends only on the coordinates of the second electron. That is,

$$\psi^0(x_1, y_1, z_1, x_2, y_2, z_2) = \varphi_1(x_1, y_1, z_1)\varphi_2(x_2, y_2, z_2),$$ (12–11)

or more simply

$$\psi^0 = \varphi_1\varphi_2.$$ (12–12)

Substitution of Eq. (12–12) into Eq. (12–8), followed by division of both sides by $\varphi_1\varphi_2$, yields

$$\left[-\frac{1}{\varphi_1}\frac{\hbar^2}{2m_e}\nabla_1^2\varphi_1 - \frac{2e^2}{r_1}\right] + \left[-\frac{1}{\varphi_2}\frac{\hbar^2}{2m_e}\nabla_2^2\varphi_2 - \frac{2e^2}{r_2}\right] = E^0.$$ (12–13)

But the value of the first term in brackets in Eq. (12–13) depends only on the coordinates of the first electron and is independent of the coordinates of the second electron, and the value of the second term in Eq. (12–13) depends only on the coordinates of the second electron and is independent of the coordinates of the first electron. Since the sum of both terms is a constant and, since the two terms are independent of one another, each term must separately be equal to a constant, that is,

$$-\frac{1}{\varphi_1}\frac{\hbar^2}{2m_e}\nabla_1^2\varphi_1 - \frac{2e^2}{r_1} = E_1,$$ (12–14)

and

$$-\frac{1}{\varphi_2}\frac{\hbar^2}{2m_e}\nabla_2^2\varphi_2 - \frac{2e^2}{r_2} = E_2, \tag{12-15}$$

where

$$E^0 = E_1 + E_2. \tag{12-16}$$

Equations (12–14) and (12–15) may each be rearranged to the more familiar forms of the Schrödinger amplitude equations:

$$-\frac{\hbar^2}{2m_e}\nabla_1^2\varphi_1 - \frac{2e^2\varphi_1}{r_1} = E_1\varphi_1, \tag{12-17}$$

and

$$-\frac{\hbar^2}{2m_e}\nabla_2^2\varphi_2 - \frac{2e^2\varphi_2}{r_2} = E_2\varphi_2. \tag{12-18}$$

Equations (12–17) and (12–18) are *each* recognized as being identical in form to Eq. (10–12) for the hydrogen-like ion, where $Z = 2$, that is, where the ion is He^+. Thus by ignoring electron repulsion we have generated a model of the helium atom which is, in effect, a superposition of the models of two He^+ ions (with only one nucleus, however).

In Chapter 10 we solved an equation identical in form to Eqs. (12–17) and (12–18), and the resultant stationary-state energies of relative motion are given by Eq. (10–79). Through identical solution, where now m_e is substituted for μ,

$$E_1 = -2\pi^4 m_e Z^2 e^4 / n_1^2 h^2, \tag{12-19}$$

and

$$E_2 = -2\pi^2 m_e Z^2 e^4 / n_2^2 h^2. \tag{12-20}$$

But we have shown previously that E_H, the energy of the hydrogen atom in the *ground state*, is given (Eq. 10–80) as $-2\pi^2\mu e^4 / h^2$. We'll now assume that within the accuracy of our present approximation, μ may be replaced by m_e without significant error, so that

$$E_H = -2\pi^2 m_e e^4 / h^2, \tag{12-21}$$

and since $Z = 2$ for the He^+ ion, E_1 and E_2, as given by Eqs. (12–19) and (12–20), may be expressed as

$$E_1 = 4E_H / n_1^2 \quad \text{and} \quad E_2 = 4E_H / n_2^2,$$

so that E^0, from Eq. (12–16), becomes

$$E^0 = 4E_H\left(\frac{1}{n_1^2} + \frac{1}{n_2^2}\right). \tag{12-22}$$

If each of the electrons in the nonrepulsion model of the helium atom is in its ground state, that is, if $n_1 = 1$ and $n_2 = 1$, the ground-state energy of the helium atom (assuming no electron repulsion) is

$$E^0 = 2E_{He^+} = 8E_H, \tag{12-23}$$

where E_{He^+} is the *ground-state* energy of the helium *ion*. But E_H was shown Eq. (10–80) to be -2.18×10^{-11} erg, which is equivalent to -13.6 eV. Thus E^0, the ground-state energy of the helium *atom* in our first crude approximation is

$$E^0 = 8(-13.6 \text{ eV}) = -108.8 \text{ eV}. \qquad (12\text{–}24)$$

The best *experimental* value for the energy E_{He} of the helium atom in the ground state is given in terms of the *first ionization potential I*, which is the *minimum* energy absorbed in the reaction in which the first electron leaves the ground state of the helium atom to become a free electron:

$$\text{He (ground state)} \rightarrow \text{He}^+ + \text{e}. \qquad (12\text{–}25)$$

Since the minimum energy corresponds to an ejected electron having no translational kinetic energy, we may write the energy balance for Eq. (12–25) as

$$E_{He} + I = E_{He^+}. \qquad (12\text{–}26)$$

But we are able to solve the Schrödinger amplitude equation for E_{He^+} directly and have shown (Eq. 12–23) that

$$E_{He^+} = E^0/2 = -54.4 \text{ eV},$$

and I is experimentally determined to be 24.6 eV. Thus, from Eq. (12–26), the experimental value for the ground-state energy of the helium atom is given as

$$E_{He} = -54.4 - 24.6 = -79.0 \text{ eV}. \qquad (12\text{–}27)$$

A comparison of E_{He} (experimental) from Eq. (12–27) with E^0 (calculated) in Eq. (12–24) indicates that in our first approximation method, the calculated value of the ground-state energy of the helium atom is 38% lower (algebraically) than the experimental value. Such a relatively large error indicates that electron repulsion cannot be ignored.

12-3 GROUND-STATE ENERGY OF THE HELIUM ATOM. SECOND APPROXIMATION: A FIRST-ORDER PERTURBATION METHOD

We have indicated that the correct Schrödinger wave equation for the helium atom is

$$\hat{H}\psi = E\psi, \qquad (12\text{–}2')$$

where

$$\hat{H} = \frac{-\hbar^2}{2m_e} \nabla_1^2 - \frac{\hbar^2}{2m_e} \nabla_2^2 - \frac{2e^2}{r_1} - \frac{2e^2}{r_2} + \frac{e^2}{r_{12}}. \qquad (12\text{–}28)$$

Equation (12–2') cannot be solved directly, which means that we cannot directly evaluate the *true wave functions*, ψ. We have shown, however, that if we are willing to ignore electron repulsion, we may solve *directly* the equation

$$\hat{H}^0\psi^0 = E^0\psi^0, \qquad (12\text{–}29)$$

where, from Eq. (12–9),

$$\hat{H}^0 = \frac{-\hbar^2}{2m_e} \nabla_1^2 - \frac{\hbar^2}{2m_e} \nabla_2^2 - \frac{2e^2}{r_1} - \frac{2e^2}{r_2}, \tag{12-30}$$

and where the eigenfunctions ψ^0 are the *wave functions for the helium atom in which electron repulsion is ignored.* Substitution of Eq. (12–30) into Eq. (12–28) yields

$$\hat{H} = \hat{H}^0 + \frac{e^2}{r_{12}}, \tag{12-31}$$

and substitution of Eq. (12–31) into (12–2′) yields

$$\left(\hat{H}^0 + \frac{e^2}{r_{12}} \right) \psi = E\psi, \tag{12-32}$$

an equation which cannot be directly solved. Let's now compare Eqs. (12–29) and (12–32). The difference in the operators in the two eigenvalue equations is the repulsion term, e^2/r_{12}. The smaller the value of e^2/r_{12}, the more nearly identical will the two equations be. We have seen, in fact, that if we let e^2/r_{12} equal zero (ignore electron repulsion), the two equations become identical. That is, E values are then given by E^0 values, and ψ functions are given by ψ^0 functions.

In using the *perturbation method* we'll assume that e^2/r_{12} is small enough to be considered as a minor modification or *perturbation* of the operator $\hat{H}^0$, an operator for which we may directly calculate the eigenfunctions ψ^0. We further hope that the perturbation will be small enough so that for a given state ψ^0 will not be too much different from ψ (which we cannot directly evaluate). Thus, in evaluating the ground-state energy of the helium atom in a given state, we'll combine the use of the correct Hamiltonian operator $\hat{H} = \hat{H}^0 + e^2/r_{12}$ with an incorrect wave function ψ^0 for that state which is regarded to be fairly close to what the correct wave function would be. The nonrepulsion operator $\hat{H}^0$ is said to be perturbed to the first order by the term e^2/r_{12}, and the present method of approximation is called a *first-order* perturbation method. The previous approximation method in which we ignored electron repulsion altogether is called a *zero-order* perturbation method, since we used $\hat{H}^0$ directly as the operator. That is, we added *no* perturbation terms to the nonrepulsion Hamiltonian that we used. It is important to note that for a given state ψ^0 is *not* an eigenfunction of $\hat{H}$ and therefore the operation of $\hat{H}$ on ψ^0 will not yield an exact energy eigenvalue. Rather, because of the perturbation of energy due to the repulsion term, the value of E obtained by the operation of $\hat{H}$ on ψ^0 for a given state will depend on the relative positions of the two electrons so that an *approximate* average value of energy, $\tilde{E}$, must be calculated through use of the mean-value postulate (Eq. 6–81) as

$$\tilde{E} = \frac{\int_{-\infty}^{\infty} \psi^{0^*} \hat{H} \psi^0 \, d\tau}{\int_{-\infty}^{\infty} \psi^{0^*} \psi^0 \, d\tau} = \frac{\int_{-\infty}^{\infty} \psi^{0^*} (\hat{H}^0 \psi^0 + e^2 \psi^0 / r_{12}) \, d\tau}{\int_{-\infty}^{\infty} \psi^{0^*} \psi^0 \, d\tau}. \tag{12-33}$$

Substitution of Eq. (12–29) into Eq. (12–33) yields

$$\tilde{E} = \frac{\int_{-\infty}^{\infty} \psi^{0^*} E^0 \psi^0 \, d\tau + \int_{-\infty}^{\infty} \psi^{0^*} (e^2/r_{12}) \psi^0 \, d\tau}{\int_{-\infty}^{\infty} \psi^{0^*} \psi^0 \, d\tau}, \tag{12–34}$$

but, since for any given state E^0 is constant,

$$\tilde{E} = E^0 + \frac{\int_{-\infty}^{\infty} \psi^{0^*} (e^2/r_{12}) \psi^0 \, d\tau}{\int_{-\infty}^{\infty} \psi^{0^*} \psi^0 \, d\tau}. \tag{12–35}$$

According to Eq. (12–35), the approximate *first-order* energy $\tilde{E}$ for the helium atom in a given state is given by the sum of E^0, the energy of the *nonperturbed* atom in that same state in which electron repulsion is ignored (the *zero-order energy*), and a term which is equivalent to the approximate average potential energy of electron repulsion over all space. Note that the second term on the right-hand side of Eq. (12–35) would be, according to the mean-value postulate (Eq. 6–81), the mean value of e^2/r_{12} if ψ^0 were used as the correct wave function, since the quantum-mechanical operator for e^2/r_{12} is also e^2/r_{12}. The second term on the right-hand side of Eq. (12–35) is called the *perturbation energy E'*.

We may therefore write

$$\tilde{E} = E^0 + E', \tag{12–36}$$

where the *perturbation energy* is

$$E' = \frac{\int_{-\infty}^{\infty} \psi^{0^*} (e^2/r_{12}) \psi^0 \, d\tau}{\int_{-\infty}^{\infty} \psi^{0^*} \psi^0 \, d\tau}. \tag{12–37}$$

Thus, in using the perturbation method we think of the energy $\tilde{E}$ of a given state of the helium atom as consisting of the total of E^0, the exact sum of the kinetic energy and the potential energy of attraction in that same state calculated by ignoring repulsion, and E', the approximate mean potential energy of repulsion, which in this case is a perturbation energy.

Since we desire to evaluate E' for the *ground state* of the helium atom ($n_1 = 1$ and $n_2 = 1$), we must use 1s orbitals for each of the electrons. Thus, from Table 11–1, where $Z = 2$,

$$\varphi_1 = \varphi_1^* = \left(\frac{1}{\pi}\right)^{1/2} \left(\frac{2}{a_0}\right)^{3/2} e^{-2r_1/a_0}, \tag{12–38}$$

and

$$\varphi_2 = \varphi_2^* = \left(\frac{1}{\pi}\right)^{1/2} \left(\frac{2}{a_0}\right)^{3/2} e^{-2r_2/a_0}. \tag{12–39}$$

Since $\psi^0 = \varphi_1 \varphi_2$ and $\psi^{0^*} = \varphi_1^* \varphi_2^*$,

$$\psi^0 = \psi^{0^*} = \varphi_1 \varphi_2 = \left(\frac{1}{\pi}\right) \left(\frac{2}{a_0}\right)^3 e^{-2r_1/a_0} e^{-2r_2/a_0}. \tag{12–40}$$

But φ_1 and φ_2, as given by Table 11–1, are normalized, so that

$$\int_{-\infty}^{\infty} \varphi_1^* \varphi_1 \, d\tau_1 = 1,$$

$$\int_{-\infty}^{\infty} \varphi_2^* \varphi_2 \, d\tau_2 = 1,$$

and

$$\int_{-\infty}^{\infty} \psi^{0*} \psi^0 \, d\tau = \int_{-\infty}^{\infty} \int_{\infty}^{\infty} \varphi_1^* \varphi_1 \cdot \varphi_2^* \varphi_2 \, d\tau_1 \, d\tau_2 = 1. \qquad (12\text{–}41)$$

Substitution of Eqs. (12–40) and (12–41) into Eq. (12–37) yields

$$E' = \left(\frac{8}{\pi a_0^3}\right)^2 \int_{-\infty}^{\infty} \frac{e^2}{r_{12}} (e^{-2r_1/a_0} e^{-2r_2/a_0})^2 \, d\tau \qquad (12\text{–}42)$$

or, in terms of the coordinates of each of the particles,

$$E' = \left(\frac{8}{\pi a_0^3}\right)^2 \int_{-\infty}^{\infty} \int_{-\infty}^{\infty} \frac{e^2}{r_{12}} (e^{-2r_1/a_0} e^{-2r_2/a_0})^2 \, d\tau_1 \, d\tau_2. \qquad (12\text{–}43)$$

The integral in Eq. (12–43) can be handled more easily if we convert the r coordinates to a new set in terms of dimensionless *atomic units*, that is,

$$R_1 = r_1/a_0, \quad R_2 = r_2/a_0, \quad R_{12} = r_{12}/a_0,$$

so that

$$d\tau_1 \, d\tau_2 = a_0^6 R_1^2 R_2^2 \sin\theta_1 \sin\theta_2 \, dR_1 \, dR_2 \, d\theta_1 \, d\theta_2 \, d\phi_1 \, d\phi_2,$$

and Eq. (12–43) becomes

$$E' = \frac{64e^2}{\pi^2 a_0} \int_0^\infty \int_0^\infty \int_0^\pi \int_0^\pi \int_0^{2\pi} \int_0^{2\pi}$$

$$\frac{e^{-4R_1} e^{-4R_2}}{R_{12}} R_1^2 R_2^2 \sin\theta_1 \sin\theta_2 \, dR_1 \, dR_2 \, d\theta_1 \, d\theta_2 \, d\phi_1 \, d\phi_2 \qquad (12\text{–}44)$$

Let's first consider holding the coordinates of electron 1 fixed as we integrate over the coordinates of electron 2. Since we must integrate the coordinates of both electrons over all space, the initial fixed position of electron 1 is arbitrary. However, for convenience we'll initially place electron 1 along the z-axis, as shown in Fig. 12–2. Then, by the law of cosines,

$$R_{12} = \left(R_1^2 + R_2^2 - 2R_1 R_2 \cos\theta_2 \right)^{1/2},$$

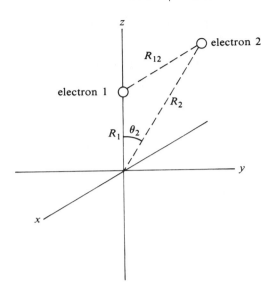

FIG. 12-2. Relation of R coordinates when the position of electron 1 is fixed on the z axis.

and Eq. (12–44) can be written

$$E' = \frac{64e^2}{\pi^2 a_0} \int_0^\infty R_1^2 e^{-4R_1}\, dR_1 \int_0^\infty R_2^2 e^{-4R_2}\, dR_2 \int_0^\pi \sin\theta_1\, d\theta_1$$

$$\times \int_0^\pi \frac{\sin\theta_2\, d\theta_2}{(R_1^2 + R_2^2 - 2R_1 R_2 \cos\theta_2)^{1/2}} \int_0^{2\pi} d\phi_1 \int_0^{2\pi} d\phi_2$$

The θ_1, ϕ_1, and ϕ_2 integrals are easily evaluated as 2, 2π, and 2π respectively, so that

$$E' = \frac{512e^2}{a_0} \int_0^\infty R_1^2 e^{-4R_1}\, dR_1 \int_0^\infty R_2^2 e^{-4R_2}\, dR_2$$

$$\times \int_0^\pi \frac{\sin\theta_2\, d\theta_2}{(R_1^2 + R_2^2 - 2R_1 R_2 \cos\theta_2)^{1/2}}. \qquad (12\text{–}45)$$

The θ_2 integral is readily integrated to

$$\frac{1}{R_1 R_2}\left[(R_1^2 + R_2^2 - 2R_1 R_2 \cos\theta_2)^{1/2}\right]_0^\pi = \frac{1}{R_1 R_2}\left[R_{12}\right]_{\theta_2=0}^{\theta_2=\pi}$$

$$= \frac{1}{R_1 R_2}\left[R_{12}^\pi - R_{12}^0\right].$$

In evaluating R_{12}^π and R_{12}^0 we must note carefully that R_{12} is a vector *magnitude* and is thus inherently positive, which forces us to consider two different regions of space. For the first region of space, in which $R_2 < R_1$,

$$R_{12}^{\pi} = R_1 + R_2, \quad \text{and} \quad R_{12}^{0} = R_1 - R_2$$

as shown graphically in Fig. 12–3(a). In this first region, the θ_2 integral becomes

$$\frac{2R_2}{R_1 R_2} = \frac{2}{R_1} \quad (0 < R_2 < R_1).$$

For the second region of space, in which $R_2 > R_1$

$$R_{12}^{\pi} = R_1 + R_2, \quad \text{and} \quad R_{12}^{0} = R_2 - R_1$$

as shown in Fig. 12–3(b). In the second region, the θ_2 integral becomes

$$\frac{2R_1}{R_1 R_2} = \frac{2}{R_2} \quad (R_1 < R_2 < \infty).$$

Then, in integrating R_2 over the two separate regions of space, using the appropriate values of the θ_2 integral for each region, we write Eq. (12–45) as

$$E' = \frac{1024d^2}{a_0} \int_0^{\infty} R_1^2 e^{-4R_1} \, dR_1 \left[\int_0^{R_1} \frac{R_2^2}{R_1} e^{-4R_2} \, dR_2 + \int_{R_1}^{\infty} \frac{R_2^2}{R_2} e^{-4R_2} \, dR_2 \right]$$

With R_1 constant, the R_2 integrals are evaluated using the forms,

$$\int x^2 e^{-ax} \, dx = e^{-ax} \left[-\frac{x^2}{a} - \frac{2x}{a^2} - \frac{2}{a^3} \right],$$

and

$$\int x e^{-ax} \, dx = e^{-ax} \left[-\frac{x}{a} - \frac{1}{a^2} \right]$$

to give

$$E' = \frac{1024e^2}{a_0} \int_0^{\infty} R_1^2 e^{-4R_1} \left[e^{-4R_1} \left(-\frac{1}{32R_1} - \frac{1}{16} \right) + \frac{1}{32R_1} \right] dR_1$$

$$= \frac{1024e^2}{a_0} \int_0^{\infty} \left(-\frac{1}{32} R_1 e^{-8R_1} - \frac{1}{16} R_1^2 e^{-8R_1} + \frac{1}{32} R_1 e^{-4R_1} \right) dR_1$$

$$= \frac{1024e^2}{a_0} \left[-\frac{1}{32} \left(\frac{1}{64} \right) - \frac{1}{16} \left(\frac{1}{256} \right) + \frac{1}{32} \left(\frac{1}{16} \right) \right] = \frac{5e^2}{4a_0}$$

But, $E_H = -e^2/2a_0$, so that

$$E' = -\tfrac{5}{2} E_H = -\tfrac{5}{2}(-13.6 \text{ eV}) = +34.0 \text{ eV}.$$

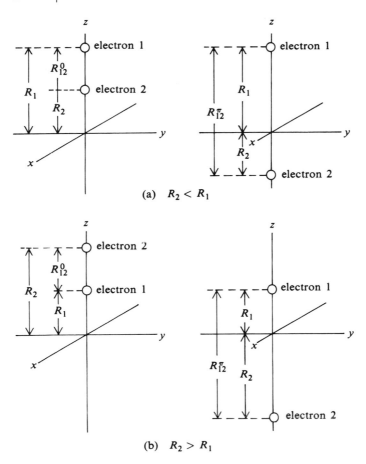

(a) $R_2 < R_1$

(b) $R_2 > R_1$

FIG. 12–3.   Relation of R coordinates at $\theta_2 = 0$ and at $\theta_2 = \pi$.

Substitution of the value for E^0 from Eq. (12–24) and E' from the above equation into Eq. (12–36) yields

$$\tilde{E} = -108.8 + 34.0 = -74.8 \text{ eV}$$

for the approximate ground-state energy of the helium atom, as calculated by a first-order perturbation method.   A comparison of $\tilde{E}$ (-74.8 eV) with the experimental value (-79.0 eV) shows that the calculated value is 5.3% too high (algebraically).   If we use the calculated $\tilde{E}$ value to compute a value for the ionization potential, we obtain

$$I_{\text{calc}} = E_{\text{He}^+} - \tilde{E}$$
$$= -54.4 - (-74.8) = 20.4 \text{ eV}.$$

As compared with the experimental value of I, which is 24.6 eV, the value calculated by the perturbation method is 17% too low. It thus appears that a comparison of calculated and experimental first ionization potentials is a more sensitive measure of the accuracy of a given approximation method than is a comparison of ground-state energies, the reason being that the ionization potential is a measure of a *difference* in energy levels. We often encounter the same situation in other atomic and molecular problems, and it is important to note that most experimental quantities of interest are dependent on *differences* in energy levels. Consequently, small errors in total energies can be very serious when we attempt to estimate quantities that are to be measured experimentally.

The accuracy of the perturbation method is dependent primarily on the accuracy with which ψ^0 reflects the form of the correct wave function, ψ. For small perturbations on $\hat{H}$, ψ is apparently close to ψ^0 as judged by those calculated results which we are able to compare with experimental measurements. We'll define the difference between the energy calculated by a first-order perturbation method and the observed experimental energy as the *charge correlation energy*.* Thus, for the helium atom, the charge correlation energy is

$$\text{(first-order energy)} - \text{(observed energy)} = \text{charge correlation energy,}$$

$$-74.8 - (-79.0) = 4.2 \text{ eV.}$$

The charge correlation energy is the *lowering* of the calculated total energy of the system which would result were we able to correct or to adjust the *wave function* to take into account the tendency of the like-charged electrons to correlate their movements in order to avoid one another.

In the same way that we have used the first-order perturbation method to calculate the energy of the helium atom, we may also calculate first-order energies of other two-electron systems, which we'll refer to as *helium-like ions*. We may obtain experimental energies for helium-like ions from measured ionization potentials and then calculate appropriate charge correlation energies as the differences between the first-order and observed energies. Resultant values for the charge correlation energies of several helium-like ions are given in Table 12–1. Evidently the charge correlation energy is not too dependent on atomic number Z.

Although the first-order perturbation method has led to much better results for the ground-state energy of the helium atom than our first zero-order approximation technique in which we ignored electron repulsion altogether, there is one

---

* As suggested by J. W. Linnett, *Wave Mechanics and Valency*, Methuen & Co., London, 1960, p. 65. The term, *charge correlation energy*, as we have defined it, is to be distinguished from the more general term *correlation energy*, which is commonly defined as the difference between the exact eigenvalue energy and its expectation value in the Hartree-Fock approximation (see Sec. 14–2).

TABLE 12–1

Charge Correlation Energies for the Ground States of Helium and Helium-Like Ions*	
Species	Charge correlation energy, eV (first-order energy − observed energy)
H⁻	4.11
He	4.19
Li⁺	4.21
Be²⁺	4.21
B³⁺	4.24
C⁴⁺	4.32
N⁵⁺	4.43
O⁶⁺	4.60
F⁷⁺	4.84

* Values converted from data of W. Kauzmann, *Quantum Chemistry*, Academic Press, New York, 1957, p. 288.

major difficulty in the approach. Unless an experimental result is available for comparison, we have no way of knowing how good or how bad our approximation has been.

12–4 THE VARIATION THEOREM

In using the perturbation method we made use of the mean-value postulate (Eq. 6–81) in order to estimate the ground-state energy of the helium atom. However, the energy which we obtained was not really equal to the quantum-mechanical expectation value because ψ^0 for the ground state was only an approximation of the true wave function. A better choice of trial wave function (better than ψ^0) would have given an energy value nearer to the true expectation value. In this section we'll present a method, called the *variation method*, which provides a systematic approach for trying and comparing different guesses at the form of the correct wave function. We may state the *variation theorem* as follows:

If we use any well-behaved approximate function $\tilde{\psi}$ in evaluating the integral in the mean-value postulate, the value of the expectation energy $\tilde{E}$ so obtained will always be algebraically equal to or greater than the true ground-state energy E_0 of the system. That is,

$$\tilde{E} = \frac{\int_{-\infty}^{\infty} \tilde{\psi}^* \hat{H} \tilde{\psi}\, d\tau}{\int_{-\infty}^{\infty} \tilde{\psi}^* \tilde{\psi}\, d\tau} \geq E_0, \qquad (12\text{–}46)$$

where $\hat{H}$ is the complete and correct Hamiltonian operator for the system of interest.

It is important to note that $\tilde{E}$ in Eq. (12–46) is *not* the true expectation energy. It is merely a trial energy evaluated from the same expression which yields the true expectation energy *when the true wave function is used*. One way of using the variation theorem would be to keep trying well-behaved $\tilde{\psi}$ functions until we are satisfied that no appreciably lower energy can be obtained, at which point the $\tilde{\psi}$ function would be very close to ψ_0, the wave function for the true ground state. Of course, the mathematics could become quite burdensome, and the advantages of using high-speed computers in such a trial and error approach are apparent.

Before proving the variation theorem, let's consider a very simple problem in order to illustrate the method. Specifically, let's assume that we are not able to calculate directly the ground-state energy for a particle of mass m in a one-dimensional box of length L. That is, assume that we cannot directly solve the Schrödinger equation

$$\hat{H}\psi = \frac{-\hbar^2}{2m}\frac{d^2\psi}{dx^2} = E\psi. \tag{12–47}$$

Instead, we'll devise a trial function, $\tilde{\psi}$, which is well-behaved and obeys the boundary conditions, and insert the trial function into Eq. (12–46). One such acceptable trial function which goes to zero at $x = 0$ and at $x = L$ is given by

$$\tilde{\psi} = x(L - x) = Lx - x^2. \tag{12–48}$$

Insertion of the correct form for $\hat{H}$ and the trial form for $\tilde{\psi}$ into Eq. (12–46) yields

$$\tilde{E} = \frac{\displaystyle\int_0^L (Lx - x^2)\left(\frac{-\hbar^2}{2m}\frac{d^2}{dx^2}\right)(Lx - x^2)\,dx}{\displaystyle\int_0^L (Lx - x^2)(Lx - x^2)\,dx}, \tag{12–49}$$

which, after simple differentiation and integration, becomes

$$\tilde{E} = 1.013\frac{h^2}{8mL^2}. \tag{12–50}$$

Recall that the ground-state energy E_0 for the particle in the one-dimensional box was given as

$$E_0 = h^2/8mL^2, \tag{12–51}$$

so that Eq. (12–50) may be written as

$$\tilde{E} = 1.013E_0.$$

Thus, in the above particularly simple case, the expectation energy calculated by the variation method is surprisingly close to and slightly larger than the true ground-state energy. The unexpected accuracy of our first approximate calculation results from the fortunate choice of a trial function which is really not too far different from the true function for the ground state. The variation theorem states, however, that *any* trial function will yield a value of $\tilde{E}$ which is equal to or larger than E_0.

Let's now prove the variation theorem. We have previously mentioned in Chapter 4 that in the absence of nonlinear distortion, the resultant waveform of a plucked violin string, no matter how complicated, can always be represented as a linear superposition of the normal modes of vibration for the stretched and fixed string (see Fig. 12–4). However, in order to apply the *principle of superposition*, we must require that the resultant wave function be well-behaved. For example, the resultant wave function certainly must be continuous, since a discontinuous wave function would imply a violin string *in*

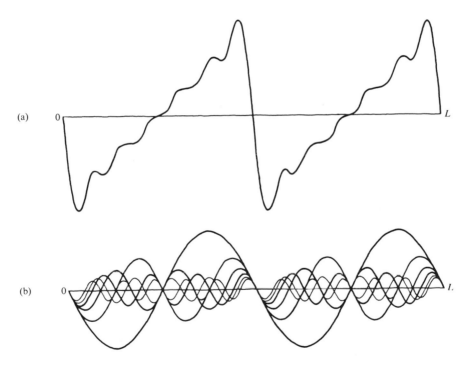

(a)   0   L

(b)   0   L

FIG. 12–4   The "sawtooth" form of the resultant wave shown in (a) may be represented by the graphical linear superposition of the six normal modes shown in (b). As more normal modes are included with appropriate amplitudes, the resultant wave may be made much smoother. According to *Fourier's theorem*, any well-behaved standing wave may be represented mathematically by a series which includes weighted terms representing each of the normal modes.

two pieces! Furthermore, the resultant wave equation must be such that the amplitude is zero at the fixed ends, since all of the normal mode wave equations go to zero at the fixed ends of the string.

In a similar manner, it can be shown* through Fourier's expansion theorem that any *well-behaved* trial function $\tilde{\psi}$ may be expressed as a linear superposition of the orthonormal eigenfunctions $\psi_0, \psi_1, \psi_2, \ldots$, of the system of interest. (Of course, we cannot always directly evaluate these eigenfunctions.) We may then write

$$\tilde{\psi} = a_0\psi_0 + a_1\psi_1 + a_2\psi_2 + \cdots, \tag{12–52}$$

where $a_0, a_1, a_2, \ldots$, are constants and

$$\tilde{\psi}^* = a_0^*\psi_0^* + a_1^*\psi_1^* + a_2^*\psi_2^* + \cdots \tag{12–53}$$

Substitution of Eqs. (12–52) and (12–53) into Eq. (12–46) yields

$$\bar{E} = \frac{\int_{-\infty}^{\infty} (a_0^*\psi_0^* + a_1^*\psi_1^* + a_2^*\psi_2^* + \cdots)\hat{H}(a_0\psi_0 + a_1\psi_1 + a_2\psi_2 + \cdots)\,d\tau}{\int_{-\infty}^{\infty} (a_0^*\psi_0^* + a_1^*\psi_1^* + a_2^*\psi_2^* + \cdots)(a_0\psi_0 + a_1\psi_1 + a_2\psi_2 + \cdots)\,d\tau}. \tag{12–54}$$

After multiplying out all the terms in Eq. (12–54) we may simplify the somewhat bulky equation by using the following relationships:

1. Since any of the functions, for example ψ_m or ψ_n, from the correct set are eigenfunctions of $\hat{H}$,

$$\int \psi_m^*\hat{H}\psi_m\,d\tau = \int \psi_m^*E_m\psi_m\,d\tau = E_m\int \psi_m^*\psi_m\,d\tau, \tag{12–55}$$

and

$$\int \psi_n^*\hat{H}\psi_n\,d\tau = E_n\int \psi_n^*\psi_n\,d\tau, \tag{12–56}$$

where E_m and E_n are the eigenvalue energies in the mth and nth states.

2. Since each of the functions in the correct set is normalized,

$$\int_{-\infty}^{\infty} \psi_m^*\psi_m\,d\tau = 1 \quad \text{and} \quad \int_{-\infty}^{\infty} \psi_n^*\psi_n\,d\tau = 1. \tag{12–57}$$

3. Since each of the functions of the correct set is orthogonal to each of the other functions,

$$\int_{-\infty}^{\infty} \psi_m^*\psi_n\,d\tau = 0. \tag{12–58}$$

---

* See, for example, W. Kauzmann, *Quantum Chemistry*, Academic Press, New York, 1957, p. 63.

The result we obtain after expanding Eq. (12–54) and then simplifying the expression through use of the three relationships above is

$$\tilde{E} = \frac{a_0^* a_0 E_0 + a_1^* a_1 E_1 + a_2^* a_2 E_2 + \cdots}{a_0^* a_0 + a_1^* a_1 + a_2^* a_2 + \cdots},$$

or

$$\tilde{E} = \frac{|a_0|^2 E_0 + |a_1|^2 E_1 + |a_2|^2 E_2 + \cdots}{|a_0|^2 + |a_1|^2 + |a_2|^2 + \cdots},$$

which may also be written as

$$\tilde{E} = \left(\frac{|a_0|^2}{|a_0|^2 + |a_1|^2 + |a_2|^2 + \cdots}\right) E_0 + \left(\frac{|a_1|^2}{|a_0|^2 + |a_1|^2 + |a_2|^2 + \cdots}\right) E_1$$

$$+ \left(\frac{|a_2|^2}{|a_0|^2 + |a_1|^2 + |a_2|^2 + \cdots}\right) E_2 + \cdots. \tag{12–59}$$

But the numerical coefficients of each of the energy terms in Eq. (12–59) may be designated by fractions, b, such that

$$b_0 = \frac{|a_0|^2}{|a_0|^2 + |a_1|^2 + |a_2|^2 + \cdots},$$

$$b_1 = \frac{|a_1|^2}{|a_0|^2 + |a_1|^2 + |a_2|^2 \cdots}, \ldots,$$

where the sum of the fractions is unity, or

$$b_0 + b_1 + b_2 + \cdots = 1.$$

Thus, Eq. (12–59) may be written as

$$\tilde{E} = b_0 E_0 + b_1 E_1 + b_2 E_2 + \cdots \tag{12–60}$$

The energy $\tilde{E}$ may then be thought of as the average of the eigenvalue energies $E_0, E_1, E_2, \ldots$, weighted according to the fractional coefficients $b_0, b_1, b_2, \ldots$. It is immediately obvious that the average energy $\tilde{E}$ must be larger than the lowest energy in the group which comprises the average, unless the lowest energy is the *only* energy, in which case $\tilde{E}$ would be equal to E_0. It is thus required that

$$\tilde{E} \geq E_0, \tag{12–61}$$

and the variation theorem is proved.

12–5 GROUND-STATE ENERGY OF THE HELIUM ATOM. THIRD APPROXIMATION: A VARIATION METHOD

We'll now use the variation theorem to approximate the ground-state energy of the helium atom. As a first step, we must choose a good trial function for helium. The unperturbed function for the ground state of helium was shown to be

$$\psi^0 = \varphi_1 \varphi_2 = \left(\frac{1}{\pi}\right)\left(\frac{2}{a_0}\right)^3 e^{-2r_1/a_0}e^{-2r_2/a_0}, \tag{12-40'}$$

and we'll guess that the correct eigenfunction is not too different in form. Note that the first-order perturbation method we have used for the helium atom in Section 12–3 is equivalent to the variation method in which ψ^0 itself is used as the trial function. That is, Eq. (12–46) is identical to Eq. (12–33) if ψ^0 is substituted for $\tilde{\psi}$. In order to construct a trial function more accurate than ψ^0, we note that the negative charge cloud of the first electron results in an effective screening of the positive nuclear charge from the second electron. That is, because of the presence of the first electron, the second electron in effect "sees" a nucleus whose positive charge is somewhat less than $+2e$. The effect of the second electron on the first electron is considered to be the same. Therefore, a promising trial modification of Eq. (12–40') would result from the reduction of the nuclear charge from $+2e$ to an *effective value* of $+Z'e$ so that the trial function is written, through substitution of Z' for 2 in Eq. (12–40'), as

$$\tilde{\psi} = \left(\frac{1}{\pi}\right)\left(\frac{Z'}{a_0}\right)^3 e^{-Z'r_1/a_0}e^{-Z'r_2/a_0}. \tag{12-62}$$

We'll now proceed to leave Z' undetermined and insert $\tilde{\psi}$ from Eq. (12–62) into Eq. (12–46) in order to obtain a value for $\tilde{E}$ in terms of Z'. We'll then select the value of Z' which minimizes $\tilde{E}$, that is, which brings $\tilde{E}$ as close as possible to the true ground-state energy. Since the $\tilde{\psi}$ function given by Eq. (12–62) may be varied by choice of value for Z', it is called a *variation function*. Because the variation function is not complex and is already normalized, Eq. (12–46) may be written as

$$\tilde{E} = \int_{-\infty}^{\infty} \tilde{\psi} \hat{H} \tilde{\psi} \, d\tau. \tag{12-63}$$

Substitution of Eq. (12–62) into Eq. (12–63) yields

$$\tilde{E} = \left(\frac{1}{\pi}\right)^2 \left(\frac{Z'}{a_0}\right)^6 \int_{-\infty}^{\infty}\int_{-\infty}^{\infty} (e^{-Z'r_1/a_0}e^{-Z'r_2/a_0})\hat{H}(e^{-Z'r_1/a_0}e^{-Z'r_2/a_0}) \, d\tau_1 \, d\tau_2, \tag{12-64}$$

where

$$\hat{H} = \frac{-\hbar^2}{2m_e}(\nabla_1^2 + \nabla_2^2) - \frac{2e^2}{r_1} - \frac{2e^2}{r_2} + \frac{e^2}{r_{12}}. \tag{12-65}$$

The correct solution* of Eq. (12–64) yields

$$\tilde{E} = [-2(Z')^2 + \tfrac{27}{4}Z']\left[\frac{-2\pi^2 m_e e^4}{h^2}\right],$$

---

* Davis, J. C., *Advanced Physical Chemistry*, Ronald Press, New York, 1965, pp. 221 ff.

or, through substitution of Eq. (12–21),

$$\tilde{E} = [-2(Z')^2 + \tfrac{27}{4} Z'] E_H \tag{12–66}$$

The best value for Z' will be that value for which $\tilde{E}$ is minimum, that is, that value of Z' for which $d\tilde{E}/dZ' = 0$. The first derivative of Eq. (12–66) equated to zero is

$$\frac{d\tilde{E}}{dZ'} = (-4Z' + \tfrac{27}{4})E_H = 0,$$

from which

$$Z' = \tfrac{27}{16}.$$

That this value of Z' represents a minimum rather than a maximum in $\tilde{E}$ can be verified by noting that $d^2\tilde{E}/dZ'^2 = -4E_H$ is positive, since E_H is negative. We may now calculate $\tilde{E}$ by inserting the value $\tfrac{27}{16}$ for Z' in Eq. (12–66):

$$\tilde{E} = 5.70E_H = 5.70(-13.6 \text{ eV}) = -77.5 \text{ eV}. \tag{12–67}$$

Recall that the experimental value for the ground-state energy of the helium atom is -79.0 eV. In comparison, then, the value -77.5 eV yielded by the variation theorem method is 1.9% too high (algebraically). The ionization potential calculated by the variation method is

$$I = E_{He^+} - \tilde{E} = -54.4 - (-77.5) = 23.1 \text{ eV}.$$

Since the experimental value of I is 24.6 eV, the calculated result is 6.1% too low.

Of the three approximation methods we have thus far considered, the variation method provides calculated values for the ground-state energy and ionization potential in best agreement with experimental values. It is possible to further improve the results of the variation method if we use a trial function which takes into account the tendency of the two electrons to avoid one another. That is, in addition to having mutual nuclear screening effects on one another, the two like-charged electrons in the helium atom correlate their movements and positions in such a way as to remain as far apart as possible, and the trial function should take such *electron correlation* into account. Specifically, we might introduce a second parameter c into the trial function given by Eq. (12–62) such that

$$\tilde{\psi} = \left(\frac{1}{\pi}\right)\left(\frac{Z'}{a_0}\right)^3 (1 + cr_{12})e^{-Z'r_1/a_0}e^{-Z'r_2/a_0}. \tag{12–68}$$

The incorporation of the additional polynomial term $(1 + cr_{12})$ causes $\tilde{\psi}$ to be larger when r_{12} is larger (if $c > 0$). Thus the probability $\tilde{\psi}^2$ for a particular distribution will also be greater as r_{12} increases. That is, distributions in which the electrons are farther apart become more probable. Using Eq. (12–68) in the variation method, we are able to calculate a function for $\tilde{E}$ in terms of c and Z'. Minimization of $\tilde{E}$ in terms of both c and Z' then leads to parameter values of 0.364 and 1.849 respectively, and to a ground-state energy for the helium atom which is only 0.34 eV larger than the experimental value.

Apparently the incorporation of more parameters into the variation function allows one to more successfully minimize $\tilde{E}$ and thus arrive at an energy value more closely in agreement with the experimental value. Hylleraas* in 1930 used a function of the general form of Eq. (12–68) in which the polynomial contained 14 terms, each term containing an adjustable constant. Agreement with the experimental value was obtained to within 0.003 eV. More recently Kinoshita** has used linear combinations of up to 80 terms and Pekeris† has used up to 1078 functional terms to obtain a ground-state energy for helium within experimental error.

Things on a very small scale behave like nothing that you have any direct experience about. They do not behave like waves, they do not behave like particles, they do not behave like clouds, or billiard balls, or weights on springs, or like anything that you have ever seen.

— RICHARD P. FEYNMAN,   *The Feynman Lectures on Physics.*‡

PROBLEMS

12–1 a) Write the complete Schrödinger amplitude equation for the lithium atom.

  b) Ignoring the potential energy terms for electron repulsion, separate the total wave equation into three equations, each involving only the coordinates of one of the electrons.

  c) Ignoring repulsion energy, approximate the ground-state energy of the Li atom.

12–2 Set up the expression for the first-order perturbation energy E', for the ground state of the Be^{2+} ion.

12–3 Calculate zero-order energies E^0 in the ground states of the following helium-like ions: Li^+, Be^{2+}, B^{3+}.

12–4 Show that Eq. (12–48) is not a satisfactory solution of Eq. (12–47), that is, that $Lx - x^2$ is not an eigenfunction of $\hat{H}$ given in Eq. (12–47).

12–5 Complete in detail the integration of Eq. (12–49) to obtain the value of $\tilde{E}$ given by Eq. (12–50).

12–6 Devise an acceptable trial function for the particle in the one-dimensional box other than $\tilde{\psi} = Lx - x^2$. Use the function you have devised in the variation theorem

---

* E. Hylleraas, *Z. Physik*, **65**, 209 (1930).

** T. Kinoshita, *Phys. Rev.*, **115**, 366 (1959).

† C. L. Pekeris, *Phys. Rev.*, **115**, 1216 (1959).

‡ R. P. Feynman, R. B. Leighton, and M. Sands, *The Feynman Lectures on Physics*, Vol. III, *Quantum Mechanics*, Addison-Wesley, Reading, Mass., 1965.

method in order to evaluate $\tilde{E}$, the expectation energy. Compare to E_0, the true ground-state energy.

12–7 Assume that the ground-state energy for the particle in the three-dimensional box cannot be directly solved.

a) Postulate a satisfactory well-behaved trial function which may be used in the variation method, and which does not involve either a trigonometric or exponential function.

b) Write the expression for the evaluation of $\tilde{E}$ according to the variation method, using the above trial function and the appropriate Hamiltonian operator.

12–8 A suitable trial function for the first excited state ($n = 2$) in the variation method for the particle in a one-dimensional box is $\psi = x(2x - L)(x - L)$. That is, ψ goes to zero at $x = 0$, $x = L/2$, and $x = L$.

a) Normalize the above function in the range $0 \leqslant x \leqslant L$.

b) Calculate the ratio $\tilde{E}/E_2$, where E_2 is the true energy in the first excited state ($n = 2$). Comment on the result.

THE HELIUM ATOM II.
EXCITED STATES AND SOME INTERESTING
ASPECTS OF TWO-ELECTRON SYSTEMS

We have used the perturbation method to calculate the ground-state energy of the helium atom, for which each electron was considered to be in a 1s hydrogen-like orbital. As a first approximation, we totally ignored the electron repulsion term (e^2/r_{12}) in the Hamiltonian operator and were then able to directly solve the Schrödinger equation in order to calculate the zero-order energy E^0 for the unperturbed system in the ground state. In obtaining the solution, we showed that if the unperturbed wave function is expressed as the *product* of wave functions for each of the electrons, that is, if $\psi^0 = \varphi_1 \varphi_2$, we are able to separate the Schrödinger equation into two hydrogen-like eigenvalue equations for which solutions may be obtained directly.

13–1 EXCITED STATES:
THE INDISTINGUISHABILITY OF ELECTRONS

We'll now consider the application of a first-order perturbation method to the calculation of the energy of an excited state of helium in which one of the electrons is promoted from the 1s orbital to some higher-energy orbital, for example, the 2s or 2p orbital. As we have done in the ground-state calculation, we'll initially ignore the effects of electron repulsion. That is, we'll initially consider the unperturbed system. A satisfactory solution for the unperturbed Schrödinger equation (Eq. 12–8) may once again be written as a product.

$$\psi^0_{1,2} = \psi_A(1)\psi_B(2), \tag{13–1}$$

where $\psi_A(1)$ signifies that electron 1 is in the ψ_A orbital and $\psi_B(2)$ signifies that electron 2 is in the ψ_B orbital. For example, ψ_A may be the 1s orbital and ψ_B may be the 2s orbital. If E_A is the energy of the first electron in the ψ_A orbital and E_B is the energy of the second electron in the ψ_B orbital, then the total zero-order energy E^0 of the system ignoring repulsion would be $E_A + E_B$.

However, it is apparent that the function

$$\psi_{2,1}^0 = \psi_A(2)\psi_B(1) \tag{13-2}$$

is an *equally satisfactory* solution to the unperturbed Schrödinger equation. In writing $\psi_{2,1}^0$ we assume that electron 2 is in the ψ_A orbital and electron 1 is in the ψ_B orbital. But it is experimentally impossible to distinguish between the two different distributions represented by $\psi_{1,2}^0$ and $\psi_{2,1}^0$. That is, *we cannot determine which electron is in which orbital*.

Thus far we have been concerned only with ensuring that the wave function for a given system satisfy the Schrödinger equation and the imposed boundary and continuity conditions. However, *if the wave function is to completely define a system containing two electrons, it must also account for the indistinguishability of the electrons*. In order for a wave function ψ for a two-electron system to account satisfactorily for electron indistinguishability, it is specifically necessary that the resultant electron probability density ψ^2 remain the same when the coordinates of the two electrons are interchanged, since either of the two indistinguishable configurations must lead to the same electron probability distribution. This means that the interchange of electron coordinates between the two orbitals must yield either the original function ψ or the negative of the original function $(-\psi)$ since the square of either is ψ^2.

Let's now determine whether $\psi_{1,2}^0$ as given by Eq. (13–1) is a satisfactory function for the system in terms of accounting for electron indistinguishability. The interchange of electron coordinates (if ψ_A and ψ_B were spherically symmetrical, this would involve exchanging r_1 and r_2 between the two orbitals) must yield either $\psi_{1,2}^0$ or $(-\psi_{1,2}^0)$. Let's define the *interchange operator* $\rightleftarrows$ as an operator which interchanges coordinates between the two electrons. Then

$$\rightleftarrows \psi_{1,2}^0 = \psi_A(2)\psi_B(1),$$

which is equal to neither $\psi_{1,2}^0$ nor $(-\psi_{1,2}^0)$. Thus $\psi_{1,2}^0$ as given by Eq. (13–1) is *not* a satisfactory representation of the state in which the electrons are indistinguishable and each is in a different orbital. By the same argument, neither is $\psi_{2,1}^0$ as given by Eq. (13–2) satisfactory.

As a comparison, note that if the two electrons were in the *same* orbital, as in the previous ground-state calculation for the helium atom, the wave function

$$\psi_{2,1}^0 = \psi_{1s}(2)\psi_{1s}(1)$$

is identical to the interchanged function

$$\psi_{1,2}^0 = \psi_{1s}(1)\psi_{1s}(2).$$

By virtue of their being in the same orbital, the two electrons in the ground state are automatically indistinguishable in terms of space coordinates.

But let's again return to the excited state. Even though neither the $\psi_{1,2}^0$ function as given by Eq. (13–1) nor the $\psi_{2,1}^0$ function as given by Eq. (13–2) is in itself a satisfactory function in terms of electron indistinguishability, we may create two new satisfactory functions through linear combination of Eqs. (13–1) and (13–2). Because the two states represented by $\psi_{1,2}^0$ and $\psi_{2,1}^0$ correspond to the same eigenvalue energy, any linear combination of $\psi_{1,2}^0$ and $\psi_{2,1}^0$ is a satisfactory solution to the unperturbed Schrödinger equation [review Eq. (7–4) ff.]. The two linear combinations which are *also* satisfactory with respect to the criterion of *electron indistinguishability* are the sum and the difference of Eq. (13–1) and Eq. (13–2):

$$\psi_s = (1/\sqrt{2})[\psi_A(1)\psi_B(2) + \psi_A(2)\psi_B(1)], \qquad (13\text{–}3)$$

and

$$\psi_a = (1/\sqrt{2})[\psi_A(1)\psi_B(2) - \psi_A(2)\psi_B(1)]. \qquad (13\text{–}4)$$

Assuming ψ_A orthogonal to ψ_B, the coefficient, $1/\sqrt{2}$, is required to ensure that ψ_s and ψ_a are normalized. The interchange of electron coordinates in ψ_s yields

$$\rightleftarrows \psi_s = (1/\sqrt{2})\left[\psi_A(2)\psi_B(1) + \psi_A(1)\psi_B(2)\right] = \psi_s,$$

which is identical to the original function. The interchange of electron coordinates in ψ_a yields

$$\rightleftarrows \psi_A = (1/\sqrt{2})\left[\psi_A(2)\psi_B(1) - \psi_A(1)\psi_B(2)\right] = -\psi_a,$$

which is the negative of the original function. Because the sign of ψ_s does *not* change when the coordinates of the electrons are interchanged, it is said to be *symmetric*; hence the subscript s. Because the sign of ψ_a changes when the coordinates of the electrons are interchanged, it is said to be *antisymmetric*, hence the subscript a.

Either of the functions ψ_s or ψ_a is a satisfactory wave function for the unperturbed two-electron system in which one electron is in the ψ_A orbital and the other electron is in the ψ_B orbital because:

1. Either ψ_s or ψ_a is a satisfactory solution to the unperturbed Schrödinger equation, and meets the required boundary and continuity conditions.

2. Either ψ_s or ψ_a satisfactorily accounts for the indistinguishability of electrons.

In the unperturbed state, ψ_s and ψ_a correspond to degenerate states; however the degeneracy will be seen to disappear when we perturb the system by bringing electron repulsion into play.

Let's now examine in more detail the lowest excited state for the helium atom, in which one electron is in the 1s orbital and the second electron is in the 2s orbital. We'll begin with the unperturbed system (ignore electron repulsion) and assume that ψ_A and ψ_B are given by the 1s and 2s orbitals for hydrogen-like ions. Thus, from Table 11–1, where $Z = 2$,

$$\psi_A = 1s = \left(\frac{1}{\pi}\right)^{1/2}\left(\frac{2}{a_0}\right)^{3/2} e^{-2r/a_0} \tag{13–5}$$

and

$$\psi_B = 2s = \left(\frac{1}{\pi}\right)^{1/2}\left(\frac{1}{a_0}\right)^{3/2}\left(1 - \frac{r}{a_0}\right)e^{-r/a_0}. \tag{13–6}$$

The symmetric and the antisymmetric wave functions for the 1s2s excited state in the helium atom are then obtained by substituting Eqs. (13–5) and (13–6) into Eqs. (13–3) and (13–4), noting that r_1 must be substituted for r in orbitals of electron 1, and r_2 must be substituted for r in orbitals for electron 2. Thus

$$\psi_s = \frac{1}{\sqrt{2}}\left[\left(\frac{1}{\pi}\right)^{1/2}\left(\frac{2}{a_0}\right)^{3/2} e^{-2r_1/a_0}\left(\frac{1}{\pi}\right)^{1/2}\left(\frac{1}{a_0}\right)^{3/2}\left(1 - \frac{r_2}{a_0}\right)e^{-r_2/a_0}\right.$$
$$\left. + \left(\frac{1}{\pi}\right)^{1/2}\left(\frac{2}{a_0}\right)^{3/2} e^{-2r_2/a_0}\left(\frac{1}{\pi}\right)^{1/2}\left(\frac{1}{a_0}\right)^{3/2}\left(1 - \frac{r_1}{a_0}\right)e^{-r_1/a_0}\right], \tag{13–7}$$

or

$$\psi_s = \frac{2}{\pi a_0^3}\left[e^{-2r_1/a_0}\left(1 - \frac{r_2}{a_0}\right)e^{-r_2/a_0} + e^{-2r_2/a_0}\left(1 - \frac{r_1}{a_0}\right)e^{-r_1/a_0}\right]. \tag{13–8}$$

Similarly, for the antisymmetric function,

$$\psi_a = \frac{2}{\pi a_0^3}\left[e^{-2r_1/a_0}\left(1 - \frac{r_2}{a_0}\right)e^{-r_2/a_0} - e^{-2r_2/a_0}\left(1 - \frac{r_1}{a_0}\right)e^{-r_1/a_0}\right]. \tag{13–9}$$

For the symmetric state ψ_s, the probability per unit radius of simultaneously finding electron 1 at radius r_1 in *any* direction and electron 2 at radius r_2 in *any* direction is given as

$$D(r_1, r_2)_s = 4\pi r_1^2 \cdot 4\pi r_2^2 \cdot \psi_s^2, \tag{13–10}$$

and for the antisymmetric state, ψ_a, the corresponding probability is

$$D(r_1 r_2)_a = 4\pi r_1^2 \cdot 4\pi r_2^2 \cdot \psi_a^2. \tag{13–11}$$

Contour plots for both $D(r_1 r_2)_s$ and $D(r_1 r_2)_a$ are given as functions of r_1 and r_2 in Fig. 13–1. The electron distributions in the two states are quite different. For the symmetric ψ_s state, there is a maximum probability of finding both electrons at the same radius of about $0.5a_0$, where a_0 is the radius of the first Bohr orbit. In addition, there is also a lower but finite probability of finding both electrons simultaneously at a radius of about $1.6a_0$.

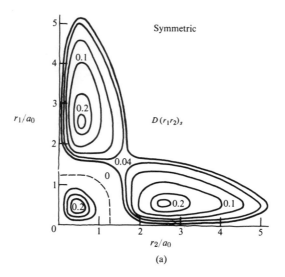

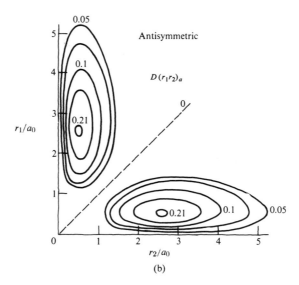

FIG. 13–1   Contour plots showing the probability per unit radius that electron 1 is at radius r_1 at the same time that electron 2 is at radius r_2. (a) The symmetric 1s2s excited state of the helium atom: $D(r_1 r_2)_s = 4\pi r_1^2 \times 4\pi r_2^2 \psi_s^2$. (b) The antisymmetric 1s2s excited state of the helium atom: $D(r_1 r_2)_a = 4\pi r_1^2 \times 4\pi r_2^2 \psi_a^2$. (Adapted from J. W. Linnett, *Wave Mechanics and Valency*, Methuen & Co., London; Wiley and Sons, New York, 1960, pp. 74, 75.)

For the antisymmetric ψ_a state, however, there is *zero probability* of finding the two electrons at the same radius for *any* value of r. This situation is also clearly evident from a consideration of Eq. (13–9). If $r_1 = r_2$ in Eq. (13–9), that is, if both electrons are at the same radius, $\psi_a = 0$.

Then, in terms of the crude zero-order non-repulsion model for the 1s2s state of helium, we are led to conclude that the electrons are on the average farther apart in the antisymmetric state than in the symmetric state. It must be emphasized that the electron interactions or correlations which appear in Fig. 13–1 do not arise from direct forces between the electrons. Recall that up to now we have been dealing with an unperturbed system in which we have ignored electron repulsion. For such a system we might well have predicted that the position of either one of the electrons in its orbital should have been entirely independent of the position of the second electron in its orbital. Apparently, the unexpected correlation of electron positions which is shown in Fig. 13–1 arises entirely from the additional requirement that the wave function account for the indistinguishability of the two electrons, that is, from the requirement that the space coordinates of the pair of electrons be interchangeable without any effect on the resultant probability distribution. It is important to note that the zero-order model is only a crude approximation. More refined calculations of excited states of helium lead, in fact, to the opposite conclusion that the electrons are on the average *farther apart* in the symmetric state than in the antisymmetric state.

Before we return to a further consideration of the energies of the two allowed excited 1s2s states, we must first consider the additional significance of the intrinsic spins of the two electrons.

13–2 INDISTINGUISHABILITY OF ELECTRON SPINS

In order that the wave function ψ might satisfactorily account for the indistinguishability of electrons with respect to coordinate positions, we have shown that it is necessary that we be able to interchange the coordinates of any pair of electrons without changing the value of ψ^2. *But electrons must also be indistinguishable with respect to the orientations of intrinsic spin.* That is, if an atom contains two electrons, one with a spin momentum component of $+\hbar/2$ (spin *up*) with respect to a specified z-axis and the other with a spin momentum component of $-\hbar/2$ (spin *down*) with respect to a specified z-axis, it is possible experimentally to determine only that the *net spin angular momentum is zero* with respect to the specified axis. *We cannot determine which of the two electrons has which spin.* In order that the wave function ψ might also satisfactorily account for the indistinguishability of electrons with respect to their intrinsic spins, it is necessary that we be able to interchange the spins of any pair of electrons without changing the value of ψ^2.

In Section 11–10 we showed that the total wave function, including spin, may be expressed as the product of the spatial function $R\Theta\Phi$ times either of two spin functions α or β (Eqs. 11–52 and 11–53). We arbitrarily called α the function for *spin up*, in which $m_s = +\frac{1}{2}$, and β the function for *spin down*, in which $m_s = -\frac{1}{2}$. Thus, for an electron in the ψ_A orbital, the total wave function including electron spin may be either $\psi_A\alpha$ or $\psi_A\beta$, and for an electron in the ψ_B orbital, the total wave function may be either $\psi_B\alpha$ or $\psi_B\beta$.

If for an unperturbed *two-electron* system we assume that the interaction between spin and orbital motion is negligible, we may consider the complete wave function for the system to be the product of a suitable spatial function for the two electrons times a suitable spin function for the two electrons. We have already shown that when one electron is in the ψ_A hydrogen-like space orbital and the other electron is in the ψ_B hydrogen-like space orbital, the only two-electron spatial functions which are satisfactory are given by the symmetric function ψ_s (Eq. 13–3), and the antisymmetric function ψ_a (Eq. 13–4). By a similar argument concerning spin indistinguishability, in order for the spin function to be a suitable representation for the pair of electrons, it must either remain the same or at the most change signs when the spins of the two electrons are interchanged. It is easily shown that the only spin functions for a pair of electrons which satisfactorily account for the indistinguishability of intrinsic spins are

$$\alpha(1)\alpha(2) \qquad\qquad\qquad \text{(symmetric)}, \qquad\qquad (13\text{–}12)$$

$$\beta(1)\beta(2) \qquad\qquad\qquad \text{(symmetric)}, \qquad\qquad (13\text{–}13)$$

$$(1/\sqrt{2})[\alpha(1)\beta(2) + \alpha(2)\beta(1)] \quad \text{(symmetric)}, \qquad\qquad (13\text{–}14)$$

$$(1/\sqrt{2})[\alpha(1)\beta(2) - \alpha(2)\beta(1)] \quad \text{(antisymmetric)}, \qquad (13\text{–}15)$$

where $1/\sqrt{2}$ is once more used as a normalization factor.

Interchange of electrons in the first three of the above functions reproduces the original function. That is, they are symmetric spin functions. The interchange of electrons in Eq. (13–15), however, reverses the sign of the spin function. Hence, the last spin function is an antisymmetric spin function.

It is now apparent that there are eight suitable complete wave functions for the pair of electrons, which are given by the eight possible products obtained when each of the two satisfactory spatial functions, Eqs. (13–3) and (13–4), is multiplied by each of the four suitable spin functions, Eqs. (13–12), (13–13), (13–14), and (13–15). Furthermore, the product of a symmetric spatial function and a symmetric spin function yields a symmetric complete function. The product of an antisymmetric spatial function and an antisymmetric spin function also yields a symmetric complete function. However, when either the spatial function or the spin function is symmetric and the other is antisymmetric, the complete function is antisymmetric. The eight complete wave functions which satisfy the unperturbed Schrödinger equation and which *also* satisfy the criterion of indistinguishability of electron coordinates *and* electron spins are

$$\frac{1}{\sqrt{2}}[\psi_A(1)\psi_B(2) + \psi_A(2)\psi_B(1)][\alpha(1)\alpha(2)] \qquad \text{(symmetric)}, \qquad (13\text{--}16)$$

$$\frac{1}{\sqrt{2}}[\psi_A(1)\psi_B(2) + \psi_A(2)\psi_B(1)][\beta(1)\beta(2)] \qquad \text{(symmetric)}, \qquad (13\text{--}17)$$

$$\frac{1}{\sqrt{2}}[\psi_A(1)\psi_B(2) + \psi_A(2)\psi_B(1)]\frac{1}{\sqrt{2}}[\alpha(1)\beta(2) + \alpha(2)\beta(1)] \text{ (symmetric)}, \quad (13\text{--}18)$$

$$\frac{1}{\sqrt{2}}[\psi_A(1)\psi_B(2) \\ + \psi_A(2)\psi_B(1)]\frac{1}{\sqrt{2}}[\alpha(1)\beta(2) - \alpha(2)\beta(1)] \quad \text{(antisymmetric)}, \quad (13\text{--}19)$$

$$\frac{1}{\sqrt{2}}[\psi_A(1)\psi_B(2) - \psi_A(2)\psi_B(1)][\alpha(1)\alpha(2)] \qquad \text{(antisymmetric)}, \quad (13\text{--}20)$$

$$\frac{1}{\sqrt{2}}[\psi_A(1)\psi_B(2) - \psi_A(2)\psi_B(1)][\beta(1)\beta(2)] \qquad \text{(antisymmetric)}, \quad (13\text{--}21)$$

$$\frac{1}{\sqrt{2}}[\psi_A(1)\psi_B(2) \\ - \psi_A(2)\psi_B(1)]\frac{1}{\sqrt{2}}[\alpha(1)\beta(2) + \alpha(2)\beta(1)] \quad \text{(antisymmetric)}, \quad (13\text{--}22)$$

$$\frac{1}{\sqrt{2}}[\psi_A(1)\psi_B(2) \\ - \psi_A(2)\psi_B(1)]\frac{1}{\sqrt{2}}[\alpha(1)\beta(2) - \alpha(2)\beta(1)] \quad \text{(symmetric)}. \qquad (13\text{--}23)$$

13-3 THE PAULI EXCLUSION PRINCIPLE

Even though each of the above eight wave functions for the electron pair is suitable with respect to the criteria we have thus far established, a great amount of experimental evidence necessitates the postulation of an additional principle which further restricts the forms of allowed functions which describe the system. The postulate, first enunciated by Wolfgang Pauli in 1924, is called the *Pauli exclusion principle* and may be stated in its quantum-mechanical form as follows:

To be acceptable, the total wave function for a system of electrons must be anti-symmetric to the simultaneous exchange of coordinates and spins between any pair of electrons.

It should be noted that nature also requires that descriptions of systems of *other* identical particles, such as protons or neutrons, be restricted to antisymmetric functions. In general, identical particles whose systems must be described by antisymmetric functions are called *Fermi particles* or *fermions* (after Enrico Fermi). On the other hand, identical particles, such as deuterons or helium nuclei, whose systems must be described by *symmetric* wave functions are called

Bose particles or *bosons* (after Satyandra Nath Bose). The Pauli postulate, like other basic postulates of quantum mechanics, cannot be proved except by the success of the conclusions to which it leads.

According to the Pauli principle, only the *antisymmetric* complete functions given above are experimentally satisfactory for the two-electron system in which one electron is in the ψ_A orbital and one electron is in the ψ_B orbital. That is, the only satisfactory total functions are (13–19), (13–20), (13–21), and (13–22).

If one of the electrons is in the 1s hydrogen-like orbital and the other electron is in the 2s orbital, the four satisfactory total wave functions may be written as

$$\frac{1}{\sqrt{2}}\,[1s(1)2s(2) + 1s(2)2s(1)]\,\frac{1}{\sqrt{2}}\,[\alpha(1)\beta(2) - \alpha(2)\beta(1)], \qquad (13\text{–}24)$$

which contains the *symmetric spatial* function, and

$$\frac{1}{\sqrt{2}}\,[1s(1)2s(2) - 1s(2)2s(1)][\alpha(1)\alpha(2)], \qquad (13\text{–}25)$$

$$\frac{1}{\sqrt{2}}\,[1s(1)2s(2) - 1s(2)2s(1)][\beta(1)\beta(2)], \qquad (13\text{–}26)$$

$$\frac{1}{\sqrt{2}}\,[1s(1)2s(2) - 1s(2)2s(1)]\,\frac{1}{\sqrt{2}}\,[\alpha(1)\beta(2) + \alpha(2)\beta(1)], \qquad (13\text{–}27)$$

each of which contains the *antisymmetric spatial* function. Note that the one allowed function (Eq. 13–24) which contains the symmetric spatial part is also associated with two electrons which have opposed spins. We say that the spins of such electrons are *paired*.

For the first of the total functions (Eq. 13–25) which contains the antisymmetric spatial part, note that the electron spins are unpaired and *up*. The net z-component of total spin angular momentum is then

$$+ \frac{\hbar}{2} + \frac{\hbar}{2} = +\hbar$$

For the second total function (Eq. 13–26) which contains the antisymmetric spatial part, we see that the electron spins are again unpaired but oriented *down*. The z-component of spin angular momentum is then

$$- \frac{\hbar}{2} - \frac{\hbar}{2} = -\hbar$$

For the third total function (Eq. 13–27) which contains the antisymmetric spatial part, the net z-component of spin angular momentum is zero. This may be pictured as resulting when the electron spins are coupled with a vectorial sum perpendicular to the z-axis, as shown in Fig. 13–2.

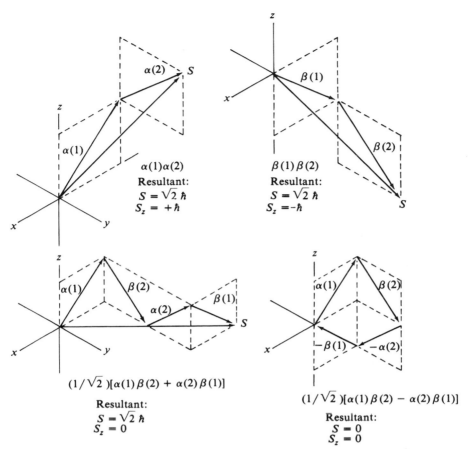

FIG. 13-2 Vector addition of the spins of two electrons. To construct the models, the spin vectors of the two electrons are drawn in planes perpendicular to one another. The resultant S vectors are to scale and in the plane of the drawings.

We thus associate the symmetric space function with two electrons having opposed spins and the antisymmetric space function with two electrons having parallel spins.

Finally, we may state the Pauli exclusion principle in a more familiar form. We have noted that when two electrons in the same atom are in different spatial orbitals, that is, when any one of their space quantum numbers n, l, or m is different, they may have either the same spins or different spins (their m_s value may be the same or different). However, when both electrons are in the same spatial orbital, that is, when they have the same value for the quantum numbers n, l, and m, the only way in which the complete wave function may be antisymmetric and thus conform to the Pauli principle is for the spin function to be antisymmetric, that is

$$(1/\sqrt{2})[\alpha(1)\beta(2) - \alpha(2)\beta(1)]. \tag{13-15'}$$

In such a case the complete wave function would be

$$\psi_A(1)\psi_A(2)(1/\sqrt{2})[\alpha(1)\beta(2) - \alpha(2)\beta(1)], \tag{13-28}$$

and the electrons would thus be required to have opposed spins or different values for the spin quantum number m_s. The *Pauli exclusion principle* is often stated as follows:

No two electrons in a given atom may have all four quantum numbers the same.

The above statement of the Pauli principle will be especially helpful when we extend the assignment of quantum numbers and energy levels to electrons in more complex atoms.

13-4 SINGLET AND TRIPLET STATES

The allowed complete wave functions for the two electrons in the 1s2s excited state of the helium atom are given by Eqs. (13–24), (13–25), (13–26), and (13–27). When the spatial function is symmetric, the only complete wave function which is satisfactory is the one (Eq. 13–24) in which the electron spins are opposed or paired. Since the symmetric spatial function appears in only one of the allowed states, it is said to be a *singlet* state. The singlet state is nondegenerate and has only one spin arrangement for the pair of electrons.

When the spatial function is antisymmetric, however, the complete wave functions given by Eqs. (13–25), (13–26), and (13–27) represent three degenerate states in which three different spin arrangements are possible. The three-fold-degenerate state which involves the antisymmetric space function is called a *triplet* state. The three states are degenerate (have identical energies) in our present approximation because we are assuming no interaction between spin and orbital magnetic moments. Actually, there are small magnetic interactions between spin and orbital motions in the helium atom which split the triplet state into three closely spaced energy levels (we'll not, however, be concerned here with the calculation of these energies).

It is interesting to note that whenever the spatial function for a pair of electrons in *any* two hydrogen-like orbitals is antisymmetric, any one of the three symmetric spin functions given by Eqs. (13–12), (13–13), and (13–14) may be chosen in conformity with the Pauli principle. Thus, antisymmetric two-electron spatial functions are always triplet states in which the electrons are unpaired. Symmetric two-electron spatial functions, on the other hand, always represent singlet states in which the electrons are paired. Since there are only two allowed spatial functions for the two-electron system, the symmetric function given by Eq. (13–3) and the antisymmetric function given

TABLE 13–1

Some Allowed States in the Unperturbed Helium Atom

State		Symmetry of spatial function	Electron spins	Degeneracy
$1s^2$	singlet	Symmetric	Paired	1
$1s2s$	singlet	Symmetric	Paired	1
	triplet	Antisymmetric	Unpaired	3
$1s2p_x$	singlet	Symmetric	Paired	1
	triplet	Antisymmetric	Unpaired	3
$1s2p_y$	singlet	Symmetric	Paired	1
	triplet	Antisymmetric	Unpaired	3
$1s2p_z$	singlet	Symmetric	Paired	1
	triplet	Antisymmetric	Unpaired	3
$2s^2$	singlet	Symmetric	Paired	1

by Eq. (13–4), we'll always expect a singlet state and a triplet state for each two-electron combination in which the hydrogen-like orbitals are different. If the two electrons are in the same orbital, the antisymmetric spatial function given by Eq. (13–4) vanishes to zero and the only allowed state is a nondegenerate singlet state. Some of the allowed states in the *unperturbed* helium atom are given in Table 13–1. These same allowed states apply to other *unperturbed* two-electron systems such as Li^+, Be^{2+}, B^{3+}, C^{4+}, etc.

13–5 AN APPROXIMATION OF ENERGY LEVELS IN THE 1s2s EXCITED STATE OF THE HELIUM ATOM

Let's now use a first-order perturbation method, such as we have previously used for the ground state of the helium atom, to approximate the energy levels in the 1s2s excited state. We'll once more consider the perturbation in the Hamiltonian operator to be due to the repulsion of the two electrons. The total energy of the atom, $\tilde{E}$, will again be given by the sum of the zero-order energy E^0, which is calculated for the atom in which electron repulsion is ignored, and the perturbation energy E', which is given by Eq. (12–37) as

$$E' = \frac{\int_{-\infty}^{\infty} \psi^{0*}(e^2/r_{12})\psi^0 \, d\tau}{\int_{-\infty}^{\infty} \psi^{0*}\psi^0 \, d\tau}. \tag{12–37'}$$

For the helium atom in the 1s2s state, E^0 is given as the sum of the energies of an electron in the 1s orbital of He^+ and an electron in the 2s orbital of He^+. This sum is given by Eq. (11–22), where $n_1 = 1$ and $n_2 = 2$, as

$$E^0 = 4E_H(\tfrac{1}{1} + \tfrac{1}{4}) = 5E_H$$

$$= 5(-13.6 \text{ eV}) = -68.0 \text{ eV}. \tag{13–29}$$

In computing the perturbation energies for the 1s2s state by separately inserting each of the four complete functions given by Eqs. (13–24), (13–25), (13–26), and (13–27) into Eq. (12–37′), we note immediately that the spin functions cancel out completely since they are independent of space coordinates. Thus, in the present treatment in which we assume no interaction between spin and orbital motions of electrons, the energy of any of the four 1s2s states is determined entirely by the form of the spatial function.

Substitution of the normalized and unperturbed *spatial* functions given for the singlet and triplet states as part of Eqs. (13–24) and (13–25) into Eq. (12–37′) yields the combined form for the perturbation energy,

$$E' = \int_{-\infty}^{\infty} \int_{-\infty}^{\infty} \frac{e^2}{r_{12}} \left(\tfrac{1}{2}\right) [1s(1)2s(2) \pm 1s(2)2s(1)]^2 \, d\tau_1 \, d\tau_2, \qquad (13\text{–}30)$$

where the (+) sign refers to the perturbation energy of the singlet state and the (−) sign refers to the perturbation energy of the triplet state. Since the limits on τ_1 and τ_2 are identical,

$$\int_{-\infty}^{\infty} \int_{-\infty}^{\infty} \frac{e^2}{r_{12}} [1s(1)2s(2)]^2 \, d\tau_1 \, d\tau_2 = \int_{-\infty}^{\infty} \int_{-\infty}^{\infty} \frac{e^2}{r_{12}} [1s(2)2s(1)]^2 \, d\tau_1 \, d\tau_2, \qquad (12\text{–}31)$$

so that expansion of Eq. (13–30) yields

$$E' = \int_{-\infty}^{\infty} \int_{-\infty}^{\infty} \frac{e^2}{r_{12}} [1s(1)2s(2)]^2 \, d\tau_1 \, d\tau_2$$

$$\pm \int_{-\infty}^{\infty} \int_{-\infty}^{\infty} \frac{e^2}{r_{12}} [1s(1)2s(1)1s(2)2s(2)] \, d\tau_1 \, d\tau_2 \qquad (13\text{–}32)$$

or

$$E' = J \pm K, \qquad (13\text{–}33)$$

where J, the *coulombic energy*, is given by the *coulombic integral* as

$$J = \int_{-\infty}^{\infty} \int_{-\infty}^{\infty} \frac{e^2}{r_{12}} [1s(1)2s(2)]^2 \, d\tau_1 \, d\tau_2, \qquad (13\text{–}34)$$

and K, the *exchange energy*, is given by the *exchange integral* as

$$K = \int_{-\infty}^{\infty} \int_{-\infty}^{\infty} \frac{e^2}{r_{12}} [1s(1)2s(1)1s(2)2s(2)] \, d\tau_1 \, d\tau_2. \qquad (13\text{–}35)$$

The *coulombic energy J* is a measure of the mean repulsion of the two electrons, each in a different orbital, and is identical in form to the perturbation term we have previously used in our calculation of the ground-state energy of the helium atom. It is a measure of the approximate mean potential energy of repulsion which would exist if there were no effects arising from the indistinguishability of electrons and the Pauli principle.

The *exchange energy K* is a measure of the change in energy which results from the tendency of two electrons to correlate their positions in such a way as to keep apart for parallel spins and come together for opposed spins. The exchange energy may also be considered as the change in energy resulting from a required exchangeability of coordinate positions.

Substitution of Eq. (13–33) into Eq. (12–36) yields the total energy of the helium atom in the 1s2s state as

$$\tilde{E} = E^0 + J \pm K, \tag{13–36}$$

where the $(+)$ sign refers to the singlet state and the $(-)$ sign refers to the triplet state.

Substitution of the 1s and 2s hydrogen-like orbitals from Table 11–1 into Eq. (13–34), followed by integration,* yields for the coulombic energy,

$$J = -0.839E_{\mathrm{H}} = -0.839(-13.6 \text{ cV}) = +11.4 \text{ eV.} \tag{13–37}$$

Substitution of the 1s and 2s hydrogen-like orbitals into Eq. (13–35), followed by integration,* yields for the exchange energy,

$$K = -0.088E_{\mathrm{H}} = -0.088(-13.6 \text{ eV}) = 1.2 \text{ eV.} \tag{13–38}$$

The energy of the singlet state is then calculated to be

$$\tilde{E}_{\mathrm{singlet}} = E^0 + J + K$$
$$= -68.0 + 11.4 + 1.2 = -55.4 \text{ eV,} \tag{13–39}$$

and the energy of the triplet state is

$$\tilde{E}_{\mathrm{triplet}} = E^0 + J - K$$
$$= -68.0 + 11.4 - 1.2 = -57.8 \text{ eV.} \tag{13–40}$$

The experimentally observed energies for the 1s2s singlet and 1s2s triplet states are -58.4 eV and -59.2 eV respectively. In each case the observed energy is below the corresponding calculated energy. For the most part, these discrepancies may be attributed to the inadequacy of the zero-order orbital functions we have used in the calculation of the perturbation energies. As a first approximation, we have simply used hydrogen-like orbitals which do not at all take into account the effects of nuclear screening. In addition, the effects of instantaneous electron repulsion are not included in the forms of the orbital equations. In Section 12–3, we have defined the charge correlation energy as the difference between the energy calculated by a first-order perturbation method and the experimentally observed energy.

---

* A method for the integration of coulombic and exchange integrals is given in W. Kauzmann, *Quantum Chemistry*, Academic Press, New York, 1957, p. 302.

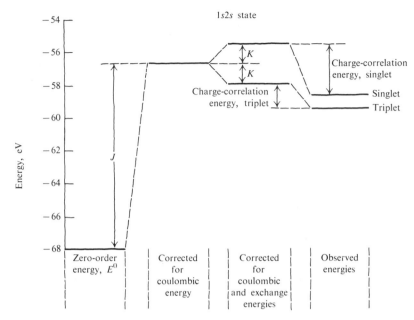

FIG. 13–3 Energy relationships in the 1s2s excited state of the helium atom.

Thus, for the 1s2s singlet state, the charge correlation energy is

$$-55.4 - (-58.4) = 3.0 \text{ eV},$$

and for the 1s2s triplet state the charge correlation energy is

$$-57.8 - (-59.2) = 1.4 \text{ eV}.$$

The higher charge correlation energy for the singlet state as compared to the triplet state is expected, since the non-repulsion function for the singlet state requires that the electrons be closer together; thus the wave function would have to be modified to a greater extent in order to take instantaneous electron repulsion into account. The relationship among the various energy terms is shown graphically in Fig. 13–3.

In a similar fashion one may readily calculate coulombic energies, exchange energies, and charge correlation energies for other excited states of the helium atom. In Table 13–2 calculated and experimental energies for any of the three possible excited 1s2p states ($1s2p_x$, $1s2p_y$, $1s2p_z$) are compared to those for the 1s2s state. An energy correlation diagram for the excited 1s2p state of the helium atom is presented in Fig. 13–4. In terms of the first-order model, the main reason that the 1s2s energies are lower than the 1s2p energies is that the coulombic energy J is larger for a 1s2p state than for the 1s2s state, which in turn results from the fact that the average distance between the electrons is

TABLE 13–2

A Comparison of Calculated and Experimental
Energies for the 1s2s and 1s2p Excited States of the Helium Atom

State		E^0, eV	J, eV	K, eV	Calculated first-order energy $\tilde{E}$ eV	Observed energy, eV	Charge correlation energy, eV
1s2s	Singlet	−68.0	11.4	1.2	−55.4	58.4	3.0
	Triplet	−68.0	11.4	1.2	−57.8	−59.2	1.4
1s2p	Singlet	−68.0	13.2	0.9	−53.9	−57.8	3.9
	Triplet	−68.0	13.2	0.9	−55.7	−58.0	2.3

Values calculated from data given by J. W. Linnett, *Wave Mechanics and Valency*, Methuen & Co., London, 1960, p. 82.

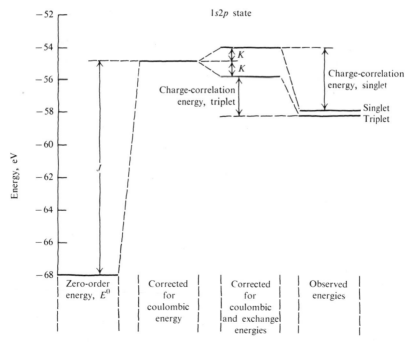

FIG. 13–4 Energy relationships in the 1s2p excited state of the helium atom.

less in a 1s2p state than in the 1s2s state. In addition, larger charge correlation energies are expected to accompany the larger coulombic energies for a 1s2p state, since the wave function would have to be corrected to a greater extent in order to accommodate the increased electron repulsion.

It must be emphasized that the first-order perturbation model is oversimplified. The charge-correlation energies we have calculated are no more than mathematical devices to accommodate inadequacies in the wave functions we have chosen. If we use more refined methods with better wave functions which more adequately account for nuclear screening and electron correlation, we not only obtain better agreement between calculated and experimental energies, but we are forced to reverse our concept of the relative positions of the two electrons. That is, refined calculations* of excited states of helium indicate that on the average the electrons in the triplet state are not only closer to each other than in the singlet state, but are also *closer to the nucleus*, and that the lower energy of the triplet state as compared to the singlet state is due to greater nuclear attraction rather than to lower electronic repulsion.

13–6 HUND'S MULTIPLICITY RULE

In calculating energies of the 1s2s and 1s2p states of the helium atom, we have found that, when the electrons are unpaired as in the triplet state, the energy of the system is lower than when the electrons are paired, as in the singlet state. In general, the concept can be extended to multielectron atoms. That is, the greater the number of electrons having unpaired (parallel) spins, the lower will be the energy and the higher will be the stability of the atom.

These conclusions are stated in *Hund's multiplicity rule*, which is

Other things being equal, the state having the highest multiplicity or the greatest number of unpaired electrons is the most stable.

Because Hund's multiplicity rule follows rather directly from the Pauli principle, it has very few exceptions, and we will find it to be most helpful in the assignment of electrons to different orbitals when we develop the electronic configuration of more complex atoms.

It was as if the electrons were politely told they might not enter and meekly obeyed; somewhat as if, instead of having the police force to prevent overcrowding, one should hang out a sign saying MEASLES or MUMPS.

—BANESH HOFFMANN, *The Strange Story of the Quantum.*[†]

---

* R. P. Messmer and F. W. Birss, *J. Phys. Chem.*, **73**, 2085 (1969)
[†] From *The Strange Story of the Quantum.* Dover, New York, 1959. Reprinted through permission of the publisher.

PROBLEMS

13-1 Show that if $\psi_{1,2}$ and $\psi_{2,1}$ as given by Eqs. (13–1) and (13–2), each individually satisfies the unperturbed Schrödinger equation, then the linear combinations ψ_s and ψ_a as given by Eqs. (13–3) and (13–4) also satisfy the unperturbed Schrödinger equation. Is it necessary that $\psi_{1,2}$ and $\psi_{2,1}$ be degenerate in order for the above relationship to hold?

13-2 Prove that the antisymmetric spatial function ψ_a as given by Eq. (13–9) vanishes whenever the two electrons in the excited helium atom are at the same radius.

13-3 Which of the following wave functions satisfy the principle of indistinguishability of electrons? Which satisfy the Pauli principle?

a) $1s(1)2s(2)\alpha(1)\beta(2)$
b) $\frac{1}{2}[1s(1)2s(2) - 2s(1)1s(2)][\alpha(1)\beta(2) + \beta(1)\alpha(2)]$
c) $1/\sqrt{2}[1s(1)2s(2)\alpha(1)\beta(2) - 2s(1)1s(2)\beta(1)\alpha(2)]$
d) $1/\sqrt{2}[1s(1)2s(2) + 2s(1)1s(2)]\alpha(1)\gamma(?)$

13-4 Write the 12 satisfactory complete wave functions corresponding to the excited 1s2p state of the helium atom. Indicate the singlet and triplet states.

13-5 Show that the spin functions α and β cancel out when the total wave function given by Eq. (13–24) is substituted into Eq. (12–37′) in order to evaluate E'.

In Chapter 10 we calculated directly the allowed energy levels for the hydrogen atom and noted that the energy of a given state is dependent entirely on the value of n, the principal quantum number. The lowest energy state, or ground state, is that state in which the lone electron is in the 1s orbital. The next highest energy level for the hydrogen atom may be represented by either the 2s or 2p state, since these two states are degenerate in hydrogen. Next in energy come the 3s, 3p, and 3d states, which are again degenerate in the hydrogen atom.

However, when we considered the energies of different states of the two-electron *helium* atom in Chapters 12 and 13, we noted that the 1s2s and 2s2p states were not degenerate, primarily because of the difference in coulombic energies due to electron repulsion.

14–1 RELATIVE ENERGIES OF s, p, d AND f SUBLEVELS

A summary of some of the energies which we have calculated for the helium atom through first-order perturbation theory is shown in Fig. 14–1. The energies shown are corrected for coulombic energies but not for exchange energies. This is equivalent to taking the simple (not weighted) mean of the energies of the singlet and triplet states for the 1s2s or 1s2p state. That is, the simple mean of the calculated energies for the singlet and triplet states is given as

$$\frac{(E^0 + J + K) + (E^0 + J - K)}{2} = E^0 + J.$$

As a result of the difference in the magnitude of electron repulsion energy, the 1s2s state of the helium atom is a more stable state (a lower energy state) than the 1s2p state. If we were to extend a first-order perturbation treatment to the

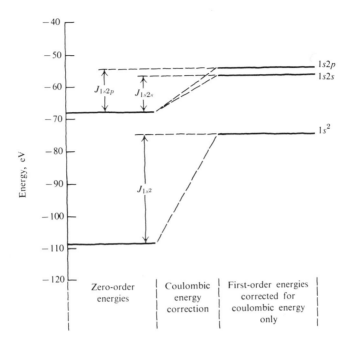

FIG. 14-1 Energies for three states of the helium atom, as calculated by first-order perturbation theory. No corrections are included for exchange energies for the 1s2s and 1s2p states; that is, the energies of the singlet and triplet states are averaged.

1s3s, 1s3p, and 1s3d excited states of helium, we would find that the mean energies of the singlet and triplet states would follow the order

$$1s^2 < 1s2s < 1s2p < 1s3s < 1s3p < 1s3d.$$
$$\text{increasing energy} \rightarrow$$

That is, if the first electron is in the 1s state, the states of increasing energy for the second electron in the helium atom are

$$1s < 2s < 2p < 3s < 3p < 3d.$$

Of course, as we consider more complex atoms containing larger numbers of electrons, the coulombic and exchange interactions become more complex and the relative positions of the energies of the various one-electron orbitals will change. Nevertheless, we'll find that within a given major energy level, that is, for a given value of the principal quantum number n, the order of increasing energy within sublevels is always given as

$$s < p < d < f.$$

14-2 SELF-CONSISTENT FIELD APPROXIMATION FOR POLYELECTRONIC ATOMS

Although it is a simple matter to *write* the Schrödinger equation for any given polyelectronic atom, a direct solution is impossible, even in the relatively simple case of the helium atom. The specific terms which make separation of variables impossible are the electronic repulsion terms in the potential-energy portion of the Hamiltonian operator.

Because of the lack of a direct formal solution in finite terms, it is necessary to resort to approximation methods for complex atoms. But even if we could solve the Schrödinger equation exactly for polyelectronic atoms, there are severe practical limitations. Hartree* has summarized the problem as follows:

> "It has been said that the tabulation of a function of one variable requires a page, of two variables a volume, and of three variables a library; but the full specification of a single wave function of neutral Fe is a function of seventy-eight variables. It would be rather crude to restrict to ten the number of values of each variable at which to tabulate this function, but even so, full tabulation of it would require 10^{78} entries, and even if this number could be reduced somewhat from consideration of symmetry, there would still not be enough atoms in the whole solar system to provide the material for printing such a table."

It is obvious that approximation methods are required, if only for practical reasons. Hartree† in 1928 proposed a successive approximation method known as the *self-consistent field (SCF) method* for solving the Schrödinger equation for a polyelectronic atom, which we'll now outline in very general terms. According to the Hartree method we first ignore interelectron repulsion and define a tentative normalized wave function φ for each electron. We then assume that the total nonperturbed wave function for an atom containing n electrons may be expressed as

$$\psi = \varphi_1(1)\varphi_2(2)\varphi_3(3) \cdots \varphi_n(n), \tag{14-1}$$

where electron 1 is in the φ_1 orbital, electron 2 is in the φ_2 orbital, etc. This, of course, is the same type of function we used in approximating the ground-state energy of the unperturbed helium atom. (Compare with Eq. 12–12.)

In order to select reasonably accurate *tentative* one-electron orbitals φ_i to use in Eq. (14–1), we initially assume that each of the electrons moves in the *same* central *spherically symmetric* potential-energy field $V(r)$, so that we may write the same one-electron Schrödinger equation for each electron. That is,

---

* D. R. Hartree, *Reports on the Prog. of Physics.* (Phys. Soc. of London), **11**, 1B42 (1947).
† D. R. Hartree, *Proc. Cambridge Phil Soc.*, **24**, 89, 111, 426 (1928).

assuming a stationary nucleus, we write

$$\frac{-\hbar^2}{2m_e} \nabla^2 \varphi + V(r)\varphi = E\varphi, \qquad (14\text{--}2)$$

in which $V(r)$ is a first rough approximation of the net potential energy field in the polyelectronic atom. Recall that in the Schrödinger equation for the single electron in the hydrogen-like ion, $V(r)$ was given as $-Ze^2/r$ (Eq. 10–1). However, in constructing $V(r)$ for use in Eq. (14–2), we attempt to take into account the spherically averaged effect of interaction with other electrons, using *numerical* methods, so that the resultant $V(r)$ is usually presented as a table or as a plot.* Thus, Eq. (14–2) usually must be integrated by numerical methods. After obtaining the set of allowed φ functions and the corresponding eigenvalue energies by numerical solution of Eq. (14–2), we assign tentative orbitals to each of the n individual electrons in order of increasing energy, always conforming to the Pauli exclusion principle which limits the number of electrons in each spatial orbital to two (one with spin up and one with spin down). At this point in the calculation, the forms of the individual one-electron functions φ_i and the corresponding energies E_i are only rough approximations, due primarily to the inaccuracy of the initial potential energy term $V(r)$ which was used in Eq. (14–2).

In order to improve the accuracy of the potential-energy term we now assume that for any *one selected electron*, the entire potential energy of interaction with the nucleus and with all of the remaining electrons may be expressed in terms of a *spherically symmetric* potential-energy function V_i *different for each electron*, which at any point in space represents the sum of the potential energy of attraction to the nucleus and the potential energies of repulsion between the selected electron and each of the remaining electrons. The spherical symmetry of the potential energy term V_i is important because the Schrödinger equation may be solved directly *only* when V_i is a function of r_i alone, where r_i is the distance between the nucleus and the electron under consideration. Furthermore, the assumption of spherical symmetry is reasonably accurate because the potential energy of interaction between a given electron and the nucleus is spherically symmetric and because the total charge distribution for any *filled* l level (closed s, p, d, or f shell) may be shown to be spherically symmetric; that is, charge distributions in configurations such as 1s or $1s2s^22p^6$ are spherically symmetric (see Problem 14–1). In the Hartree method we further average the remaining nonspherical charge distributions over all angles to ensure spherical symmetry for the entire potential-energy function and assume that the resultant error will be small.

---

* A good review and several examples of central-field plots are given by R. Latter, *Phys. Rev.*, **99**, 510–512 (1955).

We may now write the independent Schrödinger equation for electron 1 as

$$\frac{-\hbar^2}{2m_e} \nabla_1^2 \varphi_1 + V_1 \varphi_1 = E_1 \varphi_1, \tag{14-3}$$

where V_1 depends only on r_1. For the ith electron we may write

$$\frac{-\hbar^2}{2m_e} \nabla_i^2 \varphi_i + V_i \varphi_i = E_i \varphi_i, \tag{14-4}$$

where V_i depends only on r_i. The total energy of the atom is given as

$$\bar{E} = \sum_{i=1}^{n} E_i. \tag{14-5}$$

However, in order to solve Eq. (14–4), we must determine how we'll formulate the spherically symmetric potential-energy term V_i. To do this, let's first consider the interaction between a selected electron, electron 1 for example, and one of the other remaining electrons, say electron 2. We'll imagine that electron 2 is smeared out into a charge distribution such that at any point in space the fraction of the electron which is contained in a unit volume is given by $|\varphi_2|^2$ and the charge per unit volume is $e|\varphi_2|^2$. Then the charge in the differential volume element $d\tau_2$ is given as

$$e|\varphi_2|^2 \, d\tau_2 = e|\varphi_2|^2 \, dx_2 \, dy_2 \, dz_2. \tag{14-6}$$

The total potential energy of repulsion between electron 1 at a given point (x_1, y_1, z_1) and electron 2 is obtained by integrating the interaction of the charge e of electron 1 with each differential charge of electron 2 (given by Eq. 14–6) over all space, and is written as

$$\int\int\int \frac{e \cdot e |\varphi_2|^2 \, dx_2 \, dy_2 \, dz_2}{r_{12}} = e^2 \int\int\int \frac{|\varphi_2|^2}{r_{12}} \, dx_2 \, dy_2 \, dz_2. \tag{14-7}$$

The complete potential energy term for electron 1 due to attraction to the nucleus of charge Z and to repulsion of *all of the other electrons* is then given as

$$V_1(x_1, y_1, z_1) = \frac{-Ze^2}{r_1} + e^2 \sum_{j \neq 1} \int\int\int \frac{|\varphi_j|^2}{r_{1j}} \, dx_j \, dy_j \, dz_j. \tag{14-8}$$

The solution (usually numerical) of Eq. (14–8) for all closed shell systems leads to a spherically symmetric function for V_1. For atoms containing electrons in open shells, the summation term of Eq. (14–8) must be averaged over all angles, so that V_1 will be spherically symmetric. In either event, once a spherically symmetric form for V_1 is obtained, it is then used in solving Eq. (14–3) to evaluate a first value for E_1 and a first-improved version of φ_1.

   An identical calculation is then repeated for each of the other electrons. That is, for each of the other electrons a potential energy field V_i is calculated from the expression

$$V_i(x_i, y_i, z_i) = \frac{-Ze^2}{r_i} + e^2 \sum_{j \neq i} \int \int \int \frac{|\varphi_j|^2}{r_{ij}} \, dx_j \, dy_j \, dz_j, \qquad (14\text{–}9)$$

which is then used in Eq. (14–4) in order to evaluate a first energy E_i and a first-improved version of φ_i. At this point, we would have a complete set of first-improved wave functions, one for each electron.

The next step is to repeat each of the above calculations, using the set of first-improved φ functions, in order to calculate a new set of energies and a new set of second-improved wave functions. We then again repeat the process a third time, this time using the second-improved φ functions in order to calculate third-improved energies and third-improved φ functions. This iterative process is repeated until there are no further appreciable changes in the E_i values, at which point the φ functions are said to be *self-consistent*, and the energy of the atom is then given by Eq. (14–5).

Since the Schrödinger equation, Eq. (14–4), for each of the one-electron functions is identical in form to the Schrödinger equation for the single electron in the hydrogen-like ion (Eq. 10–12), the mechanism of solution of Eq. (14–4) is the same as that for the hydrogen-like ion. The two Schrödinger equations differ only with respect to the forms of the spherically symmetric potential-energy term V_i. Furthermore, since V_i appears *only* in the equation for the radial solution, *the solutions for* $\Phi(\phi_i)$ *and* $\Theta(\theta_i)$ *are exactly the same as for the hydrogen-like ion*, which means that the angular shapes and probability distributions for the one-electron φ functions are identical to those for the single electron in the hydrogen-like ion. That is, the individual electrons in a poly-electronic atom may be represented by the same angular distributions for s, p, d, and f orbitals which we have already developed in Chapter 11.

On the other hand, the radial functions for the individual electrons $R'(r_i)$ are *not* the same as those for the single electron in a hydrogen-like ion, nor are the corresponding eigenvalue energies E_i, which are obtained from the solution of the one-electron radial equations, the same.

For each of the one-electron functions, we may write

$$\varphi_i = R'_{nl}(r_i)\Theta_{lm}(\theta_i)\Phi_m(\phi_i),$$

where Θ_{lm} and Φ_m are identical to the corresponding functions for the hydrogen-like ion and are given in Tables 10–2 and 10–1, respectively, and where R'_{nl} is different for each electron and is obtained numerically as a table or plot, rather than as an explicit equation. The final wave equation ψ for the entire atom of n electrons is then given according to Eq. (14–1) as the product of n individual φ functions as

$$\psi = \prod_{i=1}^{i=n} \varphi_i = \prod_{i=1}^{i=n} [R'_{nl}(r_i)\Theta_{lm}(\theta_i)\Phi_m(\phi_i)].$$

The Hartree self-consistent field method, which is simple in concept but bulky and cumbersome in computation, has four inherent shortcomings.

First of all, the expression of ψ as a product of φ functions for each of the electrons ignores the problem of electron indistinguishability. Note that the integrals in Eqs. (14–7), (14–8), and (14–9) are *coulombic* integrals of the form given by Eq. (13–34). To be more accurate we should write φ in a form which ensures antisymmetry. Such an antisymmetric wave function (which is usually written as a determinant) introduces *exchange* integrals and energies, as might be expected from our previous calculations for the excited helium atom. Calculations based on antisymmetric determinant forms* of ψ yield somewhat improved energy levels and wave functions. Secondly, the method does not consider the energies involved in spin-orbital interactions, which are important when one deals with atomic spectra. Thirdly, many atoms contain electrons which are not part of closed l subshells, so that the potential-energy field is not truly spherically symmetrical and a resultant error is introduced. Finally, no provision is made to include the energies associated with the instantaneous correlation of relative electron positions.

14–3 RADIAL DISTRIBUTION OF ELECTRONS IN COMPLEX ATOMS

With appropriate corrections, the SCF method yields remarkable agreement with experimental data for most atoms. Consider, for example, the comparison in Fig. 14–2 of radial charge distribution plots, $D(r)$ versus r, for argon as

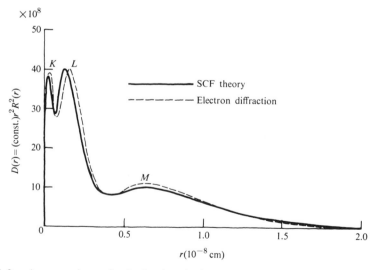

FIG. 14–2   A comparison of calculated and observed radial probability distributions for electrons in the argon atom. The constant in the $D(r)$ expression is chosen so that the total area under the curve is equal to the number of electrons in the atom (18). [L. S. Bartell and L. O. Brockway, *Phys. Rev.*, **90**, 833 (1953).]

---

* V. Fock, *Z. Physik*, **61**, 126 (1930).

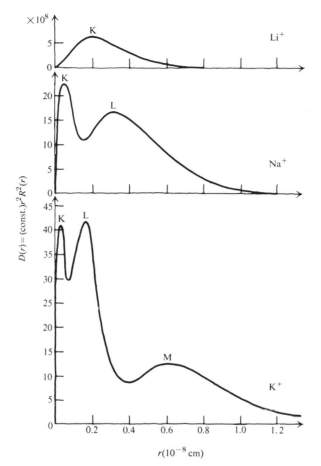

FIG. 14-3   Radial electron distributions for the ground states of several ions by the Hartree method. The $D(r)$ functions in each case are normalized so that the total area under the curve is equal to the number of electrons in the respective ion. (After C. A. Coulson, *Valence*, 2nd ed., Oxford University Press, London, 1961, p. 38.)

calculated by the SCF method and as determined by electron diffraction measurements. The electronic configuration of argon (element 18) is $1s^2 2s^2 2p^6 3s^2 3p^6$, which means that the first two principal energy levels (K and L shells) are filled and that there is a stable octet of eight electrons in the third, or M, shell. The most probable radius for each of the three shells is shown clearly in Fig. 14-2. We further note that the shells become less clearly defined, that is, more smeared out, as their radii increase.

Radial electron distributions for Li^+, Na^+, and K^+ as calculated by the Hartree SCF method are shown in Fig. 14-3. Note that as the nuclear charge

TABLE 14–1

Radii of Maximum Radial Charge Density for Various Wave Functions of Selected
Light Atoms (Radii are given in Angstroms)*

Element	Z	1s	2s	2p	3s	3p	3d	4s	4p
H	1	0.53							
He	2	0.30							
Be	4	0.143	1.19						
C	6	0.090	0.67	0.66					
O	8	0.069	0.48	0.45					
Ne	10	0.055	0.37	0.32					
Mg	12	0.046	0.30	0.25	1.32				
Si	14	0.040	0.24	0.21	0.98	1.06			
S	16	0.035	0.21	0.18	0.78	0.82			
Ar	18	0.031	0.19	0.155	0.66	0.67			
Ca	20	0.028	0.16	0.133	0.55	0.58		2.03	
Ti	22	0.025	0.150	0.122	0.48	0.50	0.55	1.66	
Cr	24	0.023	0.138	0.112	0.43	0.44	0.45	1.41	
Fe	26	0.021	0.127	0.101	0.39	0.39	0.39	1.22	
Ni	28	0.019	0.117	0.090	0.35	0.36	0.34	1.07	
Zn	30	0.018	0.106	0.081	0.32	0.32	0.30	0.97	
Ge	32	0.017	0.100	0.076	0.30	0.30	0.27	0.88	1.06
Se	34	0.016	0.095	0.071	0.28	0.28	0.24	0.81	0.95
Kr	36	0.015	0.090	0.067	0.25	0.25	0.22	0.74	0.86

* From *Quantum Theory of Atomic Structure*, Vol. I, by J. C. Slater. McGraw-Hill,
New York, 1960, p. 210. Used by permission of McGraw-Hill Book Company.

increases, the radius of equivalent shells decreases. Although we realize that
the electronic charge is actually distributed over a large radial distance, it is
often convenient to think of orbital radii in complex atoms in terms of radii of
maximum charge density, which are the radii at the peaks in the radial charge
density curves. Radii of maximum charge density for several wave functions of
selected light atoms are given in Table 14–1. Values have been calculated by the
self-consistent field method. It is immediately apparent that the radial distribution of a particular orbital is compressed as the nuclear charge increases. Such a
reduction in size is certainly to be expected according to simple Bohr theory,
since the size of a Bohr electron orbit is inversely proportional to the nuclear
charge Z. We find that the actual values for the most probable radii of orbitals
in heavier elements are somewhat *larger* than those calculated by dividing the
corresponding hydrogen-like orbital radii by Z. This is due to nuclear screening
effects which tend to reduce the effective nuclear charge.

Two other interesting observations may be made with respect to Table 14–1.
First, for each of the atoms listed, the most probable radius for 2p electrons is

smaller than that for 2s electrons, even though the 2s electrons are more tightly bound (have lower energies). This is due to the difference in shapes of the two orbitals. The lower energy of the 2s orbital is explained in terms of the penetration hump which is very near to the nucleus (see Fig. 11–10). Second, in the particular case of the elements from scandium (element 21) to nickel (element 28) the most probable radii for 3d electrons are much smaller than those for 4s electrons, even though the energies of the two states are approximately the same. As a result, even though d electrons may be easily removed from transition metal atoms, they may act as inner electrons and may easily undergo transitions between the 3d and 4s states.

14–4 ANGULAR DISTRIBUTION OF ELECTRONS IN COMPLEX ATOMS

As shown in Section 14–2, the calculated angular parts of s, p, d, and f orbitals in complex atoms are the same as for hydrogen-like ions, since we have assumed a spherical potential-energy function for each electron, which is mathematically similar to the potential energy function in the hydrogen-like ion. Thus, the angular shapes of all p orbitals are generally the same, although the orbital size is dependent on the radial distributions discussed in the previous sections. In the same way, 3d, 4d, 5d, and 6d orbitals have the same angular shapes, as given by the hydrogen-like wave functions.

14–5 GROUND-STATE ENERGY LEVELS IN COMPLEX ATOMS

We may use the SCF procedure in order to calculate the energy for each of the electrons in a polyelectronic atom. For example, the ground state configuration of the carbon atom is $1s^2 2s^2 2p^2$ and the total energy of the atom in the ground state is given by

$$E = 2E_{1s} + 2E_{2s} + 2E_{2p}, \qquad (14\text{–}10)$$

where E_{1s}, E_{2s} and E_{2p} are the energies of the electrons in the *carbon atom* orbitals indicated by the subscripts. Conceivably, then, we might calculate the individual orbital energies for each of the elements in order to obtain an overall picture of atomic behavior. From a practical point of view, however, SCF calculations are complicated by the lengthy numerical integrations which are required to obtain the self-consistent solutions. Furthermore, straightforward iterations are not necessarily convergent so that further numerical complications may arise.

Energy calculations may be very much simplified without significant loss of accuracy if we assume that each of the electrons moves independently in the *same* central potential-energy field. Using such a *common central field* potential

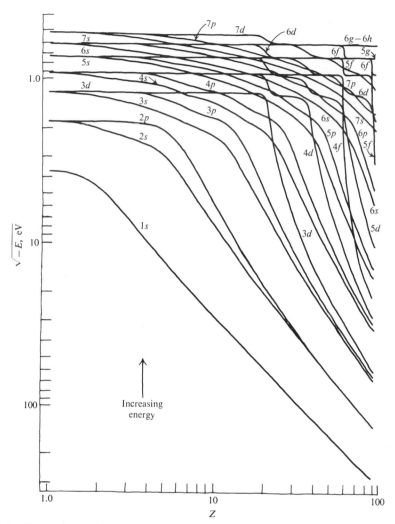

FIG. 14–4   Energy levels of atomic orbitals as calculated for a central Thomas-Fermi-Dirac statistical potential energy field. [Curves based on the data of R. Latter, *Phys. Rev.,* **99,** 517 (1955).]

energy given by a Thomas–Fermi-Dirac* statistical potential, Latter† has used modern computer methods to calculate the ground-state energies for all atoms in terms of the energies of the contributing orbitals. The resulting orbital energies, which agree in general with experimental energies as well as do those energies calculated by the more complicated SCF method, are shown in Fig.

---

* P. A. M. Dirac, *Proc. Cambridge Phil. Soc.,* **23,** 542 (1927).
† R. Latter, *Phys. Rev.,* **99,** 510–519 (1955).

14–4. As the nuclear charge Z increases, the energies of the principal levels begin to split into separate energies in each of the l sublevels. The observed energy splitting is due to nuclear shielding effects and to differences in relative penetrations among the orbitals in multi-electronic systems. For example, the 3s orbital is lower in energy than the 3p orbital because the 3s electron is more easily able to penetrate the screen of 1s and 2s electrons than is the 3p electron. In general the lower the l quantum number for an orbital, the more penetrating is the orbital. On the other hand, orbitals having a lower value of l are less effective in screening than orbitals of higher l quantum number. At intermediate Z values, the order of relative orbital energies becomes fairly complicated, and crossovers in energy levels occur which account for many of the periodic properties of the elements. At high values for Z, the l sublevels within a given n principal level again converge, apparently due to the overshadowing effect of the very high nuclear charge which tends to reduce differences in energies between l suborbitals.

14–6 ELECTRONIC DISTRIBUTIONS IN COMPLEX ATOMS

Let's now consider an imaginary process by which we construct electronic configurations for each of the elements by adding electrons to the atom one at a time, while also simultaneously adding the same number of protons to the nucleus in order to maintain electrical neutrality. Of course, neutrons must also be added to the nucleus in order to ensure nuclear stability. In constructing an atom according to such an *Aufbau* (German for *build-up*) procedure we'll follow three guidelines:

1. *Electrons must always be placed in orbitals of lowest possible energy consistent with*

2. *the Pauli principle, which requires that no two electrons may have the same set of four quantum numbers, and*

3. *Hund's rule of maximum multiplicity which requires that, other things being equal (for example, in partially filled degenerate orbitals), the state with the maximum number of unpaired electrons be lowest in energy.*

Although the relative energies of orbitals vary with Z, we may construct a chart of the energies of orbitals *in the order in which they would be filled* by successive addition of electrons, beginning with the hydrogen atom and continuing through the elements to the most complex atom. Such a chart is shown in Fig. 14–5. Each of the s orbitals may contain one pair of electrons, opposed in spin. For the p level, however, $l = 1$ and m may have the values $-1, 0, +1$, so that each p level contains three degenerate m orbitals, each of which may contain one pair of electrons. In similar fashion, for each d level, $l = 2$, and m may have the values $-2, -1, 0, +1, +2$. Each of the five degenerate m orbitals may in turn contain a pair of electrons, one with spin up and one with spin down.

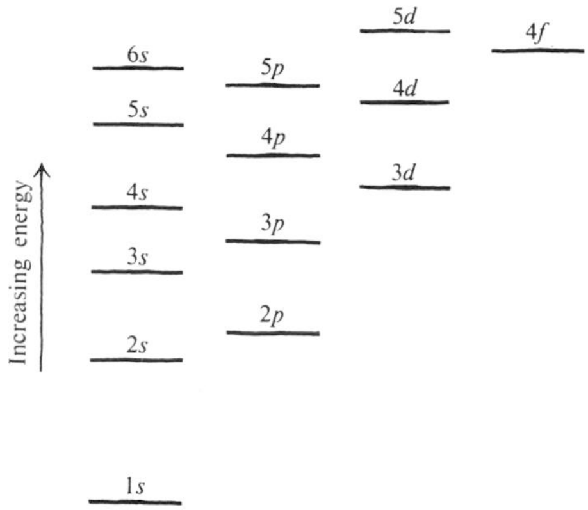

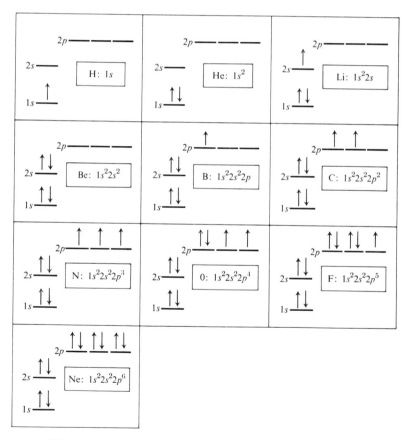

FIG. 14–5   Relative orbital energies *in the order in which atomic orbitals are filled* by successive addition of single electrons according to the Aufbau process.

FIG. 14–6   Electronic configurations for the first 10 elements.

Thus each d level may contain a total of 10 electrons, or five degenerate pairs. Similarly, each f level may contain a total of 14 electrons, or seven degenerate pairs.

Let's now use the energy chart given in Fig. 14–5 in conjunction with the Pauli principle and Hund's rule in order to construct the electronic configurations of the elements. In doing so, we'll indicate relative electron spin by arrows pointed up or down. The configurations for the first 10 elements are shown in Fig. 14–6. Note that when the configuration for carbon is formed by adding an electron to the configuration for boron, the two electrons in the 2p orbitals have parallel spins according to Hund's rule. For nitrogen, the three 2p electrons also have parallel spins. As a general rule, when p levels are filled, the first three electrons are placed with parallel spins in separate orbitals. Similarly, when d levels are filled, the first five electrons have parallel spins in separate orbitals, and when f levels are filled, the first seven electrons are placed with parallel spins in separate f orbitals.

Electronic configurations of all of the elements are presented in Table 14–2. In interpreting these configurations we should always keep Hund's rule in mind. For example, the configuration for Ru (element 44) is given as $[Kr]4d^7 5s$, so that the configuration of the 4d-level electrons may be interpreted as

$$4d \quad \uparrow\downarrow \quad \uparrow\downarrow \quad \uparrow \quad \uparrow \quad \uparrow$$

That is, three of the 4d electrons are unpaired and have parallel spins. Note also that the 5s electron in Ru is unpaired. All of the remaining electrons in the [Kr] kernel are paired.

Although the observed configurations agree fairly well with predictions, there are a few notable exceptions. Specifically, let's compare predicted configurations with observed configurations, as shown in Table 14–3, for the group of elements from Sc through Zn. These are the elements whose configurations are constructed as we fill in the 10 electrons in the 3d orbital. This group of elements is known as the *first transition series*. Note that the configurations of both Cr and Cu are apparent exceptions to the general rule. For each of these atoms, one 4s electron has unpaired and entered the higher 3d energy level in order to either exactly half-fill or completely fill the 3d energy level, with an apparent increase in stabilization (lowering of energy) of the atom. In each case, the net lowering of energy may be explained in terms of the exchange energy, K, which we have previously defined for a set of *two* electrons of parallel spin. Recall that the exchange energy is the lowering of energy for a set of two electrons of parallel spin which arises as a result of the requirement of indistinguishability or exchangeability of the electron coordinates. Two electrons with parallel spins remain farther apart than two electrons with opposed spins and consequently the coulombic repulsion energy is lower. In a system containing more than two electrons with parallel spins, the lowering of energy in the ground state due to exchange (the total *exchange energy stabilization*) is

TABLE 14–2

Electronic Configurations of Gaseous Atoms in the Ground State

Atomic number	Element	Electronic configuration	Atomic number	Element	Electronic configuration
1	H	1s	53	I	$-4d^{10}5s^25p^5$
2	He	$1s^2$	54	Xe	$-4d^{10}5s^25p^6$
3	Li	[He] 2s	55	Cs	[Xe] 6s
4	Be	$-2s^2$	56	Ba	$-6s^2$
5	B	$-2s^22p$	57	La	$-5d6s^2$
6	C	$-2s^22p^2$	58	Ce	$-4f^26s^2$
7	N	$-2s^22p^3$	59	Pr	$-4f^36s^2$
8	O	$-2s^22p^4$	60	Nd	$-4f^46s^2$
9	F	$-2s^22p^5$	61	Pm	$-4f^56s^2$
10	Ne	$-2s^22p^6$	62	Sm	$-4f^66s^2$
11	Na	[Ne] 3s	63	Eu	$-4f^76s^2$
12	Mg	$-3s^2$	64	Gd	$-4f^75d6s^2$
13	Al	$-3s^23p$	65	Tb	$-4f^96s^2$
14	Si	$-3s^23p^2$	66	Dy	$-4f^{10}6s^2$
15	P	$-3s^23p^3$	67	Ho	$-4f^{11}6s^2$
16	S	$-3s^23p^4$	68	Er	$-4f^{12}6s^2$
17	Cl	$-3s^23p^5$	69	Tm	$-4f^{13}6s^2$
18	Ar	$-3s^23p^6$	70	Yb	$-4f^{14}6s^2$
19	K	[Ar] 4s	71	Lu	$-4f^{14}5d6s^2$
20	Ca	$-4s^2$	72	Hf	$-4f^{14}5d^26s^2$
21	Sc	$-3d4s^2$	73	Ta	$-4f^{14}5d^36s^2$
22	Ti	$-3d^24s^2$	74	W	$-4f^{14}5d^46s^2$
23	V	$-3d^34s^2$	75	Re	$-4f^{14}5d^56s^2$
24	Cr	$-3d^54s$	76	Os	$-4f^{14}5d^66s^2$
25	Mn	$-3d^54s^2$	77	Ir	$-4f^{14}5d^76s^2$
26	Fe	$-3d^64s^2$	78	Pt	$-4f^{14}5d^96s$
27	Co	$-3d^74s^2$	79	Au	[ ] 6s
28	Ni	$-3d^84s^2$	80	Hg	$-6s^2$
29	Cu	$-3d^{10}4s$	81	Tl	$-6s^26p$
30	Zn	$-3d^{10}4s^2$	82	Pb	$-6s^26p^2$
31	Ga	$-3d^{10}4s^24p$	83	Bi	$-6s^26p^3$
32	Ge	$-3d^{10}4s^24p^2$	84	Po	$-6s^26p^4$
33	As	$-3d^{10}4s^24p^3$	85	At	$-6s^26p^5$
34	Se	$-3d^{10}4s^24p^4$	86	Rn	$-6s^26p^6$
35	Br	$-3d^{10}4s^24p^5$	87	Fr	[Rn] 7s
36	Kr	$-3d^{10}4s^24p^6$	88	Ra	$-7s^2$
37	Rb	[Kr] 5s	89	Ac	$-6d7s^2$
38	Sr	$-5s^2$	90	Th	$-6d^27s^2$
39	Y	$-4d5s^2$	91	Pa	$-5f^26d7s^2$
40	Zr	$-4d^25s^2$	92	U	$-5f^36d7s^2$
41	Nb	$-4d^45s$	93	Np	$-5f^46d7s^2$
42	Mo	$-4d^55s$	94	Pu	$-3f^67s^2$
43	Tc	$-4d^55s^2$	95	Am	$-5f^77s^2$
44	Ru	$-4d^75s$	96	Cm	$-5f^76d7s^2$
45	Rh	$-4d^85s$	97	Bk	$-5f^86d7s^2$
46	Pd	$-4d^{10}$	98	Cf	$-5f^{10}7s^2$
47	Ag	$-4d^{10}5s$	99	Es	$-5f^{11}7s^2$
48	Cd	$-4d^{10}5s^2$	100	Fm	$-5f^{12}7s^2$
49	In	$-4d^{10}5s^25p$	101	Md	$-5f^{13}7s^2$
50	Sn	$-4d^{10}5s^25p^2$	102	No	$-5f^{14}7s^2$
51	Sb	$-4d^{10}5s^25p^3$	103	Lr	$-5f^{14}6d7s^2$
52	Te	$-4d^{10}5s^25p^1$		Ku	$-5f^{14}6d^27s^2$
				Ha	$-5f^{14}6d^37s^2$

TABLE 14-3

Predicted and Observed Ground-State
Configurations for the Elements in the
First Transition Series

Element	Electron configuration	
	Predicted	Observed
Sc	[Ar]3d4s^2	[Ar]3d4s^2
Ti	—3d^{2}4s^2	—3d^{2}4s^2
V	—3d^{3}4s^2	—3d^{3}4s^2
Cr	(—3d^{4}4s^2)*	—3d^{5}4s
Mn	—3d^{5}4s^2	—3d^{5}4s^2
Fe	—3d^{6}4s^2	—3d^{6}4s^2
Co	—3d^{7}4s^2	—3d^{7}4s^2
Ni	—3d^{8}4s^2	—3d^{8}4s^2
Cu	(—3d^{9}4s^2)*	—3d^{10}4s
Zn	—3d^{10}4s^2	—3d^{10}4s^2

* Exceptions.

given by the product of the total number of possible exchanges of coordinates between sets of two electrons of parallel spin times the average exchange energy K per set of electrons of parallel spin. For example, for a system containing three electrons, 1, 2, and 3, all having parallel spins, there are three possible exchanges of coordinates as follows:

1 exchanges with 2,

1 exchanges with 3,

2 exchanges with 3.

Thus the total exchange energy stabilization is $3K$. For a system containing four electrons, 1, 2, 3, and 4, all having parallel spins, there are six possible exchanges of coordinate positions among electrons as follows:

1 exchanges with 2,

1 exchanges with 3,

1 exchanges with 4,

2 exchanges with 3,

2 exchanges with 4,

3 exchanges with 4,

and the total exchange energy stabilization is $6K$. In general, it may be shown that for a system containing n electrons of parallel spin, the number of possible sets, each containing two electrons of parallel spin, is given as

$$n!/2(n-2)!.$$

Recognizing that each set of two electrons of parallel spin contributes an average energy K to the total energy stabilization, we may write the total exchange energy stabilization for each of the d level configurations as shown in Table 14–4.

In order to explain the unexpected configuration for the chromium atom, we note that in the absence of exchange effects, the $3d^4 4s^2$ state would be slightly lower in energy (slightly more stable) than the $3d^5 4s$ state. However, because of the greater possibility for electron exchange in the $3d^5 4s$ state, its energy is lowered by $15K$ (6 parallel spins), whereas the lowering of energy due to electron exchange in the $3d^4 4s^2$ state is only $10K$ (5 parallel spins). The net result is that the observed energy level of the $3d^5 4s$ state is *lower* than that of the $3d^4 4s^2$ state. The calculation is based on spins of 3d and 4s electrons only. The 18 electrons in the [Ar] kernel have no effect on the *difference* in energies between the two configurations. The unexpected stability of the $3d^{10} 4s$ configuration for the copper atom may be explained in the same way. As would be expected, it can be observed that exactly half-filled or completely filled p shells and f shells also show unusual stability, because of exchange effects.

TABLE 14–4

Total Exchange Energy Stabilization for d Level Configurations

Configuration	Number of possible sets of electrons of parallel spins	Total exchange-energy stabilization
d^1: ↑ _ _ _ _	0	0
d^2: ↑ ↑ _ _ _	1	K
d^3: ↑ ↑ ↑ _ _	3	$3K$
d^4: ↑ ↑ ↑ ↑ _	6	$6K$
d^5: ↑ ↑ ↑ ↑ ↑	10	$10K$
d^6: ↑↓ ↑ ↑ ↑ ↑	10	$10K$
d^7: ↑↓ ↑↓ ↑ ↑ ↑	11 (10 up, 1 down)	$11K$
d^8: ↑↓ ↑↓ ↑↓ ↑ ↑	13 (10 up, 3 down)	$13K$
d^9: ↑↓ ↑↓ ↑↓ ↑↓ ↑	16 (10 up, 6 down)	$16K$
d^{10}: ↑↓ ↑↓ ↑↓ ↑↓ ↑↓	20 (10 up, 10 down)	$20K$

14–7 EXPERIMENTAL ENERGIES: IONIZATION POTENTIALS

The most convenient experimental indications of the relative stabilities of various electronic configurations are the *ionization potentials* which may be calculated from spectroscopic data. The ionization potential I is defined as the *amount of energy required at absolute zero to remove the most loosely bound electron from a single atom or ion*. The measurement is referred to absolute zero in order to ensure that no translational energy changes are involved. Specifically, the free electron has no kinetic energy, so that all of the energy is absorbed by the atom as potential energy required to remove the electron.

Ionization potentials for most of the elements are presented in Table 14–5. In addition to listing the ionization potential for removal of the first electron from a given atom, ionization potentials are also listed for removal of the second, third, and fourth electrons. Consider, for example, the four ionization potentials for the beryllium atom. An energy of 9.320 eV is required to remove the first electron, that is,

$$Be(g) \xrightarrow{0°K} Be^+(g) + e^-, \qquad I_1 = 9.320 \text{ eV}.$$

The second electron, however, is considerably more difficult to remove since the effective nuclear charge has increased:

$$Be^+(g) \xrightarrow{0°K} Be^{2+}(g) + e^-, \qquad I_2 = 18.206 \text{ eV}.$$

In removing the third and fourth electrons, we break into the completed 1s shell and even larger energies are required:

$$Be^{2+}(g) \xrightarrow{0°K} Be^{3+}(g) + e^-, \qquad I_3 = 153.850 \text{ eV},$$

$$Be^{3+}(g) \xrightarrow{0°K} Be^{4+}(g) + e^-, \qquad I_4 = 217.7 \text{ eV}.$$

The *total electronic energy* of the beryllium atom is given as the sum of the four ionization potentials, or 399.1 eV, which is the energy required to strip all four of the electrons from an isolated beryllium nucleus at 0°K.

Let's now examine the periodic variation in first ionization potential as a function of nuclear charge (Fig. 14–7). The ionization potential for helium is much larger than that for hydrogen primarily because of the doubling of nuclear charge, which in addition is accompanied by very little screening. For lithium, however, the most easily removed 2s electron is farther from the nucleus and is screened by two 1s electrons. Thus, I_1 for lithium is relatively small, as it is for all of the alkali metals. When any of the alkali metal atoms loses its first electron from its outermost s orbital, the atom is left with the very stable closed shell configuration characteristic of the noble gases. In proceeding through the group of elements from lithium to neon, we note a gradually increasing first ionization potential. This increase may be attributed to the lack of full nuclear screening by electrons in the same principal level, which results in a gradual

TABLE 14-5 Ionization Potentials for the Elements*

Atomic number	Symbol	Ionization Potential, eV/atom			
		I_1	I_2	I_3	I_4
1	H	13.595	—	—	—
2	He	24.581	54.403	—	—
3	Li	5.390	75.619	122.419	—
4	Be	9.320	18.206	153.850	217.7
5	B	8.296	25.149	37.920	259.3
6	C	11.256	24.376	47.871	64.5
7	N	14.53	29.593	47.426	77.4
8	O	13.614	35.108	54.886	77.4
9	F	17.418	34.98	62.646	87.1
10	Ne	21.559	41.07	63.5	97.0
11	Na	5.138	47.29	71.65	98.9
12	Mg	7.644	15.031	80.12	109.3
13	Al	5.984	18.823	28.44	120.0
14	Si	8.149	16.34	33.46	45.1
15	P	10.484	19.72	30.156	51.4
16	S	10.357	23.4	35.0	47.3
17	Cl	13.01	23.80	39.90	53.5
18	Ar	15.755	27.62	40.90	59.8
19	K	4.339	31.81	46	60.9
20	Ca	6.111	11.868	51.21	67
21	Sc	6.54	12.80	24.75	73.9
22	Ti	6.82	13.57	27.47	43.2
23	V	6.74	14.65	29.31	48
24	Cr	6.764	16.49	30.95	50
25	Mn	7.432	15.636	33.69	~53
26	Fe	7.87	16.18	30.643	~56
27	Co	7.86	17.05	33.49	~53
28	Ni	7.633	18.15	35.16	—
29	Cu	7.724	20.29	36.83	—
30	Zn	9.391	17.96	39.70	—
31	Ga	6.00	20.51	30.70	64.2
32	Ge	7.88	15.93	34.21	45.7
33	As	9.81	18.63	28.34	50.1
34	Se	9.75	21.5	32	43
35	Br	11.84	21.6	35.9	47.3
36	Kr	13.996	24.56	36.9	—
37	Rb	4.176	27.5	40	—
38	Sr	5.692	11.027	25	57
39	Y	6.38	12.23	20.5	—
40	Zr	6.84	13.13	22.98	34.3
41	Nb	6.88	14.32	25.04	38.3
42	Mo	7.10	16.15	27.13	46.4
43	Tc	7.28	15.26	—	—

* From *Electronic Structure and Chemical Bonding*, by D. K. Sebera. Blaisdell Publishing Company, a division of Ginn and Company, Waltham, Mass., 1964. Reprinted by permission of the publishers.

TABLE 14–5 (*continued*)

Atomic number	Symbol	Ionization Potential, eV/atom			
		I_1	I_2	I_3	I_4
44	Ru	7.364	16.76	—	—
45	Rh	7.461	18.07	31.05	—
46	Pd	8.33	19.42	32.92	—
47	Ag	7.574	21.48	34.82	--
48	Cd	8.991	16.904	37.47	—
49	In	5.785	18.86	28.03	54.4
50	Sn	7.342	14.628	30.49	40.7
51	Sb	8.639	16.5	25.3	44.1
52	Te	9.01	18.6	31	38
53	I	10.454	19.09	—	—
54	Xe	12.127	21.2	32.1	—
55	Cs	3.893	25.1	—	—
56	Ba	5.210	10.001	—	—
57	La	5.61	11.43	19.17	—
58	Ce	6.91	14.8	—	—
59	Pr	5.76	—	—	—
60	Nd	6.3	—	—	—
61	Pm	—	—	—	—
62	Sm	5.6	11.2	—	—
63	Eu	5.67	11.24	—	—
64	Gd	6.16	12	—	—
65	Tb	6.74	—	—	—
66	Dy	6.82	—	—	—
67	Ho	—	—	—	—
68	Er	—	—	—	—
69	Tm	—	—	—	—
70	Yb	6.22	12.10	—	—
71	Lu	6.15	14.7	—	—
72	Hf	7	14.9	—	—
73	Ta	7.88	16.2	—	—
74	W	7.98	17.7	—	—
75	Re	7.87	16.6	—	—
76	Os	8.7	17	—	—
77	Ir	9	—	—	—
78	Pt	9.0	18.56	—	—
79	Au	9.22	20.5	—	—
80	Hg	10.43	18.751	34.2	—
81	Tl	6.106	20.42	29.8	50.7
82	Pb	7.415	15.028	31.93	42.3
83	Bi	7.287	16.68	25.56	45.3
84	Po	8.43	—	—	—
85	At	—	—	—	—
86	Rn	10.746	—	—	—
87	Fr	—	—	—	—
88	Ra	5.277	10.14	—	—
89	Ac	6.9	12.1	20	—

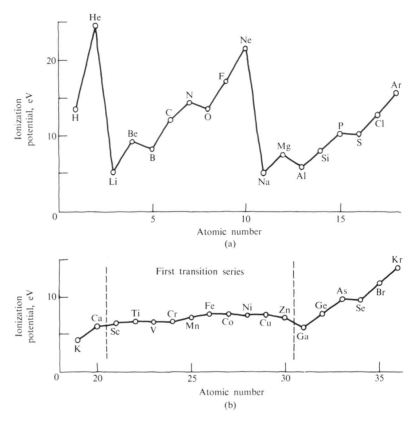

FIG. 14–7 First ionization potentials for the initial 36 elements of the periodic chart: (a) The first three periods, (b) the fourth period including the first transition series.

increase in effective nuclear charge as we pass from lithium to neon or from sodium to argon.

The drop in ionization potential observed in passing from beryllium to boron may be attributed to the fact that the electron which is removed from the boron atom is in the less penetrating p orbital and is relatively well shielded from the nuclear charge. The same type of break in the curve is observed in passing from magnesium to aluminum and is explained in the same way. Smaller breaks in the smooth curve are also observed in passing from nitrogen to oxygen and from phosphorus to sulfur. For both oxygen and sulfur, the ionization potential measures the energy required to remove a single paired electron in the highest p level, that is:

$$\text{p:} \quad \underline{\uparrow} \;\; \underline{\uparrow} \;\; \underline{\uparrow}.$$

The coulombic energy of repulsion of the pair is appreciable and there are no parallel spin electrons to lower the total energy of the atom. Thus, the electron is easily removed. By losing this electron, the atom may attain the relatively stable half-filled p configuration.

Ionization potentials for the elements in the first transition series (scandium through zinc) are fairly constant, which reflects the good screening afforded by d electrons. Apparently, it is relatively difficult to remove electrons from the noble gases. Even here, however, we note that as Z increases, electrons are more easily removed since they are farther away from the nucleus. Thus, the first ionization potential for argon is lower than that for neon or helium.

In general, the experimental data provided by the measurement of ionization potentials are in good agreement with the approximate calculations of quantum mechanics.

14–8 FURTHER STUDIES OF ATOMS

In the past three chapters, we have introduced some important basic considerations involved in the determination of electron probability distributions, energies, and spectra of complex atoms. Although our treatment has necessarily been limited, we have developed a sound basis for more detailed discussions of topics such as spin-orbit interactions, the effects of external magnetic and electric fields, and atomic spectra in general. Further studies of atoms must also involve treatments of *nonstationary* states in which the *time-dependent* Schrödinger equation is used to describe phenomena such as the emission of light by excited atoms, electron absorption spectra, and light scattering.

Finally, we recall that we have had to resort to approximation methods in the treatment of all atoms more complex than hydrogen. Although we have restricted our brief discussion to perturbation and variation methods, since these seem to be most important, we should note that other approximation methods have been developed and used successfully. Some of these methods, along with detailed treatments of the other topics mentioned above, are given in the general references cited in the Bibliography at the end of the text.

Perhaps the most spectacular application of the exclusion principle is to the "building-up" process of the elements . . . where it is also shown that different atoms owe their characteristic features to a kind of social behavior of the electrons which may be summed up by saying: One electron knows what the others are doing and acts accordingly. And this knowledge is not conveyed by forces, or dynamic interactions, of the ordinary kind. The exclusion principle introduces a correlation into the behavior of particles which, though its effects are similar to the effects of forces, has no explanation in dynamic terms.

—HENRY MARGENAU, *The Nature of Physical Reality.**

PROBLEMS

14–1 In the development of the self-consistent field method, we have stated that the total charge distribution $|\psi|^2$ for any atom with closed l shells (closed s, p, d, f shells) is spherically symmetrical, that is, independent of θ or ϕ. Prove that this is true for the neon atom $(1s^2 2s^2 2p^6)$ by assuming that the total charge distribution is the sum of the independent electron distributions, that is,

$$|\psi|^2 = |\varphi_1|^2 + |\varphi_2|^2 + |\varphi_3|^2 + \cdots + |\varphi_{10}|^2.$$

Use the orbitals given in Table 11–1 for the individual electrons.

14–2 List all the elements whose atoms have spherically symmetrical electron distributions.

14–3 What factors are responsible for Nb (element 41) having an electronic configuration of $[Kr]4d^4 5s$ rather than the expected configuration $[Kr]4d^3 5s^2$?

14–4 Prepare a table similar to Table 14–4 for the total exchange energy stabilization for each of the 14 f level configurations.

14–5 The predicted electron configuration of the copper atom is $[Ar]3d^9 4s^2$ whereas the experimentally observed configuration is $[Ar]3d^{10} 4s$.

  a) In each of the above configurations, how many electrons have spins up and how many down?
  b) Consider exchange possibilities for all 29 electrons in the atom and calculate for each of the configurations the total number of possible exchanges of coordinates between sets of electrons of parallel spin.
  c) If K is the average exchange energy per set, what is the difference in exchange energy stabilization between the two configurations?
  d) Calculate the difference in exchange energy stabilization by considering the 3d and 4s electrons only.

14–6 Calculate the energy required to strip all of the orbital electrons from a neutral lithium atom at $0°K$. Would you expect the experimentally observed energy to be higher or lower than the calculated energy? Why?

14–7 a) Calculate the first ionization potential, in electron volts, for the potassium atom, assuming that Eq. (10–79) for the hydrogen-like ion applies to the lone outer electron in the K atom.
  b) Compare the calculated value from part (a) with the observed value for I_1 given in Table 14–5. Comment on the difference in terms of nuclear screening by inner electrons.
  c) What would have to be the effective nuclear charge, Z_{eff}, experienced by the lone outer electron in the K atom for the calculation in part (a) to yield the correct ionization potential?

Up to this point we have been concerned primarily with energies and electronic distributions in atoms. Although it is not our purpose to consider molecules in any detail, a few brief introductory remarks are certainly in order. In view of the mathematical difficulties we have encountered in trying to solve the Schrödinger equation for relatively small atoms containing only *one* nucleus, it should come as no surprise that the Schrödinger amplitude equation for molecules containing *several* atomic nuclei is even more formidable. In writing the complete Schrödinger equation for a molecule, we must include one kinetic-energy operator term for each of the nuclei and one kinetic-energy operator term for each of the electrons. In addition, we must include a potential-energy term for *each* possible two-particle interaction. Thus the form of the complete time-independent molecular wave equation is

$$\left(\frac{-\hbar^2}{2} \sum_j \frac{1}{m_j} \nabla_j^2 - \frac{\hbar^2}{2m_e} \sum_i \nabla_i^2 + \sum_{i,i'} \frac{e^2}{r_{ii'}} - \sum_{i,j} \frac{Z_j e^2}{r_{ij}} + \sum_{j,j'} \frac{Z_j Z_{j'} e^2}{r_{jj'}} \right) \psi = E\psi,$$

$$(15\text{--}1)$$

where the subscripts i and i' refer to electrons and the subscripts j and j' refer to nuclei, and where the summations are taken over all the corresponding electrons and/or nuclei in the molecule. By now we have learned to expect that we cannot solve wave equations such as Eq. (15–1) without making simplifying assumptions and approximations.

15–1 THE BORN-OPPENHEIMER APPROXIMATION

As a first step toward the solution of the molecular wave equation, we assume that the motion of the very light electrons is much more rapid than that of the relatively heavy and sluggish nuclei, so that the electrons are able to adjust immediately to any change in nuclear positions. According to this assumption, which is called the *Born-Oppenheimer approximation*, we may imagine at the outset that the nuclei are fixed in space as far as electronic motion is concerned, so that the two forms of motion may be considered to be independent and the molecular wave function ψ may thus be written as the product of a wave function ψ_e which describes the electrons, and a wave function ψ_n which describes the nuclei. That is,

$$\psi = \psi_e\psi_n. \tag{15–2}$$

As a result, it may be shown* that the complete Schrödinger equation may be divided into two separate wave equations. The first of these is called the *electronic wave equation:*

$$\left(\frac{-\hbar^2}{2m_e}\sum_i \nabla_i^2 + \sum_{i,i'} \frac{e^2}{r_{ii'}} - \sum_{i,j} \frac{Z_j e^2}{r_{ij}} + \sum_{j,j'} \frac{Z_j Z_{j'} e^2}{r_{jj'}}\right)\psi_e = E_e\psi_e, \tag{15–3}$$

where ψ_e are the *electronic wave functions*, and E_e are the *electronic energy* eigenvalues. The electronic wave equation describes the motions, positions, and energies of the electrons alone under conditions in which the nuclei are fixed. The second equation is called the *nuclear wave equation:*

$$\left[-\frac{\hbar^2}{2}\sum_j \frac{1}{m_j}\nabla_j^2 + E_e(r_{jj'})\right]\psi_n = E_n\psi_n, \tag{15–4}$$

where ψ_n are the *nuclear wave functions*, and E_n are the *nuclear energy* eigenvalues. The nuclear wave equation describes the motions, positions, and energies associated with translation, rotation, and vibration of the nuclei. The electronic wave equation, Eq. (14–3), *must be solved for specific values of the internuclear distances, $r_{jj'}$,* so that, as a result, a different E_e eigenvalue spectrum is calculated for *each* selected set of $r_{jj'}$ values, which means that the electronic energy in any single eigenstate (for example, the ground state) must be expressed as a function of the internuclear separations $r_{jj'}$. In turn, each of the resultant

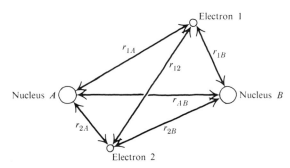

FIG. 15–1   Interparticle distances in the diatomic hydrogen molecule. Each of the nuclei is a proton.

---

* M. Born and J. R. Oppenheimer, *Ann. Physik*, **84**, 457 (1927).

$E_e(r_{jj'})$ functions, *one for each electronic-energy eigenstate*, obtained from the solutions of the electronic wave equation (15–3) for different values of $r_{jj'}$, serves as a potential-energy function in the Hamiltonian operator for a separate nuclear wave equation, Eq. (15–4). That is, *there is one nuclear wave equation for each electronic eigenstate.*

Consider, for example, the relatively simple case of the diatomic hydrogen molecule, shown schematically in Fig. 15–1. The H_2 molecule consists of two electrons, 1 and 2, and two nuclei, A and B, so that the electronic wave equation (15–3) may be written as

$$\left[-\frac{\hbar^2}{2m_e}(\nabla_1^2 + \nabla_2^2) + \frac{e^2}{r_{12}} - \frac{e^2}{r_{1A}} - \frac{e^2}{r_{1B}} - \frac{e^2}{r_{2A}} - \frac{e^2}{r_{2B}} + \frac{e^2}{r_{AB}}\right]\psi_e = E_e\psi_e,$$

$$(15\text{–}5)$$

or

$$\hat{H}_e\psi_e = E_e\psi_e.$$

15–2 APPROXIMATE SOLUTION OF THE ELECTRONIC WAVE EQUATION

In order to solve approximately equations such as Eq. (15–5), we may resort to the variation method which we outlined in Section 12–4. Trial variation functions, $\tilde{\psi}_e$, are usually constructed by simple linear combinations of atomic orbitals and contain several constants which are adjusted to minimize $\tilde{E}_e$, the variation energy, according to the variation theorem

$$\tilde{E}_e = \frac{\int \tilde{\psi}_e^* \hat{H}_e \tilde{\psi}_e\, d\tau}{\int \tilde{\psi}_e^* \tilde{\psi}_e\, d\tau} \geq (E_e)_0,$$

$$(15\text{–}6)$$

where $(E_e)_0$ is the electronic energy in the ground state. The technique is the same as that we have previously used in evaluating the ground-state energy of the helium atom. Thus, for the diatomic hydrogen molecule, we determine, *for each selected value of the internuclear distance*, r_{AB}, a minimum value for $\tilde{E}_e$ which is hopefully near (but always above) the electronic energy $(E_e)_0$ in the ground state.

The resultant plot of $\tilde{E}_e$, the approximate ground-state electronic energy for the hydrogen molecule, versus r_{AB}, the internuclear separation assumed in writing the Hamiltonian operator $\hat{H}_e$, is shown in Fig. 15–2. Such curves are relatively simple when only *one* internuclear distance is involved, but are very complicated for polyatomic molecules. A plot of the type shown in Fig. 15–2 is often called a potential energy curve because the plot of electronic energy E_e as a function of internuclear distance is also the potential energy curve which governs the motions of the nuclei. Thus the equilibrium bond distance r_e in the ground state corresponds to the minimum in the ground-state potential-energy curve shown in Fig. 15–2. The energy D_e measured from the minimum in the potential-energy curve is called the *electronic binding energy*.

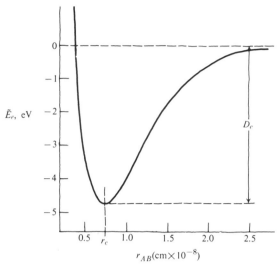

FIG. 15–2   Potential energy diagram for the ground state of the hydrogen molecule. The energy $\tilde{E}_e$ is arbitrarily defined as zero when the internuclear distance r_{AB} is infinite.

The trial electronic wave functions $\tilde{\psi}_e$ used in Eq. (15–6) are usually constructed by one of two methods:

1. the *molecular orbital* (MO) *method*, according to which *each* electron is imagined to be distributed over the entire molecule within a system of *molecular orbitals* (which are similar to the atomic orbitals we have already discussed) in such a manner that no more than two electrons occupy each space orbital, or

2. the *valence bond* (VB) *method*, in which most of the electrons generally remain associated with their original atoms and only those pairs of electrons directly involved in bonding are considered when writing the wave function.

Regardless of the method used in constructing trial electronic wave functions, we must adhere to the Pauli principle, which requires that systems of electrons be described by antisymmetric wave functions. Both the MO and the VB methods are semiquantitative at best. Although the VB method gives better results for bond energies and bond distances, the MO method is less complicated mathematically and is used more frequently for larger molecules.

15–3 THE NUCLEAR WAVE EQUATION

The nuclear wave equation, Eq. (15–4), describes the motions and energies of the nuclei associated with translation of the molecule through space, rotation of the molecule about its center of mass, and vibrations of the nuclei with respect to one another. If, as an approximation, we assume that the three types of motion are independent, we may write

$$\psi_n = \psi_{trans}\psi_{rot}\psi_{vib},$$

and

$$E_n = E_{trans} + E_{rot} + E_{vib}.$$

We must realize, however, that E_{rot} and E_{vib} are not really independent, since changes in vibrational energies change effective bond distances, which in turn affect the moment of inertia of the molecule and consequently the rotational energy. From Eq. (15–2), we may write for the molecular wave function

$$\psi = \psi_{trans}\psi_{rot}\psi_{vib}\psi_e, \tag{15–7}$$

and since

$$E = E_n + E_e,$$

we may also write for the molecular energy

$$E = E_{trans} + E_{rot} + E_{vib} + E_e. \tag{15 8}$$

Recall that the potential-energy operator term in the nuclear wave equation, Eq. (15–4), is the electronic energy in a particular eigenstate obtained from the solution of Eq. (15–3) as a function of the internuclear distances r_{ij}. For a diatomic molecule, E_e is a function of only *one* internuclear separation, r_{AB}, and Eq. (15–4) involves the six coordinates of the two nuclei. If we transform the coordinate system to the three coordinates of the center of mass of the molecule, X, Y, and Z, and to three internal coordinates of the molecule, r, θ, and ϕ, such as we have done previously in separating the wave equation for the hydrogen atom, we may further separate Eq. (15–4) into two equations, one involving translation alone, *which is identical to Eq. (8–27)*, and one involving only the internal vibrational-rotational motion and energy as a function of r, θ and ϕ.

15-4 TRANSLATIONAL ENERGY OF MOLECULES

Since the Schrödinger equation for the translational motion of the center of mass of a molecule is identical to the wave equation for the translational motion of a single particle in three dimensions, which we have solved previously, the resultant quantized translational energy levels for a molecule of mass m are given according to Eq. (8–41) as

$$E_{trans} = \frac{h^2}{8m}\left(\frac{n_x^2}{a^2} + \frac{n_y^2}{b^2} + \frac{n_z^2}{c^2}\right), \tag{8–41'}$$

where n_x, n_y, and n_z are the *translational quantum numbers* and a, b, and c are the dimensions of rectangular confinement. Because molecules are usually confined in relatively large spaces (compared to electrons confined within atoms), the spacing between adjacent translational energy levels is extremely small, so that at ordinary temperatures translational quantum numbers for molecules are very large ($\sim 10^9$). Thus, quantum jumps from one translational energy level to adjacent energy levels are not normally experimentally detectable, so that the energy levels appear to be continuous. Under these conditions the quantum-mechanical treatment of translational energy distributions for molecules is equivalent to the classical treatment, in which an energy continuum is assumed. This is another example of the *correspondence principle*, which requires that, in the limit of large quantum numbers, quantum mechanical results must be identical to classical results.

15–5   ROTATIONAL ENERGY OF MOLECULES

Let's again return to a consideration of the diatomic molecule. To a first approximation, we assume that the rotational and vibrational motions are separable, so that we may then independently write the Schrödinger equation for the two-dimensional (θ and ϕ) rotation of a rigid dumbbell-like molecule about its center of mass. Although we'll not formally write and solve this wave equation, it is interesting to note that the form of solution is exactly the same as that which we have solved previously for the spherical harmonics $\Theta(\theta)\Phi(\phi)$ of the hydrogen atom in Chapter 11. Consequently, *two* quantum numbers, J and M, arise which have the same significance with regard to the rotational angular momentum of the *molecule* as l and m have with regard to the angular momentum of the hydrogen atom. Specifically, the *rotational quantum number J* serves to quantize the rotational angular momentum L' of the molecule, so that

$$L' = \sqrt{J(J+1)}\,\hbar \qquad (15\text{–}9)$$

(Compare with Eq. 11–21), where $J = 0, 1, 2, 3, \ldots$, and the quantum number M serves to quantize the component L'_z of rotational angular momentum of the molecule along an arbitrarily prescribed z-axis, so that

$$L'_z = M\hbar$$

(Compare with Eq. 11–27), where $M = 0, \pm 1, \pm 2, \ldots, \pm J$. But the classical angular momentum of the rigid rotor is

$$L' = I\omega, \qquad (15\text{–}10)$$

where I is the moment of inertia of the molecule defined as μr^2_{AB}, μ is the reduced mass, and ω is the angular velocity. In addition the energy E_{rot} of the rigid rotor is given classically as

$$E_{\text{rot}} = \tfrac{1}{2} I\omega^2. \qquad (15\text{–}11)$$

Combination of Eqs. (15–10) and (15–11) to eliminate ω, followed by substitution of the resultant equation for L' into Eq. (15–9), yields the expression for the allowed eigenvalue energies of the rigid rotor as

$$E_{rot} = J(J + 1)\frac{\hbar^2}{2I},$$

(15–12)

where the degree of degeneracy of each level is $(2J + 1)$, since the energy of the rotor depends only on the value of the quantum number J, and since for each value of M, M may assume $(2J + 1)$ different values.

A schematic diagram of allowed rotational energy levels for a rigid diatomic molecule is shown in Fig. 15–3. The energies absorbed or emitted in transitions from one rotational energy level to an adjacent level correspond (according to $\epsilon = h\nu$) to experimentally observed frequencies in the microwave and infrared regions of the electromagnetic spectrum. Furthermore, according to the *selection rule for rotation*, only those transitions for which $\Delta J = \pm 1$ are observed. Transitions are observed in microwave and infrared spectra only for molecules which have permanent dipole moments. In *Raman* spectroscopy, however, transitions can be observed for molecules such as H_2, O_2, and CO_2, which do not have permanent dipole moments and, in this case, $\Delta J = 0, \pm 2$.

Pure rotational spectra may be observed in the microwave region and sometimes in the far infrared region, where the energies absorbed are too low to excite simultaneous vibrational transitions, and moments of inertia for

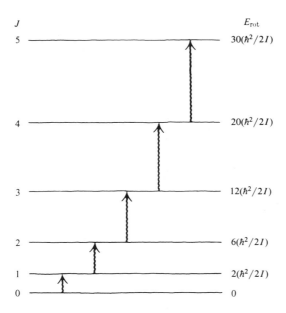

FIG. 15–3   Rotational energy levels for a rigid diatomic molecule. When energy is *absorbed*, the only allowed transitions are those for which $\Delta J = 1$ (see wavy lines in diagram).

molecules may thus be calculated.   Such calculations furnish much of our information concerning the geometry of polyatomic molecules.   Rotational transitions and resultant spectra in the infrared region are usually complicated by accompanying vibrational transitions, for which the energy changes are much larger.

15–6   VIBRATIONAL ENERGY OF MOLECULES

Although, according to the Born-Oppenheimer approximation, the solution of the nuclear wave equation for a given electronic eigenstate requires that for a diatomic molecule we use a potential-energy function $E_e(r_{AB})$ similar to that given for the ground state of hydrogen in Fig. 15–2, we may, as an initial approximation, assume that a diatomic molecule behaves as if it were a simple harmonic oscillator, such as we have previously discussed in Chapter 9. The potential energy of the simple harmonic oscillator, as given by Eq. (4–20), is

$$V = \tfrac{1}{2}\kappa y^2 = \tfrac{1}{2}\kappa(r_{AB} - r_e)^2, \tag{15–13}$$

where y is the displacement of the internuclear distance from the equilibrium position, that is, $r_{AB} - r_e$, and κ is the *force constant* which must be determined

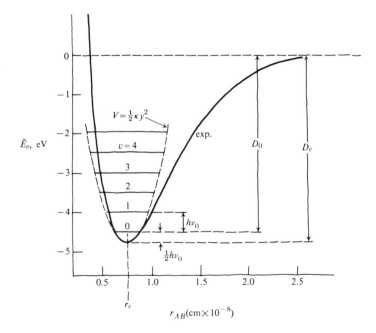

FIG. 15–4   The simple-harmonic-oscillator potential $V = \tfrac{1}{2}\kappa y^2$ as an approximation to the potential energy curve for the hydrogen molecule. Allowed idealized vibrational-energy levels are shown for the first five values of v, the vibrational quantum number.

empirically for a particular bond from observed energy levels. A plot of $V = \frac{1}{2}\kappa y^2$, which is a parabola, is shown for the hydrogen molecule in Fig. 15–4. With proper choice of κ, the approximation of a simple harmonic oscillator appears to be reasonably accurate, at least for lower energies near the bottom of the curve of $\tilde{E}_e$ versus r_{AB}.

The Schrödinger equation for a diatomic molecule of reduced mass μ, which is assumed to be a simple one-dimensional harmonic oscillator, is

$$\frac{-\hbar^2}{2\mu} \cdot \frac{d^2\psi_{\text{vib}}}{dy^2} + \left(\tfrac{1}{2}\kappa y^2\right)\psi_{\text{vib}} = E_{\text{vib}} = E_{\text{vib}}\psi_{\text{vib}}, \qquad (15\text{–}14)$$

where ψ_{vib} are the vibrational wave functions and E_{vib} are the eigenvalues for the vibrational energy. Eq. (15–14) is identical in form (with μ instead of m) to the equation for the simple harmonic oscillator which we have previously solved in detail in Chapter 9. The resultant allowed energy levels, according to Eq. (9–10) are given by

$$E_{\text{vib}} = (v + \tfrac{1}{2})h\nu_0, \qquad (15\text{–}15)$$

where ν_0 is the fundamental vibration frequency of the bond and v, the *vibrational quantum number*, is restricted to the values 0, 1, 2, 3, . . .

Several vibrational energy levels for the *idealized* linear harmonic oscillator are shown superposed on the minimum ground state electronic energy for the hydrogen molecule in Fig. 15–4. Recall that the lowest allowed vibrational energy level, or *zero-point energy*, for the linear harmonic oscillator if $\frac{1}{2}h\nu_0$ and that the spacing between adjacent energy levels is $h\nu_0$, which is in accord with Planck's original postulate concerning the quantization of oscillator energies in the blackbody radiator (see Section 2–5). The dissociation energy D_0 referred to the zero point energy level is a measure of the energy required to dissociate the molecule at absolute zero where, in addition to the electronic energy at equilibrium separation, it still has a residual vibrational energy of $\frac{1}{2}h\nu_0$. Thus D_0, often called the *spectroscopic dissociation energy*, is smaller than D_e, the electronic binding energy, by $\frac{1}{2}h\nu_0$. In addition, because the actual potential-energy curve is not really a parabola, the spacing between experimentally observed vibrational energy levels decreases quickly at higher levels, as shown in Fig. 15–5, and the levels finally merge into a continuum at the dissociation-energy level.

Transitions between adjacent vibrational energy levels appear in the infrared region of the experimental spectrum and are restricted by the *vibrational selection rule* to those for which $\Delta v = \pm 1$. In addition, the dipole moment of the molecule must change with vibration in order for an infrared band to be observed. On the other hand, *Raman* bands are once again observed for vibrations which *do not* involve changing dipoles provided the polarizability fluctuates. Since the energies required for vibrational transitions are much larger than those required for rotational transitions, both types of

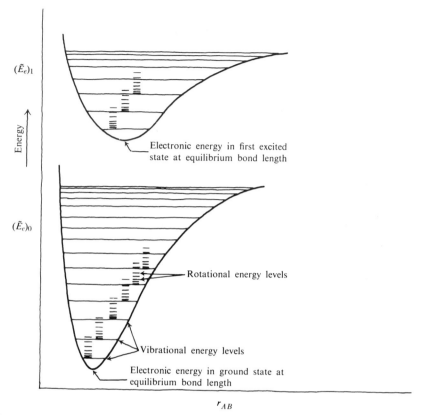

FIG. 15–5   Energy level diagram for a diatomic molecule, showing the first two electronic states, some of the more closely spaced vibrational states, and the still more closely spaced rotational states.

transitions generally occur together in infrared spectra. In addition, since ΔJ must also be ± 1, the two types of transitions are related in a fairly predictable manner, at least for diatomic molecules.

Relative differences in electronic, vibrational, and rotational energy levels for a typical diatomic molecule are shown schematically in Fig. 15–5. Only the first two electronic energy levels are shown, and it is apparent that the spacing between electronic energy levels is much larger than the spacing between vibrational energy levels, which, in turn, is much larger than the spacing between energy levels for rotational states. These relative differences account for the fact that electronic transitions are experimentally observed in the relatively energetic ultraviolet and visible regions, while vibrational transitions are observed in the less energetic infrared region, and rotational transitions are observed in both the infrared region and in the still less energetic microwave region. Because electronic transitions are always accompanied by vibrational and rotational transitions, visible and ultraviolet spectra generally appear as rather broad bands of closely spaced lines or as continuous bands.

For accurate measurements, however, because of the actual interaction of vibrational and rotational motions, the rather simple rigid rotor and linear harmonic oscillator approximations must be modified to take into account such effects as centrifugal stretching, anharmonicity of vibration, and changes in the moments of inertia for different vibrational states. Much of the study of molecular spectroscopy is concerned with the development and application of such corrections and with the rather difficult extension of basic concepts to an interpretation of polyatomic molecules.

15-7 SOME FINAL REMARKS ON BONDING

Chemical bonding was not really adequately accounted for until the advent of quantum mechanics. The basic problem of bonding must be concerned with the relationship of charge distribution to the total energy of a polynuclear species. In general, if the ground-state energy of an aggregate of nuclei or atoms is lower than the sum of the ground-state energies of the nuclei or atoms from which it is made, we say that a bond exists. Such a lowering of energy may, of course, be partially accounted for in terms of classical electrostatics, at least when we are concerned with *ionic* bonding. However, when we deal with covalent bonds, such as the bond in the hydrogen molecule, the explanation is not as simple. For such *homopolar* covalent bonds, we may attribute a large part of the lowering of energy to the fact that *in the molecule, the volume of confinement of the individual electrons is larger than it was in the original atoms.* That is, we may imagine that the electrons are now allowed to move throughout the regions of *several* atoms, rather than within the region of a single atom. Recall that the minimum energy for the single particle in a box is lowered as the size of the box increases. As a crude analogy to the lowering of energy which occurs when two electrons, each in a separate atom, are combined to form an electron pair in a covalent molecule, let's consider two particles (electrons), each in its lowest energy state in a *separate* one-dimensional box (atom) of width L, as shown in Fig. 15-6. According to Eq. (7-35), the energy of each particle in its ground state ($n = 1$) is $h^2/8mL^2$, so that the total ground-state energy of the two particles in separate boxes is

$$2(h^2/8mL^2) = 0.25(h^2/mL^2).$$

If we now *combine* the two boxes (atoms) so that both particles (electrons) are in the same new overlapped box (molecule) of *arbitrary* combined width $1.5L$ and ignore repulsion and exchange effects, the total ground-state energy of the new system is *lowered* to

$$2[h^2/8m(1.5L)^2] = 0.11(h^2/mL^2).$$

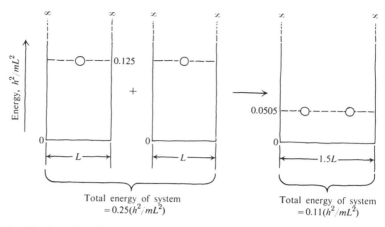

FIG. 15–6   The lowering of energy which occurs when two particles, each in its own ground state in a separate one-dimensional box of width L, are combined in the ground state of a new dimensional box of width $1.5L$.

In order to now separate the particles (electrons) in the wider box (molecule) back into the original narrow boxes (atoms), we must expend an amount of energy $0.14h^2/mL^2$, which may be thought of as the *energy of bonding*.

In a similar fashion, the bonding of *metallic* atoms may be partially accounted for through the use of a model in which we assume that to a first approximation the conducting electrons behave as if they were completely mobile particles (a *Fermi gas*) confined within a three-dimensional box the size of the conductor. Since electrons are Fermi particles whose systems must be described by anti-symmetric wave functions, we may first calculate the allowed energy levels for a single particle in the three-dimensional box according to Eq. (8–42) and then add electrons to the system in order of increasing energy, two at a time in each spatial state according to the Pauli exclusion principle, until we have placed all the electrons in the system. The procedure is analogous to the *Aufbau* principle used in constructing electronic configurations for atoms. The highest level filled is called the *Fermi level*. Because the "box" is so large and because there are such a large number of conducting electrons to place, the resultant lowering of energy for metallic bonding is appreciable.

According to similar arguments, whenever electrons are spread over the regions of *several* atoms in molecules, that is, when they are *delocalized*, the energy of the resultant configuration is lowered. In addition, the lowering of energy in bond formation must also be partially attributed to other purely quantum-mechanical effects, similar to the electron correlation effects which we discussed in terms of the excited helium atom.

In summary, our knowledge of the nature of chemical bonding is far from crystal clear. Other types of bonds, such as hydrogen bonds, van der Waals bonds, and ligand interactions provide equally stimulating challenges. Just as it is difficult in a short concluding chapter such as this to adequately describe the many problems to which the quantum-mechanical approach may be applied with advantage, it is equally difficult to conclude a short text on quantum mechanics, since so many important topics beg consideration.

The theory of the chemical bond . . . is still far from perfect. Most of the principles that have been developed are crude, and only rarely can they be used in making an accurate quantitative prediction. However, they are the best that we have, as yet, and I agree with Poincaré that "it is far better to foresee even without certainty than not to foresee at all.

—LINUS PAULING, *The Nature of the Chemical Bond.**

PROBLEMS

15–1 For the lithium hydride, LiH, molecule: a) write the complete time-independent Schrödinger wave equation; b) write the electronic wave equation; c) write the nuclear wave equation.

15–2 The value for D_e, the electronic binding energy, for H_2, as shown in Fig. 15–4, is 4.748 eV. However, the experimentally observed *spectroscopic dissociation energy D_0* is smaller because in the ground state the molecule already possesses a zero-point energy, $\frac{1}{2} h\nu_0$. Given the fundamental frequency of 1.317×10^{14} Hz for the H_2 molecule, calculate the value of the spectroscopic dissociation energy D_0.

15–3 According to statistical mechanics, the average translational energy of a molecule in each of the three coordinate directions is given as $\frac{1}{2}kT$, where k is the Boltzmann constant (see Appendix A.)

a) Using the quantum-mechanical expression for the energy of a particle in a one-dimensional box 10 cm wide, calculate the approximate value for the translational quantum number n_x of an average neon molecule at 300°K. (Atomic mass of Ne = 20.183 amu.)

b) What will be the change in the one-dimensional translational energy per molecule for a quantum jump from n_x to $(n_x + 1)$?

c) If the energy for the quantum absorption in (b) is supplied by a source of electromagnetic radiation, what would be the frequency and wavelength of the absorbed radiation?

d) Comment on the feasibility of translational spectroscopy.

15–4 a) Calculate the moment of inertia I of the HCl^{35} molecule, which has an equilibrium bond length of 1.275 Å. (The mass of the Cl^{35} atom is 34.96885 amu.)

b) Calculate the first four rotational energy levels for HCl.

15–5 According to the Boltzmann distribution, the number N_J of molecules in the Jth rotational energy level, as related to the number N_0 in the $J = 0$ rotational energy

---

level, is given as:

$$N_J = g_J N_0 e^{-E_J/kT},$$

where E_J is given by Eq. (15–12), g_J is the degeneracy of the Jth level, and k is the Boltzmann constant as given in Appendix A. Using the value of I for HCl calculated from Problem 15–4, prepare a *relative population curve*, N_J/N_0 versus J, for HCl at 25°C.

15–6 For the absorption of energy in a certain diatomic molecule, the vibrational quantum number increases from $v = 0$ to $v = 1$. Draw a schematic energy level diagram, similar to Fig. 15–3, showing the first five rotational energy levels superposed on each of the vibrational energy levels, and show, with vertical arrows, the allowed transitions from lower vibrational-rotational levels to higher vibrational-rotational levels. Use the selection rule for rotation.

15–7 The force constant κ for DCl^{35}, considering the molecule to be a simple harmonic oscillator, is 4.903×10^5 dyne/cm. The masses of the D and Cl^{35} atoms are 2.01410 amu and 34.96885 amu respectively.

  a) Using Eq. (4–16), and substituting μ, the reduced mass, for m, calculate the fundamental vibrational frequency of the DCl^{35} molecule.
  b) Using ν_0 from part (a), calculate the energies of the two vibrational states for which $v = 0$ and $v = 1$.
  c) How much energy is absorbed per molecule in the transition $(v = 0) \rightarrow (v = 1)$?
  d) If the energy is supplied as electromagnetic radiation, what is the frequency of the absorbed radiation?

APPENDIX A
PHYSICAL CONSTANTS
AND CONVERSION FACTORS

Values for the physical constants and conversion factors which follow are based on the analysis of E. R. Cohen and B. N. Taylor, *Journal of Physical and Chemical Reference Data*, **2**, No. 4, p 663 (1973). These values have been recommended for general use by CODATA (Committee on Data for Science and Technology of the International Council of Scientific Unions). Digits in parentheses correspond to one standard deviation uncertainty in the final digits of the quoted value, computed on the basis of internal consistency of the least-squares adjustment. Since the system of units used in this text is the centimeter-gram-second (cgs) system, the constants and conversion factors are given for the cgs system. Thus the unit of force ($F = ma$) is g-cm-sec^{-2}, or *dyne*, and the unit of work ($F\,dx$) is g-cm^2-sec^2, or *erg*. The electrostatic *unit* of charge (esu) is that quantity of electricity upon which there is exerted a force of 1 dyne when a second unit charge is placed 1 cm away in a vacuum.

TABLE A–1

Defined Values and Equivalents

Unit	Abbreviations	Equivalent
Meter	m	1,650,763.73 wavelengths *in vacuo* of the unperturbed transition $2p_{10} - 5d_5$ in ^{86}Kr
Kilogram	kg	Mass of the international kilogram at Sèvres, France
Second	s	the duration of $9,192,631,770$ periods of the radiation corresponding to the transition between the two hyperfine levels of the ground state of the cesium-133 atom.
Degree Kelvin	°K	Defined in the thermodynamic scale by assigning 273.16°K to the triple point of water (freezing point, 273.15°K = 0°C)
Unified atomic mass unit	u	One-twelfth the mass of an atom of ^{12}C nuclide
Mole	mole	Amount of substance containing the same number of atoms (or other units) as 12 g of pure ^{12}C
Standard acceleration of free fall	g_n	9.80665 m sec^{-2}, 980.663 cm sec^{-2}
Normal atmospheric pressure	atm	101, 325 N m^{-2}, 1,013,250 dyne cm^{-2}
Thermochemical calorie	cal$_{th}$	4.1840 J, 4.1840×10^7 erg
Liter	l	1000 cm^3
Inch	in.	0.0254 m, 2.54 cm
Micron	μ	10^{-6}m, 10^{-4} cm
Angstrom	Å	10^{-10} m, 10^{-8} cm
Pound	lb	0.45359237 kg, 453.59237 g
Hertz	Hz	s^{-1}
Newton	N	kg m s^2
Joule	J	N m
Watt	W	J s^{-1}
dyne		g cm s^{-2}
erg		dyne cm

TABLE A–2

Adjusted Values of Constants*

Constant	Symbol	Value
Speed of light in vacuum	c	$2.99792458(1.2) \times 10^{10}$ cm s^{-1}
Elementary charge	e	$4.8032424(137) \times 10^{-10}$ cm$^{3/2}$ g$^{1/2}$ s^{-1} (esu)
Avogadro constant	N_A	$6.022045(31) \times 10^{23}$ mol^{-1}
Electron rest mass	m_e	$9.109534(47) \times 10^{-28}$ g
		$5.4858026(21) \times 10^{-4}$ u
Proton rest mass	m_p	$1.6726485(86) \times 10^{-24}$ g
		$1.007276470(11) \times 10^{0}$ u
Neutron rest mass	m_n	$1.674954(30) \times 10^{-24}$ g
		$1.008665012(37) \times 10^{0}$ u
Planck constant	h	$6.626176(36) \times 10^{-27}$ erg s
	$\hbar = \dfrac{h}{2\pi}$	$1.0545887(57) \times 10^{-27}$ erg s
Rydberg constant	R_∞	$1.097373177(83) \times 10^{5}$ cm^{-1}
	R_H	$1.09677586(90) \times 10^{5}$ cm^{-1}
Bohr radius	a_0	$5.2917706(44) \times 10^{-9}$ cm
Electron radius	r_e	$2.8179380(70) \times 10^{-13}$ cm
Gas constant	R	$8.31441(26) \times 10^{7}$ erg °K^{-1} mol^{-1}
Boltzmann constant	k	$1.380662(44) \times 10^{-16}$ erg °K^{-1}

TABLE A–3

Energy Conversion Factors*

$$1 \text{ eV} = 1.6021892(46) \times 10^{-19} \text{ J}$$
$$1 \text{ eV} = 1.6021892(46) \times 10^{-12} \text{ erg}$$
$$1 \text{ cal} = 4.184 \text{ J (defined)}$$
$$1 \text{ J} = 10^{7} \text{ erg}$$

* Digits in parentheses correspond to one standard deviation uncertainty in the final digits of the quoted number.

APPENDIX B
GREEK ALPHABET

Greek Alphabet

Alpha	A	α	Nu	N	ν
Beta	B	β	Xi	Ξ	ξ
Gamma	Γ	γ	Omicron	O	o
Delta	Δ	δ	Pi	Π	π
Epsilon	E	ϵ	Rho	P	ρ
Zeta	Z	ζ	Sigma	Σ	σ, s
Eta	H	η	Tau	T	τ
Theta	Θ	θ, ϑ	Upsilon	Υ	υ
Iota	I	ι	Phi	Φ	ϕ, φ
Kappa	K	κ	Chi	X	χ
Lambda	Λ	λ	Psi	Ψ	ψ
Mu	M	μ	Omega	Ω	ω

315

BIBLIOGRAPHY

The following books provide useful references for review and further reading.

History and Philosophy

Bohr, N., *Atomic Theory and the Description of Nature*, Cambridge University Press, London, 1934.

D'Abro, A., *The Rise of the New Physics*, Vol. II, Dover, New York, 1951.

de Broglie, L., *Matter and Light: The New Physics*, W. W. Norton, New York, 1939. (Reprinted by Dover, New York.)

Gamow, G., *Mr. Tompkins in Wonderland*, Cambridge University Press, London, 1960.

Gamow, G., *Thirty Years That Shook Physics: The Story of Quantum Theory*, Doubleday & Co., New York, 1966.

Heisenberg, W., *The Physical Principles of Quantum Theory*, University of Chicago Press, Chicago, 1930. (Reprinted by Dover, New York.)

Hoffmann, B., *The Strange Story of the Quantum*, 2nd ed., Dover, New York, 1959.

Margenau, H., *The Nature of Physical Reality*, McGraw-Hill, New York, 1950.

Schrödinger, E., *Science and Human Temperament*, W. W. Norton, New York, 1935.

Mathematics Background

Margenau, H., and G. M. Murphy, *The Mathematics of Physics and Chemistry*, D. Van Nostrand, New York, 1943.

Sokolnikoff, I. S., and E. S. Sokolnikoff, *Higher Mathematics for Engineers and Physicists*, 2nd Ed. McGraw-Hill, New York, 1941.

317

Classical Physics

COULSON, C. A., *Waves, A Mathematical Account of the Common Types of Wave Motion*, Oliver and Boyd, Edinburgh, 1961.

HALIDAY, D., and R. RESNICK, *Fundamentals of Physics*, John Wiley & Sons, New York, 1974.

WALDRON, R. A., *Waves and Oscillations*, D. Van Nostrand, Princeton, 1964.

Introductory Treatments of Quantum Mechanics

COULSON, C. A., *Valence*, Oxford University Press, London, 1961.

DAVIS, J. C., *Advanced Physical Chemistry: Molecules, Structure, and Spectra*, Ronald Press, New York, 1965.

EISBERG, R. M., *Fundamentals of Modern Physics*, John Wiley & Sons, New York, 1961.

HANNA, M. W., *Quantum Mechanics in Chemistry*, Second Edition, W. A. Benjamin, New York, 1969.

HARRIS, L., and A. L. LOEB, *Introduction to Wave Mechanics*, McGraw-Hill, New York, 1963.

LINNETT, J. W., *Wave Mechanics and Valency*, Methuen & Co., London, 1960.

PAULING, L., and E. B. WILSON, *Introduction to Quantum Mechanics*, McGraw-Hill, New York, 1935.

ROJANSKY, V., *Introductory Quantum Mechanics*, Prentice-Hall, Englewood Cliffs, N.J., 1968.

SHERWIN, C. W., *Introduction to Quantum Mechanics*, Holt, Rinehart and Winston, New York, 1959.

More Complete Treatments of Quantum Mechanics: General References

DIRAC, P. A. M., *Principles of Quantum Mechanics*, 4th ed., Oxford University Press, London, 1958.

EYRING, H., J. WALTER, and G. E. KIMBALL, *Quantum Chemistry*, John Wiley & Sons, New York, 1944.

KAUZMANN, W., *Quantum Chemistry: An Introduction*, Academic Press, New York, 1957.

KRAMERS, H. A., *Quantum Mechanics*, Dover, New York, 1964.

LEVINE, I. N., *Quantum Chemistry*, Allyn and Bacon, Boston, 1970.

PILAR, F. L., *Elementary Quantum Chemistry*, McGraw-Hill, New York, 1968.

PITZER, K. S., *Quantum Chemistry*, Prentice-Hall, Englewood Cliffs, N.J., 1953.

SLATER, J. C., *Quantum Theory of Atomic Structure*, Vol. I., McGraw-Hill, New York, 1960.

ANSWERS TO PROBLEMS

Chapter 2

2-1 6230° K

2-2 a) $0.3679p_0$, $0.1353p_0$, $0.04979p_0$, $0.01831p_0$
b) $4.540 \times 10^{-5}p_0$, $2.061 \times 10^{-9}p_0$, $9.358 \times 10^{-14}p_0$, $4.248 \times 10^{-18}p_0$

2-3 a) 1.986×10^{-22} erg   b) 4.97×10^{-12} erg   c) 1.986×10^{-6} erg

2-4 7.84×10^7 cm/sec

2-5 3.4×10^{-12} erg

2-6 a) 5.03×10^{18} photons/cm^3 b) 20.8 photons/cm^3

2-7 5 sec. Much longer than observed.

2-8 414 yrs

2-9 7.9×10^{15} electrons/sec

Chapter 3

3-1 a) 911.8 Å, 8205.9 Å, 14,588.2 Å, 22,794.1 Å
b) Energy of free electrons is not quantized.

3-2 a) 9.1066×10^{-28} g   b) 109,707.5 cm^{-1} c) deuterium: 6562.9 Å,
hydrogen: 6564.7 Å   d) Sommerfeld energies must be used.

3-4 13.598 eV

3-5 $E = -2\pi^2 Z^2 m_e e^4 / n^2 h^2$, $r = n^2 h^2 / 4\pi^2 Z m_e e^2$

3-6 1640.5 Å, 1215.1 Å, 911.4 Å

3-7 a) 1.637×10^{-11} erg,   b) Yes. 1.26×10^{-11} erg

3-9 3.88×10^{-9} cm, 1.23×10^{-9} cm

3-10 a) 3.325 Å   b) 6.650 Å

Chapter 4

4-1 a) 2 cm   b) $\pi/4$ radians   c) 3π radians/sec   d) (2/3) sec/cycle
e) -1.414 cm, 13.3 cm/sec, 125 cm/sec^2

4-2 0.356 sec^{-1}, 1.13 sec^{-1}

4-3 7×10^5 dyne cm^{-1}

4-4 2.12×10^{28}, 4.72×10^{-29}

4-5 $-b^2$ **4-6** 3   **4-7** a^2

4-8 B

4-9 $b = 4a^2$, $-2a$

4-14 Does not satisfy boundary conditions

4-15 b) $n > 1$, $m > 1$, m integral multiple of n.

4-16 a) 0.5 cm   b) 1.0 sec^{-1} c) 0.57 cm

4-17 a) b,   b) $-d/c$, $-x$ direction.

4-18 a) $-(n\pi u A'/L) \sin(n\pi x/L) \sin(n\pi u t/L)$ d) zero   e) $(\mu n^2 \pi^2 u^2 / 2L^2) A'^2$

Chapter 5

5-2 $\sqrt{2}$, $\sqrt{2}$, 5, 5, π **5-4** $\frac{1}{2}$, $\frac{1}{2}$

5-5 a) $-(b^2 + c^2)$

Chapter 6

6-1 a, b, e

6-2 Only (b) is a linear homogeneous equation.

6-3 a) $(yh/i2\pi)(\partial/\partial x)$ b) $xy + (h^2/4\pi^2)(\partial/\partial x)(\partial/\partial y)$
   c) $(-h^2/8\pi^2 m)(\partial^2/\partial x^2)$

6-4 a) r b) $(-h^2 \nabla^2/8\pi^2 m) + V(x, y, z)$ c) $(h/i2\pi)[y(\partial/\partial z) - z(\partial/\partial y)]$

6-6 $a^2 b^3 c^4/24$

6-7 a) 0.103, 0.0303   b) 42.7   c) 40

6-8 b) When $E_1 = E_2$; yes

6-9 $a = \pm \sqrt{3/L^3}$

6-10 $\alpha = \pm(105/a^3 b^5 c^7)^{1/2}$

Chapter 7

7-2 $(2mE)^{1/2}$

7-3 a) $\partial^2 \psi/\partial x^2 = -8\pi^2 m(E - V_c)\psi/h^2$.
   b) $A \sin \left[8\pi^2 m(E - V_c)/h^2 \right]^{1/2} x$.
   c) $\psi = 0$ at $x = 0$ and at $x = L$
   d) $E = (n^2 n^2/8mL^2) + V_c$.

7-4 a) $A \sin \left[2\pi(2mE)^{1/2} x/h \right]$

7-6 No. Cannot satisfy boundary conditions at $x = 0$.

7-7 6.43×10^{-63} erg   **7-8** 0.00033 erg

7-9 6.02×10^{-13} erg, 2.41×10^{-12} erg, 5.42×10^{-12} erg, 9.64×10^{-12} erg
   1.1×10^{-4} cm

7-10 a) 1.13×10^{-12} erg, 4.52×10^{-12} erg, 1.017×10^{-11} erg,
   1.809×10^{-11} erg
   b) 2510 Å

7-11 $L/2$

7-12 $h^2/m^2 a^2$

7-14 a), b) 6.6256×10^{-19} cm g sec^{-1} c) h

7–15 7.27×10^8 cm/sec, 6.03×10^{-11} erg

7–16 a) Yes. $V = \kappa x^2/2 \neq f(t)$ b) $\left(\dfrac{-h^2}{8\pi^2 m} \dfrac{d^2}{dx^2} + \dfrac{\kappa x^2}{2} \right)\psi = E\psi$

7–17 0.264

Chapter 8

8–1 a) $\pm(3/L^3)^{1/2}$ b) $\pm(5/L^5)^{1/2}$ c) No

8–2 b) $1/\sqrt{2}$

8–3 b) $a^2 + b^2 = 1$, $f^2 + g^2 = 1$

8–9 a) 3.02, 6.02, 6.08, 9.02, 11.02, 11.18

b) Degeneracies: 1, 2, 1, 1, 2, 2, 1

8–11 d) $|B|^2/|A|^2$ **8–12** ~ 0.006

8–13 b) $4/(4 + \alpha^2 a^2)$

Chapter 9

9–2 $(1/96\pi^{1/4}\sqrt{5}\,)(64\xi^6 - 480\xi^4 + 720\xi^2 - 120)e^{-\xi^2/2}$

9–4 $0, \frac{1}{2}, 1/\sqrt{\pi}\,, 1$

9–5 $\bar{T} = \bar{V} = h\nu/4$

9–6 a) $\left(v_x + \frac{1}{2}\right)h\nu_x + \left(v_y + \frac{1}{2}\right)h\nu_y + \left(v_z + \frac{1}{2}\right)h\nu_z$

b)

$$\left[\frac{(\alpha_x \alpha_y \alpha_z)^{1/2}}{2^{v_x + v_y + v_z} v_x! v_y! v_z! \pi^{3/2}} \right]^{1/2} e^{-1/2(\alpha_x x^2 + \alpha_y y^2 + \alpha_z z^2)} H\left(\alpha_x^{1/2} x\right) \cdot H\left(\alpha_y^{1/2} y\right) \cdot H\left(\alpha_z^{1/2} z\right)$$

c) $\left(v_x + v_y + v_z + \frac{3}{2}\right)h\nu$

d) 1, 3, 6, 10

9–7 a) $\sim 4/\alpha^{1/2}$ b) $2(mh\nu)^{1/2}$

Chapter 10

10–1 b) $\Phi_{\sin} = (1/\sqrt{\pi}\,)\sin m\phi$, $\Phi_{\cos} = (1/\sqrt{\pi}\,)\cos m\phi$

c) $\Phi_{+1} = (1/\sqrt{2}\,)\Phi_{\cos} + (i/\sqrt{2}\,)\Phi_{\sin}$,

$\Phi_{-1} = (1/\sqrt{2}\,)\Phi_{\cos} - (i/\sqrt{2}\,)\Phi_{\sin}$.

10–2 $(35z^4/8) - (15z^2/4) + 3/8$

10–3 Yes, because $d^{|m|}z^l/dz^{|m|}$ vanishes when $|m| > l$

10–5 $(3\sqrt{14}\,/4)\left[(5/3)\cos^3\theta - \cos\theta\right]$

10–6 $24 - 96\rho + 72\rho^2 - 16\rho^3 + \rho^4$

10–8 b) $-4 + 2\rho$ **10–9** 16

10–10 a) 10^{-171} b) 0.291 c) $\sim kT$

Chapter 11

11–1 b, c, e

11–2 $h/i2\pi$, $\Delta t\,\Delta E \cong h$

11–3   No

11–7   b) $h/\pi\sqrt{2}$ c) $L_z = h/2\pi$

11–11   b) Same as Fig. 11–12, except along x-axis

11–16   a) $h/2\pi$, $-h/2\pi$

b) No. Yes. L_z is not restricted to a single discrete value in either case, but may have either of the two values, $+h/2\pi$ or $-h/2\pi$, neither of which can be specifically associated with a single orbital.

Chapter 12

12–1   c) -275.4 eV

12–2 $(4096/\pi^2 a_0^6)\int\int (e^2/r_{12})\exp\left[-8(r_1 + r_2)/a_0\right] d\tau_1\, d\tau_2$

12–3 -244.8 eV, -435.2 eV, -680.0 eV

12–7   a) Example: $xyz(a - x)(b - y)(c - z)$

12–8   a) $\sqrt{210/L^7}\,(2x^3 - 3Lx^2 + L^2 x)$;

b) 1.064

Chapter 13

13–1   Yes   **13–3**   b, c, d; b, c

13–4   For the $1s2p_x$ state, the singlet is

$$(\tfrac{1}{2})\left[1s(1)2p_x(2) + 1s(2)2p_x(1)\right]\left[\alpha(1)\beta(2) - \alpha(2)\beta(1)\right]$$

and the triplet is

$$\left\{ \begin{array}{l} (1/\sqrt{2}\,)\left[1s(1)2p_x(2) - 1s(2)2p_x(1)\right]\left[\alpha(1)\alpha(2)\right] \\ (1/\sqrt{2}\,)\left[1s(1)2p_x(2) - 1s(2)2p_x(1)\right]\left[\beta(1)\beta(2)\right] \\ (\tfrac{1}{2})\left[1s(1)2p_x(2) - 1s(2)2p_x(1)\right]\left[\alpha(1)\beta(2) + \alpha(2)\beta(1)\right] \end{array} \right\}.$$

Two similar sets of four equations each may also be written for the $1s2p_y$ state and the $1s2p_z$ state.

Chapter 14

14–3   Due to greater exchange-energy stabilization in the $4d^4 5s$ configuration

14–4   Exchange-energy stabilizations: 0K, 3K, 6K, 10K, 15K, 21K, 21K, 22K, 24K, 27K, 31K, 36K, 42K

14–5   a) 19, 10; 20, 9   b) 216, 226   c) 10K   d) 10K

14–6   203.428 eV; higher due to kinetic energies of emitted electrons

14–7   a) 306.97 eV   c) 2.259

Chapter 15

15–1   c) $(h^2/8\pi^2)\left[(\nabla_{\text{Li}}^2/m_{\text{Li}}) + (\nabla_{\text{H}}^2/m_{\text{H}})\right]\psi_n + E_{el}\psi_n = E_n\psi_n$

15–2   4.475 eV

15–3   a) 3.56×10^9 b) 1.17×10^{-23} erg   c) 1770 sec^{-1}, 1.70×10^7 cm

15–4 a) 2.69×10^{-40} g-cm^2 b) 04.14×10^{-15} erg, 12.42×10^{-15} erg, 24.84×10^{-15} erg

15–5

J:	0	1	2	3	4	6	8	10	12
N_J/N_0:	1.00	2.71	3.70	3.83	3.30	1.58	0.46	0.08	0.01

15–7 a) 6.23×10^{13} sec^{-1} b) 2.06×10^{-13} erg, 6.18×10^{-13} erg
c) 4.12×10^{-13} erg   d) 6.23×10^{13} sec^{-1}

Here is this quite beautiful theory, perhaps one of the most perfect, most accurate, and most lovely that man has discovered. We have external proof, but above all internal proof, that it has only a finite range, that it does not describe everything that it pretends to describe. The range is enormous, but internally the theory is telling us, "Do not take me absolutely or seriously. I have some relation to a world that you are not talking about when you are talking about me." This is a kind of rebuke, of course, to anyone who believes that any specialty can wholly exhaust life or its meaning.

Our knowledge is limited by the limits of our experience. It grows all the time. As we see more of the harmony, the order, and the beauty in the world, we take great pride in sharing that knowledge. This knowledge is the reward of the scientists.

—J. Robert Oppenheimer, *Physics Today*, **10**, No. 7, 12–20 (1957).